INTRODUCTION TO GROUP THEORY

INTRODUCTION TO GROUP THEORY

HANS SCHWERDTFEGER

McGill University
Montreal, Canada

NOORDHOFF INTERNATIONAL PUBLISHING
LEYDEN

A division of A. W. Sijthoff International Publishing Company B.V.
Leyden, The Netherlands

ISBN 978-90-286-0495-7 **ISBN 978-94-010-1548-6 (eBook)**
DOI 10.1007/978-94-010-1548-6

Set-in-type in the United Kingdom

To my Wife

CONTENTS

PREFACE

The present book is the outcome of a one-semester lecture course which the author has given frequently during the last three decades. The course has been gradually modified over the years in accordance with changing outlook and with the steadily increasing sophistication of the audience, third- and fourth-year honours classes in several universities in Australia and in Canada.

Out of the conviction that no branch of Mathematics can be mastered by memorizing facts and methods I have tried from the beginning to make the subject interesting for the reader. Clearly one cannot hope to please everybody in this respect. I have sought, however, to attract the reader's interest by including a number of discussions and examples which either have interested me on occasion or resulted from questions of students; some sections and examples have been taken from work done by now eminent mathematicians in an early period of their career, assuming that such selections will appeal to younger readers. So I hope that this book will be found to be a reasonably modern, although not conventional, text proposing an amount of material most of which can be dealt with in a half-year's lecture course. After studying the book the reader should be able to tackle those problems in group theory which are scattered in the problem sections of the American Mathematical Monthly and other similar periodicals.

At the end of each of the 23 sections there is a collection of examples and exercises referring to, or extending, the subject matter of the section. For some of the exercise problems hints are given in an appendix. Most of Chapter I is devoted to the introduction of a large number of special groups from various parts of mathematics which should supply motivation not only for group theory as such, but also for some of the concepts which naturally occur in the development of the theory.

References to the literature have been kept to a minimum. No list is provided of other books on group theory. But I have not hesitated to occasionally cite books and papers in the more commonly understood foreign languages although it has been objected that such references were useless for the average reader. If this is really so then, I think, it is high time for this state of affairs to be changed. We take a first step in this direction. A reader who chooses to ignore such references will still be able to understand the book. The same applies to literature of a more advanced nature which is quoted at a few places. These as well as the "foreign" references will hardly disturb the beginner; they are meant for a reader who wishes to pursue one of the special subjects initiated in the book.

If I have succeeded in my efforts to write a readable text then this is largely due to the help of two young mathematicians who have read and criticized my manuscript at different stages of its development: Drs. Jeffrey Brown and Jerrold Fischer. It was my good luck to have them with me as post-doctoral assistants under a grant from the National Research Council of Canada just at the time when I was working on the book. They provided me with valuable suggestions and corrections. Fischer, in particular, had practical influence on the content and the final formulation of some of the subsections. He also read a complete set of the proofs. Dr. Peter Lorimer of the University of Auckland, New Zealand, and Robert Seely, a student in my last lecture course on Group Theory, have read parts of an early version of the manuscript. To all these helpers I wish to express my heartfelt thanks.

The book is dedicated to my wife who helped me not only in a general way and with the proof reading, but also by discussions on some details and by occasional critical remarks based on her own experience in teaching higher algebra at the university level.

Montreal, *Hans Schwerdtfeger*
November 1975

NOTATIONS

Throughout this book the mathematical notations are those now commonly used, possibly with the one exception for an introductory book, namely that German capitals serve to designate groups (cf. the table below). These letters are now quite familiar in the more advanced literature on our subject. Moreover one should think that every mathematician will appreciate having at his disposal an additional complete alphabet which, in contrast to the Greek, admits a perfect pairing with the letters of the Latin alphabet.

Bold-faced capitals denote sets, e.g. $\mathbf{A} = \{a_1, a_2, \ldots\}$. In some places a distinction between a group $\mathfrak{G}$ and the set $\mathbf{G}$ of its elements appears to be helpful. For sets the now generally accepted inclusion notation $\mathbf{A} \subseteq \mathbf{B}$ and $\mathbf{A} \subset \mathbf{B}$ is used. By $\mathbf{A} \setminus \mathbf{B}$ we denote the part of the set $\mathbf{A}$ which is not contained in $\mathbf{B}$. The number of elements in a finite set $\mathbf{S}$ is denoted by $|\mathbf{S}|$.

In the case of groups we write $\mathfrak{H} \leqslant \mathfrak{G}$ to indicate that $\mathfrak{H}$ is a subgroup of $\mathfrak{G}$, $\mathfrak{H} < \mathfrak{G}$ that $\mathfrak{H}$ is a proper subgroup of $\mathfrak{G}$, $\mathfrak{H} \trianglelefteq \mathfrak{G}$ that $\mathfrak{H}$ is a normal subgroup of $\mathfrak{G}$, etc. However $\mathbf{H} \subset \mathfrak{G}$ indicates that $\mathbf{H}$ is a proper subset consisting of elements of $\mathfrak{G}$; clearly this is equivalent to $\mathbf{H} \subset \mathbf{G}$. Small italics $a, b, \ldots, x, y \ldots$ denote group elements: $a \in \mathfrak{G}, \ldots$. The order of a finite group $\mathfrak{G}$ is denoted by $|\mathfrak{G}|$.

The letters $m, n, d, \ldots, \mu, \nu$ denote integers; $p, q, \ldots$ are often prime numbers. Capital italics $A, B, \ldots$ represent permutations or matrices. The identity permutation is denoted by I, the unit matrix by E. Functions and mappings are denoted by $f, g, \ldots, \alpha, \beta, \ldots, \varphi, \psi$.

The symbols Α, Β, Γ, Δ, Ε, Κ, Λ, Μ, Ν, Π, Ρ, Σ, Φ, Ψ, Ω should be read as Greek capitals although some of them coincide with Roman capitals.

The German alphabet

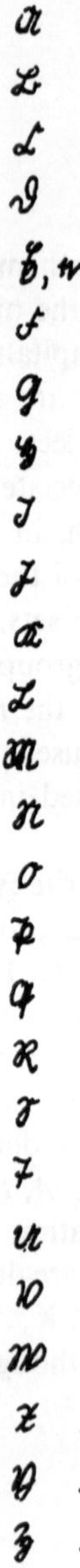

In print	*In writing*	*Roman equivalent*
𝔄		A
𝔅		B
ℭ		C
𝔇		D
𝔈, 𝔢		E, e
𝔉		F
𝔊		G
ℌ		H
ℑ		I
𝔍		J
𝔎		K
𝔏		L
𝔐		M
𝔑		N
𝔒		O
𝔓		P
𝔔		Q
ℜ		R
𝔖		S
𝔗		T
𝔘		U
𝔙		V
𝔚		W
𝔛		X
𝔜		Y
ℨ		Z

Chapter I

DEFINITION OF A GROUP AND EXAMPLES

§1 The abstract group and the notion of group isomorphism

Although it is assumed that the basic concepts of mathematics are known to the reader, it seems appropriate to outline the foundations upon which the theory of groups will be built in the present book.

a. Sets and mappings. The notions of set theory will be used in the "naive" sense.* Let **A** and **B** be two not necessarily different sets. The *cartesian product* (**A**, **B**) is defined as the set of all ordered pairs (a, b), $a \in \mathbf{A}$, $b \in \mathbf{B}$, i.e. in the standard formal description

$$(\mathbf{A}, \mathbf{B}) = \{(a, b) \mid a \in \mathbf{A},\, b \in \mathbf{B}\}.$$

A subset of the set (**A**, **B**) is called a *relation* between **A** and **B**. A relation $f \subseteq (\mathbf{A}, \mathbf{B})$ is called a *function* or a *mapping* $f\colon \mathbf{A} \to \mathbf{B}$ if

(i) for every $a \in \mathbf{A}$ there is a $b \in \mathbf{B}$ so that $(a, b) \in f$;

(ii) from $(a, b_1) \in f$ and $(a, b_2) \in f$ follows that $b_1 = b_2$.

If f is a function, $f \subseteq (\mathbf{A}, \mathbf{B})$, then as an equivalent to $(a, b) \in f$ we have the standard functional notation $b = f(a)$ which suggests the interpretation of f as an operation which to the element a of **A** associates the element b of the set **B**. The condition (ii) implies that with every element a of **A** is associated exactly one element b of **B** (cf. ex. 1).

A function $f\colon \mathbf{A} \to \mathbf{B}$ is said to be *one–one* (also *simple* or *injective*) if, whenever $f(a_1) = f(a_2) = b$, it follows that $a_1 = a_2$. In other words, the function f is one–one if the equation $f(a) = b$ has for a given $b \in \mathbf{B}$ at most one solution a in **A**. The element b is called the *image* of a under f and a is called the *pre-image* of b. A function $f\colon \mathbf{A} \to \mathbf{B}$ is said to be *onto* or *surjective* if for every b

* Cf. P. R. Halmos, Naive Set Theory, Princeton 1960.

in **B** there exists some $a \in \mathbf{A}$ (depending on the choice of b) such that $b = f(a)$, that is, this equation has for each $b \in \mathbf{B}$ at least one solution $a \in \mathbf{A}$. If a mapping $f: \mathbf{A} \to \mathbf{B}$ is not known to be onto it is said to be *into*.

A function which is one–one onto is said to be *bijective*; for a bijective function $f: \mathbf{A} \to \mathbf{B}$ the equation $f(a) = b$ has exactly one solution $a \in \mathbf{A}$ for every b in **B**. A bijective function is *invertible*; its inverse $a = f^{-1}(b)$ is a bijective function which represents, for every b in **B**, the solution of the equation $f(a) = b$. Clearly f^{-1} is a mapping of **B** onto **A** and its inverse $(f^{-1})^{-1} = f$.

b. Algebraic systems. Let **S** denote a set. A mapping f of the cartesian product (**S**, **S**) into the set **S** is a function $f(x, y)$ of the two variables $x, y \in \mathbf{S}$. This function will be called a *composition law* or a *binary operation* in the set **S**. The set **S** together with one or more composition laws $f_1, f_2, \ldots$ constitutes an *algebraic system* which is denoted by $(\mathbf{S} | f_1, f_2, \ldots)$.

The study of algebraic systems with specified composition laws is the object of various branches of Algebra. For historical as well as practical reasons the first algebraic systems in almost everybody's experience are infinite number sets having two composition laws, addition and multiplication: $f_1(x, y) = x+y$, $f_2(x, y) = xy$. Such number systems are special examples of what algebraists call rings, integral domains or fields (cf. ex. 3).

If in particular **S** is the set of all integers, the arithmetical operations of forming the greatest common divisor and the least common multiple determine another kind of algebraic system known as a distributive lattice. [Cf. Chap. IV, §1, *b.*]

In proceeding from elementary arithmetics to the so-called Elementary Theory of Numbers one usually meets for the first time an algebraic system $(\mathbf{S} | f_1, f_2)$ where the set **S** is finite, namely the finite field of the residue classes of integers (mod p), p a prime number (cf. §3).

c. Semigroups. Were it not for the lack of a primitive motivation in basic studies it would be advisable to begin the study of algebra with algebraic systems involving only one law of composition. Indeed, Group Theory is concerned with one special class of algebraic systems of this relatively simple type.

An algebraic system with exactly one composition law f is

called a *groupoid* $(\mathbf{S}|f)$. If f satisfies the functional equation of associativity, that is if for all $x, y, z \in \mathbf{S}$

(1) $f(f(x, y), z) = f(x, f(y, z))$,

the groupoid $(\mathbf{S}|f)$ is called a *semigroup*. If the function f is also symmetric, i.e. if for all $x, y \in \mathbf{S}$

$$f(x, y) = f(y, x),$$

the semigroup $(\mathbf{S}|f)$ is said to be *commutative* or *abelian*.

In the discussion of a semigroup $(\mathbf{S}|f)$, either in the general case or in a special example, the composition law f is of course fixed once for all. Its designation by a letter f is therefore unnecessary. The symbol $f(x, y)$ is usually replaced by one of a variety of symbols, such as $x \cdot y$, xy, $x+y$, (x, y) etc. Writing $x \cdot y$ or equivalently xy, suggests referring to the composition law as multiplication, $x+y$ as addition, even if x and y are not numbers or other objects for which one or the other of these operations is conventionally defined.

It is the associativity which ordinary multiplication and addition have in common with the composition law of a semigroup. Like matrix multiplication this law is in general not commutative. Convention then would have it that the multiplicative notation be used for semigroups in general whereas the additive notation is often used in the case of a semigroup which is known to be, or assumed to be abelian.

In the multiplicative notation the equation (1) appears in the form

(2) $(xy)z = x(yz)$.

A "multiplicative semigroup" will be denoted by $(\mathbf{S}|\cdot)$ and an "additive" one by $(\mathbf{S}|+)$. The adjectives "multiplicative" and "additive" refer only to the notation used to represent the composition law of a semigroup.

d. Groups. Let $\mathbb{F}$ denote a field of numbers and $\mathbf{F}$ the set of its elements; we write $\mathbb{F} = (\mathbf{F}|+, \cdot)$ to characterize the field as an algebraic system with the two composition laws $+$ and $\cdot$. We consider the system $\mathfrak{F} = (\mathbf{F}\setminus\{0\}|\cdot)$, i.e. the set of all non-zero elements of

F with multiplication as composition law. The multiplication is associative and commutative and we can state the following:

(a) There is in $\mathbf{F}\setminus\{0\}$ a unique element, the number 1, which satisfies the condition $1x = x1 = x$ for all x in $\mathbf{F}\setminus\{0\}$.

(b) With every x in $\mathbf{F}\setminus\{0\}$ there is associated a unique x' in $\mathbf{F}\setminus\{0\}$ for which $xx' = x'x = 1$.

The number 1 is the *unit element* of $\mathfrak{F}$ and x', usually denoted by $x^{-1} = 1/x$, the inverse of x in $\mathfrak{F}$. A semigroup $(\mathbf{S}|\cdot)$ which has a unique unit element as well as a definite inverse for each of its elements is called a *group*. Thus $\mathfrak{F} = (\mathbf{F}\setminus\{0\}|\cdot)$ is the multiplicative group of the field $\mathbb{F}$. The unit element of a group $(\mathbf{S}|\cdot)$ is also known as its *identity element* or its *neutral element*. In general it will be denoted by the letter e.

All elements x of a field $\mathbb{F}$ form a group $\mathbb{F}^+$ with respect to the addition in $\mathbb{F}$, i.e. the additive group $(\mathbf{F}|+)$ of $\mathbb{F}$, where 0 (zero) is the identity element and $-x$ is the inverse of x. Corresponding to the conditions (a) and (b) one has in an additive group:

(a′) $\quad 0+x = x+0 = x,$

(b′) $\quad x+(-x) = (-x)+x = 0.$

For the purposes of a systematic theory it is advisable to formulate the definition of a (multiplicative) group in a slightly weaker form.

DEFINITION. *A group $\mathfrak{G}$ consists of a set* $\mathbf{G}$ *of elements, called the elements of $\mathfrak{G}$, and a composition law defined in* $\mathbf{G}$: $\mathfrak{G} = (\mathbf{G}|\cdot)$, *subject to the following conditions:*

(i) *The algebraic system* $(\mathbf{G}|\cdot)$ *is a* ***semigroup.***

(ii) *The set* $\mathbf{G}$ *contains at least one* ***left unit element*** (*or* ***left identity***) *e which satisfies the condition $ex = x$ for all $x \in \mathbf{G}$.*

(iii) *For every $a \in \mathbf{G}$ and for every one of the left unit elements e there is an element $a' \in \mathbf{G}$ for which $a'a = e$.*

This a' is called a ***left inverse*** *of a with respect to e* (cf. ex. 11).

It will be shown that this reduced system of group axioms (i)–(iii) implies the conditions (a) and (b) stated above. We note first that the equation $ex = x$ means that the factor e can be dropped whenever it appears as left factor of an element x. Moreover, if e' denotes the left inverse of e with respect to e we have $e'e = e$; hence by the associativity (2)

$$e'ex = e'x = ex = x$$

which implies that e' is also a left unit element of $\mathfrak{G}$.

THEOREM 1.

I. *The left unit element e is unique in* $\mathfrak{G}$.

II. *The left unit element is also right unit element (and as such unique).*

III. *A left inverse of a in* $\mathfrak{G}$ *is also a right inverse of a.*

IV. *For every a in* $\mathfrak{G}$ *there is exactly one element* a^{-1} *in* $\mathfrak{G}$ *which is both a left and a right inverse of a in* $\mathfrak{G}$.

Proof. I. Let e_1 be a left identity in $\mathfrak{G}$ and e_1' its left inverse with respect to e, that is $e_1'e_1 = e$. For arbitrary $x \in \mathfrak{G}$ then $e_1'(e_1x) = e_1'x = x$. For $x = e_1$ we have $e_1'e_1 = e_1$. Hence $e_1 = e$.

II. Let $ex = x$ and $xe = y$. Multiplication by the left inverse x' of x yields $x'(xe) = x'y$; by (2) $x'(xe) = (x'x)e = ee = e$. Hence $x'x = x'y$. Denoting by x'' the left inverse of x' we have $x''x'x = x''x'y$, i.e. $ex = ey$ whence $x = y$. Thus $xe = x$.

III. By assumption $a'a = e$. Let $aa' = b$. Then $a'aa' = a'b = ea' = a'$. Thus $a'b = a'$, $a''a'b = a''a' = e$ whence $eb = b = e$, and $aa' = e$.

IV. Suppose that a_1 is another left inverse of a so that $a_1a = e = a'a$. Then $a_1aa' = (a_1a)a' = ea' = a'$ (by II.) and $a_1aa' = a_1(aa') = a_1e = a_1$. Hence $a_1 = a'$.

All that was proved for left unit and left inverse can, *mutatis mutandis*, be established for right unit and right inverse. Thus we have in $\mathfrak{G}$ the existence of a *unique unit* (*identity* or *neutral*) *element* e, and a *unique inverse* $a' = a^{-1}$, simultaneously left and right, for every element a in $\mathfrak{G}$.

The weaker form of the group axioms has the following advantage: Often one has to verify that a semigroup $(\mathbf{G}|\cdot)$ is actually a group. With regard to Theorem 1 it is sufficient to show that $(\mathbf{G}|\cdot)$ contains a left unit element and a left inverse *or* a right unit element and a right inverse. In ex. 11 we shall present a semigroup with a left unit element and a right inverse for every element, which, however, is not a group.

A group $\mathfrak{G}$ is said to be finite if the set $\mathbf{G}$ is finite; the number of its elements is denoted by $|\mathfrak{G}|$. It is called the *order* of $\mathfrak{G}$.

e. Isomorphism. We have defined what is often called an *abstract group* $\mathfrak{G}$, "abstract" with regard to the assumption that the elements of $\mathfrak{G}$ are abstract, that is, undefined. As elements of a set **G** they are assumed to be distinct from one another; they are sometimes called "marks". In older treatises the abstract group elements are called "operations".

In contrast to the idea of an abstract group we have the concept of a *group representation*, that is a group $(\mathbf{G}|\cdot)$ where **G** is a set of well defined mathematical objects: Numbers, matrices, mappings of a set or space onto itself, as e.g. rotations about a fixed point, and where the composition law is in a way natural to the set **G**. This means, for example: If **G** is a set of numbers then the composition is ordinary multiplication or addition; if **G** is a set of matrices, the composition is matrix multiplication or addition; if **G** is a set of mappings, then it is functional composition (cf. §2, *a.*).

By neglecting, or giving no attention to, the special nature of the elements of a group representation, we return to the abstract group which has the same composition law as the representation. This relation between an abstract group and a representation is a special case of *group isomorphism*. More generally this is defined as follows:

Two groups $\mathfrak{G}$ and $\mathfrak{G}'$ are said to be *isomorphic*, in symbols

$$\mathfrak{G} \simeq \mathfrak{G}',$$

if there is an invertible mapping $f: \mathfrak{G} \to \mathfrak{G}'$ which satisfies the functional equation

(3) $f(xy) = f(x)f(y), \quad x, y \in \mathfrak{G}$ (i.e. for all $x, y \in \mathfrak{G}$).

Every such mapping f will be called an *isomorphism*. (Notice that in both groups $\mathfrak{G}$ and $\mathfrak{G}'$ the composition law has been written multiplicatively: $xy \in \mathfrak{G}$, $x'y' = f(x)f(y) \in \mathfrak{G}'$.)

If x runs through all the elements of $\mathfrak{G}$, then $x' = f(x)$ runs through all the elements of $\mathfrak{G}'$. The element $f(e)$ is the unit element e' of $\mathfrak{G}'$. Indeed $f(ex) = f(e)f(x) = f(x)$, $x \in \mathfrak{G}$.

Further for all $a \in \mathfrak{G}$ we have $f(a^{-1}) = f(a)^{-1}$, because $f(aa^{-1}) = f(e) = e'$ and $f(aa^{-1}) = f(a)f(a^{-1})$; thus $f(a^{-1})$ is the inverse of $f(a)$ in $\mathfrak{G}'$.

If f is an isomorphism $\mathfrak{G} \to \mathfrak{G}'$, then the inverse mapping f^{-1} represents an isomorphism $\mathfrak{G}' \to \mathfrak{G}$. This follows from the fact that f^{-1} satisfies the condition $f^{-1}(f(x)) = x$ (cf. *a.*), for if $f(x) = x'$

and $f(y) = y'$, $x, y \in \mathfrak{G}$, we have

$$f^{-1}(x'y') = f^{-1}(f(x)\cdot f(y)) = f^{-1}(f(xy)) = xy$$
$$= f^{-1}(x')\cdot f^{-1}(y').$$

From the point of view of abstract group theory two isomorphic groups are indistinguishable; it is mainly from the computational point of view that one may be preferable to the other.

Isomorphism is an equivalence relation between groups. An equivalence class consists of all groups which are isomorphic to one of its members.

f. Cyclic groups. If $\mathfrak{G}$ is a group and $a \in \mathfrak{G}$, then it follows that all the powers of a, that is $a^0 = e$, $a^1 = a$, $a^2 = a\cdot a = aa$, $a^3 = a\cdot a\cdot a, \ldots, a^n = a\cdot a^{n-1}, \ldots$ and $a^{-1}, a^{-2} = (a^{-1})^2 = (a^2)^{-1}, \ldots$, are elements of $\mathfrak{G}$. The set $\{a^\mu \mid \mu = 0, \pm 1, \pm 2, \ldots\}$ is a group with respect to the composition law of $\mathfrak{G}$. Indeed, due to the associativity of the group multiplication in $\mathfrak{G}$ we have the ordinary power (or index) laws: For all integers μ, ν it is easy to verify that

$$(4) \qquad a^\mu a^\nu = a^{\mu+\nu}, \quad (a^\mu)^\nu = a^{\mu\nu}, \quad a^0 = e.$$

If all the elements of $\mathfrak{G}$ are powers of an element a with integral exponents, the group $\mathfrak{G}$ is said to be *cyclic*, *generated by a*. In this case we shall write $\mathfrak{G} = \langle a\rangle$, thus indicating that a is a *generator* of the group $\langle a\rangle$. With regard to (4) a cyclic group is abelian.

A cyclic group $\langle a\rangle$ may be infinite or finite (cf. ex. 3). In the latter case there must exist a least positive integer n so that a^n equals one of the earlier powers of a, i.e. $a^n = a^{n_0}$, $n_0 < n$. Thus $a^{n-n_0} = a^0 = e$ and if we put $n - n_0 = m$, we can say that m is the least positive integer for which $a^m = e$; hence in reality $n_0 = 0$ and $n = m$. Consequently the elements $e, a, a^2, \ldots, a^{m-1}$ are all distinct and they constitute the whole group $\langle a\rangle$. Indeed by (4) the product of two powers of a is a power of a, and since every integer $k = qm + r$ (with integers q, r; $0 \leqslant r < m$) we have

$$a^k = a^{qm}a^r = (a^m)^q a^r = a^r.$$

Hence the cyclic group generated by a has exactly m distinct elements. We indicate this fact by writing $\langle a\rangle = \langle a\rangle_m$. The number

m, that is *the smallest positive integer for which* $a^m = e$, *is called the order of the element* a. It is at the same time the order of the cyclic group $\langle a \rangle_m$.

Examples and exercises

1. Let **S** denote the set of all real numbers $x, y, \ldots$; the cartesian product (**S**, **S**) may be represented by the cartesian (coordinate) plane.

(a) Determine the subset **R** of (**S**, **S**) which defines the order relation in **S**, that is: $(x, y) \in \mathbf{R}$ if and only if $x < y$.

(b) If the relation is a function $f: \mathbf{S} \to \mathbf{S}$ the subset $f \subseteq (\mathbf{S}, \mathbf{S})$ is the graph of the function in the cartesian plane.

2. Let **S** denote the set of all real numbers $x, y, \ldots$. The following functions of two variables represent composition laws in **S**:

(a) $f_1(x, y) = x+y,\qquad f_2(x, y) = xy;$

(b) $f_1(x, y) = \mathrm{Min}(|x|, |y|),\qquad f_2(x, y) = \mathrm{Max}(|x|, |y|);$

(c) $f_1(x, y) = \mathrm{Min}(x, y),\qquad f_2(x, y) = \mathrm{Max}(x, y);$

(d) $f_1(x, y) = x+y,\qquad f_2(x, y) = 0;$

(e) $f_1(x, y) = x-y,\qquad f_2(x, y) = 0.$

In each of the five cases test the validity of the associative law (1), the commutative law [i.e. symmetry: $f(y, x) = f(x, y)$] and the distributive laws

$$(5)\quad \begin{cases} f_2(x, f_1(y, z)) = f_1(f_2(x, y), f_2(x, z)), \\ f_1(x, f_2(y, z)) = f_2(f_1(x, y), f_1(x, z)) \end{cases}$$

for each of the five pairs of composition laws.

3. An algebraic system $(\mathbf{S} | +, \cdot)$ where **S** denotes a set of elements $x, y, \ldots$ is called a *ring* if its elements form a commutative (abelian) group with respect to the addition, a semigroup with respect to the multiplication, and if these two operations are distributive, i.e.

$$(6)\quad x(y+z) = xy+xz,\quad (y+z)x = yx+zx.$$

The group $(\mathbf{S}|+) = \mathbf{S}^+$ is called the additive or the addition group of the ring; its neutral element is 0 and the inverse of x is $-x$. The semigroup $(\mathbf{S}|\cdot)$ is called the multiplication semigroup of the ring.

The ring $(\mathbf{S}|+, \cdot)$ is called an *integral domain* if the multiplication is commutative and if for all non-zero elements x, y the product $xy \neq 0$. The ring is a *skew field* if $(\mathbf{S}\setminus\{0\}|\cdot) = \mathfrak{S}$ is a group, the multiplication group of the skew field; it is a *field* if $\mathfrak{S}$ is an abelian group. Thus examples of groups can be derived from algebraic systems known from elementary algebra.

(a) Let $\mathbf{Z}$ denote the system of all integers. Then $(\mathbf{Z}|+, \cdot) = \mathbb{Z}$ is an integral domain, but not a field. The group $\mathbf{Z}^+ = (\mathbf{Z}|+)$ is infinite cyclic, generated by the number 1.

(b) All the multiples ("additive powers") of a fixed integer a, $a \neq 0$, form an additive cyclic group $\mathbf{Z}_a{}^+ \simeq \mathbf{Z}_1{}^+ = \mathbf{Z}^+$.

(c) Let c be a real number different from 0 and from ± 1. All powers c^n, $n \in \mathbb{Z}$, form a multiplicative infinite cyclic group $\langle c \rangle$. Show that $\langle c \rangle \simeq \mathbf{Z}^+$.

(d) For every natural number m the complex number $\epsilon = e^{2\pi i/m} = \cos(2\pi/m) + i \sin(2\pi/m)$, $i = \sqrt{(-1)}$, generates a multiplicative cyclic group of order m, the group of the mth roots of one.

4. (a) If a generates a cyclic group show that $\langle a^{-1} \rangle = \langle a \rangle$.

(b) Show that every infinite cyclic group is isomorphic to $\mathbf{Z}^+$.

(c) Show that every cyclic group of order m is isomorphic to the group of the mth roots of one.

5. Let $x, y \in \mathbb{Z}$ and

$$g_1(x, y) = (x, y) = \text{the g.c.d. of } x \text{ and } y,$$

$$g_2(x, y) = [x, y] = \text{the l.c.m. of } x \text{ and } y.$$

(a) Show that these two composition laws are associative, commutative and distributive (cf. ex. 2).

(b) Verify that the two algebraic systems $(\mathbf{Z}|g_1)$, $(\mathbf{Z}|g_2)$ fail to be groups.

6. A semigroup $\mathcal{G} = (\mathbf{G}|\cdot)$ is a group if and only if for every two elements a, b of $\mathcal{G}$ there is a unique pair $x, y \in \mathcal{G}$ which satisfies the conditions $ax = b$ and $ya = b$.

Proof. Necessity of the conditions is immediate. To prove that they are sufficient we observe that by supposition the equation $ax = a$

has a unique solution $x = e_a$. We shall show that this is in fact a right unit in $\mathcal{G}$. Indeed there is a y in $\mathcal{G}$ such that $ya = b$ is an arbitrary element of $\mathcal{G}$ and from $ae_a = a$ we conclude that $be_a = b$ for every b in $\mathcal{G}$. The inverse of a is obtained as the unique solution x of the equation $ax = e_a = e$.

7. Let S be the system of the real numbers and

$$x = (x_1, x_2, x_3),\ y = (y_1, y_2, y_3), \ldots; \quad x_i, y_i, \ldots \in \mathbf{S},$$

thus $x, y, \ldots$ are elements of the cartesian product $(\mathbf{S}, \mathbf{S}, \mathbf{S}) = \mathbf{V}_3$. Consider the algebraic system $(\mathbf{V}_3 | f_1, f_2)$ with the two composition laws

$$f_1(x, y) = x+y = (x_1+y_1, x_2+y_2, x_3+y_3),$$

$$f_2(x, y) = x\times y = (x_2y_3-y_2x_3, x_3y_1-y_3x_1, x_1y_2-y_1x_2).$$

(a) The addition f_1 in this system is associative, commutative and distributive with the cross-multiplication f_2 [cf. (6)].

(b) The cross-multiplication f_2 is neither associative nor commutative; instead one has

(7) $x\times x = 0,$

(8) $x\times(y\times z)+y\times(z\times x)+z\times(x\times y) = 0 \quad (x, y, z \in V_3),$

where 0 represents the triplet (0, 0, 0), the zero of the addition group $(\mathbf{V}_3 | f_1) = (\mathbf{V}_3 | +)$.

An algebraic system $(\mathbf{V}_3 | +, \times)$ whose elements $x, y, \ldots$ satisfy the conditions (7), (8) is called a *Lie ring*.*

(c) From (7) derive that $y\times x = -x\times y$ $(x, y \in \mathbf{V}_3)$.

8. Let a be a fixed element of a group $\mathfrak{G}$ and x run through all elements of $\mathfrak{G}$.

(a) Show that ax, xa, axa^{-1} run through all elements of $\mathfrak{G}$ whereas xax^{-1} can never exhaust the whole of $\mathfrak{G}$ if $|G| > 1$.

* Named after the Norwegian mathematician S. Lie who introduced such algebraic systems in his theory of continuous transformation groups.

(b) Show that the mapping $x \to axa^{-1}$ is an isomorphism of $\mathfrak{G}$ onto itself. Is the mapping $x \to ax$ also an isomorphism?

9. If $x, y, \ldots$ are arbitrary elements of a group $\mathfrak{G}$ show that

$$(xy)^{-1} = y^{-1}x^{-1}, (xyz)^{-1} = z^{-1}y^{-1}x^{-1}, \ldots$$

and $(xy)^2 = x^2y^2$ if and only if $\mathfrak{G}$ is abelian.

10. (a) Let **S** represent the set of the real numbers x, $-1 < x < 1$, and

$$f(x, y) = x \circ y = \frac{x+y}{1+xy}.$$

Then $(\mathbf{S}|f) = (\mathbf{S}|\circ)$ is a group. Find the unit (neutral) element and for each x find its inverse within this group.

(b) Show that the group $(\mathbf{S}|\circ)$ is ismorphic to the addition group of the field of all real numbers ξ: $-\infty < \xi < \infty$.

Remark. For a given field (or some other algebraic system) $(S|+, \cdot)$ as for instance the field of the real numbers, it is a meaningful problem to determine solutions, or all solutions, f of the functional equation (1) and thus to determine the corresponding semigroups $(S|f)$. Cf. J. Aczél, *Lectures on Functional Equations*, Academic Press, New York 1966, Chap. 6.2.

11. One might wonder what results when in the definition of a group (cf. subsection *d.*) the axiom (iii) is replaced by

(iii)′ For every $a \in \mathbf{G}$ and for every one of the left unit elements e there is an element $\hat{a} \in \mathbf{G}$ (a right inverse of a with respect to e) for which $a\hat{a} = e$.

Every group $(\mathbf{G}|\cdot)$ satisfies the conditions (i), (ii), (iii)′, but one cannot prove that a semigroup $(\mathbf{G}|\cdot)$ which satisfies these conditions is a group. Indeed, here is a semigroup $(\mathbf{G}|\cdot)$ which is not a group. Let $n \geqslant 2$ and $\mathbf{G} = \{e_1, \ldots, e_n\}$ with the multiplication law

$$e_\mu e_\nu = e_\nu \quad (\mu, \nu = 1, \ldots, n).$$

Then $(\mathbf{G}|\cdot)$ is evidently not a group, but each e_μ is a left identity for every e_ν and has itself as a right inverse within respect to e_μ $(e_\mu e_\mu = e_\mu)$.

This example is due to H. B. Mann "On certain systems which are almost groups", *Bull. Amer. Math. Soc.*, **50** (1944). Cf. also A. H. Clifford, "A system arising from a weakened set of group postulates",

Annals of Math., **34** (1933), 865–871. Both papers develop a complete theory of semigroups satisfying the axioms (ii) and (iii)′.

12. Let $(\mathbf{S}|\cdot)$ be a semigroup where every element a of $\mathbf{S}$ possesses a unique pseudo-inverse, i.e. an element a^* of $\mathbf{S}$ which satisfies the condition $aa^*a = a$. Show that $(\mathbf{S}|\cdot)$ is a group.

§2 Groups of mappings. Permutations. Cayley's theorem

a. Composition of mappings. Let $\mathbf{S}$ be a set and let f and g be two mappings of $\mathbf{S}$ into itself. We define the *product* gf (first f, then g) as the mapping

(1) $(gf)(x) = g(f(x))$.

Two mappings $f, g: \mathbf{S} \to \mathbf{S}$ are said to be equal, $f = g$, if $f(x) = g(x)$ for all $x \in \mathbf{S}$. The mapping $\epsilon: \mathbf{S} \to \mathbf{S}$ which maps every x in $\mathbf{S}$ onto itself is called the *identity mapping*; we have $\epsilon(x) = x$, $x \in \mathbf{S}$.

If the mapping f is invertible, then $f^{-1}(f(x)) = f(f^{-1}(x)) = x$, that is

$$f^{-1}f = ff^{-1} = \epsilon.$$

In the case that the function f has no inverse, the equation $f(x) = x'$ either has no solution x for some, possibly all, $x' \in \mathbf{S}$, or it has more than one solution $x \in \mathbf{S}$ for some, possibly all, elements x' in $\mathbf{S}$. Let us denote by $\mathbf{X}$ the set of all those x in $\mathbf{S}$ for which $f(x) = x'$; then

$$f^{-1}(x') = \mathbf{X}.$$

The set $\mathbf{X}$ may be empty; this means that there is no x in $\mathbf{S}$ for which $f(x) = x'$. We express this by writing $\mathbf{X} = \varnothing$.

If $\mathbf{X}$ is not empty, $\mathbf{X} \neq \varnothing$, we have $f(\mathbf{X}) = x'$. Hence the relation $ff^{-1} = \epsilon$, that is

$$f(f^{-1}(x')) = f(\mathbf{X}) = x'$$

remains true, but

$$f^{-1}(f(x)) = f^{-1}(x') = \mathbf{X}$$

which implies that $f^{-1}f$ maps x onto a subset of **S** which contains x; thus $f^{-1}f$ may not be the identity ϵ.

THEOREM 1. *All mappings of a set* **S** *into itself form a semigroup with a unit element* ϵ, *the identity mapping. The multiplication in this semigroup is the composition defined in* (1).

One has to prove that the composition of mappings $\mathbf{S} \to \mathbf{S}$ is associative; indeed

$$(hg)(f(x)) = h(g(f(x))) = h((gf)(x)).$$

COROLLARY. *All invertible mappings of a set* **S** *onto itself form a group* $\mathfrak{M}_\mathbf{S}$.

The set **S** is called the "*space* in which the group $\mathfrak{M}_\mathbf{S}$ operates".

Remark. We write functional composition as well as multiplication in groups from right to left. This is the established form in all work with linear transformations represented by matrices. It also appears to be what most readers of mathematics at the present level are familiar with: One writes $f(x)$ and $g(f(x))$, not xf and xfg in analysis. Standard texts, as van der Waerden's "Algebra", have preserved the right-to-left writing into the most recent editions. The author acknowledges certain advantages of the left-to-right writing as adopted in many recent books on algebra; he finds, however, that the fashion has changed roughly every 20–30 years; so there is hope that we shall be up-to-date in the near future.

b. Permutations are invertible mappings of a finite set **N** onto itself. The name of permutation is sometimes extended to an invertible mapping of an infinite set onto itself and attempts have been made to develop a theory of groups of permutations of an infinity of symbols, generalizing the classical finite theory. Such infinite permutations will not be dealt with in the present book.

The elements of the finite set **N** will be represented by the number symbols $1, 2, \ldots, n$. The permutation which changes $\nu \to p_\nu$ will be represented in the form

$$P = \begin{pmatrix} 1 & 2 & \ldots & n \\ p_1 & p_2 & \ldots & p_n \end{pmatrix} = \begin{pmatrix} \nu \\ p_\nu \end{pmatrix} \quad (\nu = 1, 2, \ldots, n)$$

where $p_1, p_2, \ldots, p_n$ are the symbols $1, 2, \ldots, n$ in a certain order. We call P a *permutation of degree n*.

We note that the symbols in the first row of the permutation P can be written in any order if those in the second row are changed accordingly. Thus if

$$Q = \begin{pmatrix} 1 & 2 & \dots & n \\ q_1 & q_2 & \dots & q_n \end{pmatrix} = \begin{pmatrix} \nu \\ q_\nu \end{pmatrix}$$

is a second permutation, then also

$$P = \begin{pmatrix} q_1 & q_2 & \dots & q_n \\ p_{q_1} & p_{q_2} & \dots & p_{q_n} \end{pmatrix} = \begin{pmatrix} q_\nu \\ p_{q_\nu} \end{pmatrix}.$$

This enables us immediately to define the product of two permutations [in conformity with (1)] by

$$\text{(2)} \quad PQ = \begin{pmatrix} \nu \\ p_\nu \end{pmatrix}\begin{pmatrix} \nu \\ q_\nu \end{pmatrix} = \begin{pmatrix} q_\nu \\ p_{q_\nu} \end{pmatrix}\begin{pmatrix} \nu \\ q_\nu \end{pmatrix} = \begin{pmatrix} \nu \\ p_{q_\nu} \end{pmatrix}$$

In accordance with the remark at the end of subsection *a.* the product PQ is obtained by carrying out first Q, then P: $\nu \to q_\nu$ followed by $q_\nu \to p_{q_\nu}$ yields $\nu \to p_{q_\nu}$.

The identity permutation will be denoted by

$$I = \begin{pmatrix} 1 & 2 & \dots & n \\ 1 & 2 & \dots & n \end{pmatrix} = \begin{pmatrix} \nu \\ \nu \end{pmatrix}.$$

Further every permutation P has a unique inverse

$$P^{-1} = \begin{pmatrix} p_1 & p_2 & \dots & p_n \\ 1 & 2 & \dots & n \end{pmatrix} = \begin{pmatrix} p_\nu \\ \nu \end{pmatrix}$$

which by the law (2) satisfies the condition $P^{-1}P = I = PP^{-1}$. We have therefore (cf. Theorem 1, Corollary) the following result:

THEOREM 2. *All permutations of degree n form a group with multiplication defined by* (2).

This group is called the *symmetric group* of degree n; it is denoted by $\mathfrak{S}_n$. Its order (i.e. the number of elements it contains) is $n!$.

c. Cycles. In elementary number theory it is shown that a natural number can be represented in one and only one way as a product of prime numbers. We shall now show that a permutation P can be represented in one and only one way as a product of a special kind of permutations, called *cycles.* A cycle C is a permutation which effects a cyclic interchange of its symbols:

$$C = \begin{pmatrix} \nu_1 & \nu_2 & \dots & \nu_{m-1} & \nu_m \\ \nu_2 & \nu_3 & \dots & \nu_m & \nu_1 \end{pmatrix} \quad (m \leqslant n).$$

This is called an m-cycle; one uses for it the abbreviated notation

$$C = (\nu_1 \, \nu_2 \, \dots \, \nu_m) = (\nu_2 \, \nu_3 \, \dots \, \nu_m \, \nu_1) = \dots .$$

The most adequate representation of an m-cycle would be the graphical symbol

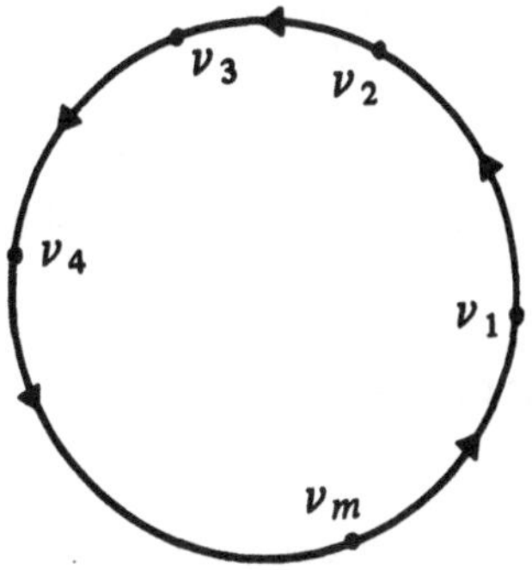

We shall say that a permutation P affects the symbols $\nu_1, \dots, \nu_m$ $(m \leqslant n)$ if $p_{\nu_1} \neq \nu_1, \dots, p_{\nu_m} \neq \nu_m$, and P does not affect the symbols $\nu_{m+1}, \dots, \nu_n$ if $p_{\nu_{m+1}} = \nu_{m+1}, \dots, p_{\nu_n} = \nu_n$. Two permutations P, Q are said to be disjoint if the set of symbols affected by P is disjoint from the set of symbols affected by Q.

LEMMA. *Two disjoint permutations P, Q commute, i.e. $QP = PQ$.*

Remark. The condition of disjointness is sufficient, but by no means necessary for the commutativity of two permutations. In fact, every permutation P commutes with its powers $P^2, P^3, \ldots$, and these cannot be disjoint from P.

Proof of the lemma follows immediately from (2): Look at the explicit form of two disjoint permutations:

$$P = \begin{pmatrix} \nu_1 & \nu_2 & \ldots & \nu_m & \nu_{m+1} & \ldots & \nu_n \\ p_{\nu_1} & p_{\nu_2} & \ldots & p_{\nu_m} & \nu_{m+1} & \ldots & \nu_n \end{pmatrix},$$

$$Q = \begin{pmatrix} \nu_1 & \ldots & \nu_m & \nu_{m+1} & \ldots & \nu_n \\ \nu_1 & \ldots & \nu_m & q_{\nu_{m+1}} & \ldots & q_{\nu_n} \end{pmatrix}$$

and since the q_ν are left fixed by P, the p_ν by Q we have

$$PQ = \begin{pmatrix} \nu_1 & \ldots & \nu_m & \nu_{m+1} & \ldots & \nu_n \\ p_{\nu_1} & \ldots & p_{\nu_m} & q_{\nu_{m+1}} & \ldots & q_{\nu_n} \end{pmatrix} = QP.$$

THEOREM 3. *Every permutation P of degree n is a product of uniquely defined disjoint cycle factors. These cycles commute.*

Proof (by induction). The permutation P maps the symbol 1 into p_1, the symbol p_1 into p_{p_1}, the symbol p_{p_1} into $p_{p_{p_1}}$ etc. finally the symbol

$$p_{p_{\cdot_{\cdot_{\cdot_{p_1}}}}} \text{ into } 1; \text{ hence } P = \left(1\ p_1\ p_{p_1} \cdots p_{p_{\cdot_{\cdot_{\cdot_{p_1}}}}}\right) Q$$

where Q is a permutation of degree n which does not affect the symbols affected by the uniquely defined cycle

$$\left(1\ p_1\ p_{p_1} \cdots p_{p_{\cdot_{\cdot_{\cdot_{p_1}}}}}\right)$$

(Q is of degree $n-1$ if $p_1 = 1$). We assume the theorem to be correct for all permutations of degree $<n$; hence it is correct also for the permutation P of degree n.

d. So it appears that the cycle permutations play the role of prime elements in the group of all permutations of degree n. However the analogy to number theory is not complete. The cycles can be further factorized.

A 2-cycle $(\nu_1\nu_2)$ is called a *transposition*.

THEOREM 4. *Every permutation P is a product of transpositions.*

This becomes evident if one realizes that every permutation of a finite number of material objects can be effected with two hands by carrying out a finite number of interchanges of two objects.

It follows also from Theorem 3 by observing that it is true for cycles:

(3) $(1\ 2\ 3\ \ldots\ m-1\ m) = (1\ m)(1\ m-1)\ \ldots\ (1\ 3)(1\ 2)$

and thus for products of cycles.

It should be noted, however, that the transposition factors of P are neither necessarily disjoint, nor uniquely defined by P. Since $(\mu\nu)(\mu\nu) = I$ one can insert into a product of transpositions an arbitrary number of products of two equal transpositions without changing the value of the given product. This might suggest the following definition:

A permutation P is said to be *even* if it can be represented as a product of an even number of transpositions; *odd* if it can be represented as a product of an odd number of transpositions.

According to (3) we would thus say that an m-cycle is an even permutation if m is odd, an odd permutation if m is even.

However we have to justify the definition, that is we have to prove that the evenness or oddness of the number of transposition factors of a permutation P is uniquely defined by P itself.

For this purpose let the symbols $1, 2, \ldots, n$ designate the columns of an $n \times n$-matrix M with number elements whose determinant $\det M$ is different from zero. By applying to M a transposition [e.g. (1 2), effecting interchange of the first and the second column of M] the determinant $\det M$ is changed into $-\det M$. Thus by applying to M an even number of transpositions in succession, $\det M$ remains unchanged; by applying an odd number of transpositions $\det M$ is turned into $-\det M$. On the other hand, the result of applying a permutation P to the columns of M, either $+\det M$ or $-\det M$, is uniquely defined by P. Hence the oddness or evenness of P is independent of any special representation of P as a product of transpositions.

Clearly the product of two even or of two odd permutations is an even permutation; the unit element I, i.e. the identity mapping, is an even permutation, and so is P^{-1} if P is even. The product of an even and an odd permutation is odd. If Q is a fixed odd permutation, e.g. the transposition (1 2), and X runs through all even permutations, then the product QX runs through all odd permutations. Indeed, if X, Y are two different even permutations then $QX \neq QY$, and for every odd permutation R there is an even permutation X so that $QX = R$. Hence

THEOREM 5. *All the even permutations of n symbols form a group* $\mathfrak{A}_n$. *Its order equals* $\frac{1}{2}n!$

The group $\mathfrak{A}_n$ is called the *alternating group* of degree n.

e. Subgroups. The alternating group $\mathfrak{A}_n$ is a subgroup of the symmetric group $\mathfrak{S}_n$. We shall now introduce the subgroup concept in a more general way.

Let $\mathfrak{G} = (\mathbf{G}|\cdot)$ be a group and $\mathbf{H}$ a subset of $\mathbf{G}$ so that $(\mathbf{H}|\cdot)$ is itself a group. We call $\mathfrak{H} = (\mathbf{H}|\cdot)$ a *subgroup* of $\mathfrak{G}$ and we write $\mathfrak{H} \leqslant \mathfrak{G}$. Not every subset $\mathbf{H}$ of $\mathbf{G}$ yields a subgroup of $\mathfrak{G}$. The symbol $(\mathbf{H}|\cdot)$ implies that for the elements of the set $\mathbf{H}$ the group axioms (cf. §1, *d.*) are satisfied with respect to the multiplication "$\cdot$" in $\mathfrak{G}$. That is: If $x, y \in \mathfrak{G}$ and $x, y \in \mathfrak{H}$ then $xy \in \mathfrak{H}$ and $x^{-1} \in \mathfrak{H}$, hence automatically $e \in \mathfrak{H}$.

The whole group $\mathfrak{G}$ is a subgroup of $\mathfrak{G}$. All other subgroups $\mathfrak{H}$ are *proper subgroups*: $\mathfrak{H} < \mathfrak{G}$. Distinguished among the proper subgroups is the *trivial subgroup*, consisting of the identity e only; it will be denoted by the symbol $\mathfrak{e}$. Thus $\mathfrak{e} \leqslant \mathfrak{G}$, but $e \in \mathfrak{G}$.

A good deal of group theory is concerned with properties of subgroups of a given group as will be seen from Chap. II on.

f. Cayley's theorem. The importance of permutations for finite group theory can be estimated from the following theorem.

THEOREM 6. (Cayley 1878). *Every finite group* $\mathfrak{G}$ *of order* $|\mathfrak{G}| = g > 1$ *is isomorphic to a subgroup of the symmetric group* $\mathfrak{S}_g$.

Proof. Let a denote a fixed element of $\mathfrak{G}$, $x_1 = e$ and let

$$x_1, \; x_2, \; \ldots, x_g$$

be the g elements of $\mathfrak{G}$. If we form the g products

$$ax_1, ax_2, \ldots, ax_g$$

we obtain again the g different elements of $\mathfrak{G}$ (cf. §1, ex. 6) in a definite order depending on a. Thus

$$(4)\quad A = \begin{pmatrix} x_1 & x_2 & \dots & x_g \\ ax_1 & ax_2 & \dots & ax_g \end{pmatrix} = \begin{pmatrix} x \\ ax \end{pmatrix} \quad (x \in \mathfrak{G})$$

is a permutation of the group elements which is defined by the group element a. It changes the element $x_1 = e$ into the element $ax_1 = a$.

Hence there is a function $f: \mathfrak{G} \to \mathfrak{S}_g$ defined by $f(a) = A$, $a \in \mathfrak{G}$, $A \in \mathfrak{S}_g$ as given by (4). Let us denote by $\mathbf{G}^*$ the set of all the permutations A in $\mathfrak{S}_g$ which appear as images of elements a in $\mathfrak{G}$. We have to show:

(i) If the dot denotes the multiplication (composition) of permutations then $(\mathbf{G}^* | \cdot)$ is a group $\mathfrak{G}^*$, namely $\mathfrak{G}^* < \mathfrak{S}_g$.

(ii) The mapping $f: \mathfrak{G} \to \mathfrak{G}^*$ is onto and one–one.

(iii) The mapping f satisfies the condition $f(ba) = f(b)f(a)$ for all a, b in $\mathfrak{G}$.

(i) For the elements a and b of $\mathfrak{G}$ we have $f(a) = A$ and $f(b) = B = \begin{pmatrix} x \\ bx \end{pmatrix}$ in $\mathfrak{G}^*$ and

$$(5)\quad f(b)f(a) = BA = \begin{pmatrix} x \\ bx \end{pmatrix}\begin{pmatrix} x \\ ax \end{pmatrix} = \begin{pmatrix} ax \\ bax \end{pmatrix}\begin{pmatrix} x \\ ax \end{pmatrix} = \begin{pmatrix} x \\ bax \end{pmatrix} = f(ba) \in \mathfrak{G}^*,$$

i.e. $BA \in \mathfrak{G}^*$. The unit element e of $\mathfrak{G}$ is mapped onto

$$f(e) = \begin{pmatrix} x \\ ex \end{pmatrix} = \begin{pmatrix} x \\ x \end{pmatrix} = I,$$

the identity permutation. The inverse a^{-1} of a is mapped onto

$$\begin{pmatrix} x \\ a^{-1}x \end{pmatrix} = \begin{pmatrix} ax \\ x \end{pmatrix} = A^{-1}.$$

Hence $A^{-1} \in \mathfrak{G}^*$. Thus $(\mathbf{G}^* | \cdot) = \mathfrak{G}^*$ is a group.

(ii) Obviously f is onto. It is also one–one: Indeed the permutation A changes e into $ae = a$ and B changes e into $be = b$; thus if $A = B$ it follows that $a = b$.

(iii) The functional equation of the isomorphism has been established in (5).

Thus the permutations $A, B, \ldots$ form a group $\mathfrak{G}^*$ which is isomorphic to the given group $\mathfrak{G}$ (cf. ex. 13).

In connection with Cayley's theorem we mention a device which under certain circumstances facilitates operation in a finite abstract group $\mathfrak{G}$: *The Group Table* (Cayley 1854). Along the upper horizontal and the left vertical side of a square we line up the elements $x_1 = e$, $x_2, \ldots, x_g$ of $\mathfrak{G}$. At the point where the horizontal line through the element x_i meets the vertical line through the element x_j we write down the product $x_i x_j$. It is readily seen that in each row and in each column of the group table every element of the group $\mathfrak{G}$ occurs

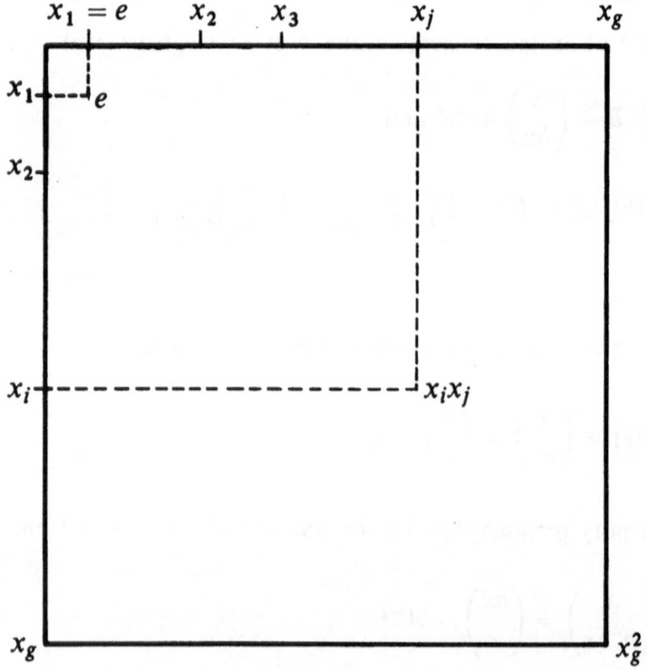

Fig. 1

exactly once. Indeed, the elements in a horizontal row of the group table are those in the second line of a permutation symbol

$$f(x_i) = \begin{pmatrix} x \\ x_i x \end{pmatrix}, \; x \in \mathfrak{G},$$

associated with the group element x_i under the isomorphism constituted by Cayley's theorem.

The group table is a valuable tool for the discussion of groups of small orders. In general, however, it does not reflect the essential properties of a group. Indeed a group $\mathfrak{G}$ of order g admits $g!^2$ different group tables, depending on the order of succession of the group elements on the sides of the square.

Examples and exercises

1. Let **S** be the interior of the unit circle in the complex number plane:

$$\mathbf{S} = \{z \,|\, |z| < 1\}.$$

Show that if a, b are constant complex numbers for which $a\bar{a} - b\bar{b} > 0$ (i.e. $|b| < |a|$) the mappings

$$z \rightarrow \frac{az+b}{\bar{b}z+\bar{a}}$$

form a group of mappings $\mathbf{S} \rightarrow \mathbf{S}$.

2. (a) $P = \begin{pmatrix} 1\,2\,3\,4 \\ 4\,3\,2\,1 \end{pmatrix}, \quad Q = \begin{pmatrix} 1\,2\,3\,4 \\ 2\,4\,1\,3 \end{pmatrix} \Rightarrow PQ = QP.$

(b) $P = \begin{pmatrix} 1\,2\,3\,4\,5 \\ 3\,5\,1\,4\,2 \end{pmatrix}, \quad Q = \begin{pmatrix} 1\,2\,3\,4\,5 \\ 5\,4\,3\,2\,1 \end{pmatrix} \Rightarrow PQ \neq QP.$

3. Verify that the permutation

$$\begin{pmatrix} 1\,2\,3\,4\,5\,6\,7\,8\,9\,\alpha\,\beta\,\gamma \\ 3\,\alpha\,5\,\beta\,7\,\gamma\,9\,8\,6\,4\,1\,2 \end{pmatrix}$$

is an 11-cycle.

4. (a) For the permutation

$$\begin{pmatrix} 1\,2\,3\,4\,5\,6\,7\,8\,9\,\alpha\,\beta\,\gamma \\ \alpha\,1\,2\,6\,3\,\gamma\,8\,7\,\beta\,5\,9\,4 \end{pmatrix}$$

find the factorization into disjoint cycles.

(b) Let P be a permutation of degree n, represented as a product of r disjoint cycles. Show that $r-1$ transpositions T_ρ can be found so that $PT_1T_2 \ldots T_{r-1} = C$ is an n-cycle.

5. Every permutation has a certain finite order m, i.e. $P^m = I$, $P^s \neq I$ if $0 < s < m$ (cf. §1, *f*.).

(a) Show that an m-cycle, e.g. $C = (1\ 2 \ldots m)$ has order m.

(b) Show that if $C_1, C_2, \ldots, C_l$ are disjoint cycles, C_λ of order m_λ, $\lambda = 1, \ldots, l$, then the permutation $C_1C_2 \ldots C_l$ has order $[m_1, m_2, \ldots, m_l]$, i.e. the l.c.m. of the numbers m_λ (cf. §3, *a*. IV′).

6. For the permutations given in exx. 2–4 decide their parity, i.e. whether they are even or odd.

7. By setting up all possible 4×4-group tables show that there are exactly two abstract groups of order 4. Note that a given abstract group can lead to several different group tables.

8. Set up a group table of the symmetric group $\mathfrak{S}_3$.

9. A group of eight elements can be defined as follows: Let i, j be two elements of order 4 subject to the conditions ("relations")

$$i^2 = j^2, \quad iji = j$$

and introduce the eight symbols

$$(6) \quad e, i, j, k = ij, \quad e' = i^2, \quad i' = e'i, \quad j' = e'j, \quad k' = e'k.$$

(a) Making use of the associative law set up the multiplication table for the eight elements (6) and thus show that they form a group.

With regard to its origin in the system of the quaternions this group is called the *quaternion group*; it is denoted by $\mathfrak{Q}$. The elements e and e' are often denoted by 1 and -1 respectively.

(b) Each of the eight elements (6) can be represented as a product of powers of i and j.

(c) Referring to Cayley's theorem find a permutation group of degree 8 which is isomorphic to $\mathfrak{Q}$. It is called an isomorphic or faithful representation of $\mathfrak{Q}$ by permutations.

10. The number of symbols affected by a cycle C is called the length of C. A permutation $P = C_1C_2\ldots C_l$ is called *regular* if all its disjoint cycle factors C_λ have the same length.

(a) Let $\mathfrak{G}$ be a finite group and $\mathfrak{G}^*$ the permutation group isomorphic to $\mathfrak{G}$, obtained according to Cayley's theorem. Then all elements of $\mathfrak{G}^*$ are regular permutations. The group $\mathfrak{G}^*$ is therefore called the *regular representation* of $\mathfrak{G}$.

(b) Show that the identity $I = \begin{pmatrix} x \\ x \end{pmatrix}$ is the only permutation in the regular representation $\mathfrak{G}^*$ which leaves a symbol fixed.

(c) A permutation P of degree n is regular if and only if it is a power of an n-cycle.

11. Let a, b, c, d be four elements of a group $\mathfrak{G}$ which in a group table of $\mathfrak{G}$ appear as vertices of a rectangle:

$$x_i x_h = a, \quad x_i x_k = b$$

$$x_j x_h = c, \quad x_j x_k = d.$$

Show that all the other rectangles in the group table which have a, b, c as vertices will have d as their fourth vertex.

12. Show that the six functions of the complex variable z

$$z \to z, \quad \frac{1}{z}, \quad 1-z, \quad 1-\frac{1}{z}, \quad \frac{1}{1-z}, \quad \frac{z}{z-1}$$

form a group with respect to functional composition. This group is isomorphic to $\mathfrak{S}_3$.

13. Cayley's theorem can be extended to arbitrary groups:

THEOREM. *Every group* $\mathfrak{G} = (\mathbf{G} \mid \cdot)$ *is isomorphic to a group* $\mathfrak{G}^*$ *of invertible mappings of the set* $\mathbf{G}$ *onto itself.*

14. Let $\mathfrak{G}$ be an abelian group and n a fixed natural number.

(a) All elements z which can be written in the form $z = x^n$, $x \in \mathfrak{G}$, form a subgroup $\mathfrak{H}_n$ of $\mathfrak{G}$.

(b) All elements y of $\mathfrak{G}$ which satisfy the condition $y^n = e$, form a subgroup $\mathfrak{L}_n$ of $\mathfrak{G}$.

15. Let $\mathfrak{G}$ be a group. Following K. Menger (1931) we introduce for every $a \in \mathfrak{G}$ the "modulus" $|a|$ as the couple $(a, a^{-1}) = (a^{-1}, a)$. (If $\mathfrak{G}$ is an additive group this becomes $|a| = (a, -a) = (-a, a)$ and has then the typical property $|-a| = |a|$ of the modulus in the additive group of the real numbers.) For two elements $x, y \in \mathfrak{G}$ the modulus $|x \cdot y^{-1}|$ is called the (left) distance of x to y in $\mathfrak{G}$. Prove the following result:

THEOREM. *The distance* $|xy^{-1}|$ *in* $\mathfrak{G}$ *is invariant under the mapping* $x \to x^{-1}$, $y \to y^{-1}$, $x, y \in \mathfrak{G}$, *if and only if either* $\mathfrak{G}$ *is abelian or every pair of non-commuting elements of* $\mathfrak{G}$ *generates a subgroup isomorphic to the quaternion group* (cf. ex. 9).*

§3 Arithmetical groups

A large number of special groups will have to be considered in order to illustrate certain aspects of the general theory. There are, however, examples of groups whose knowledge is necessary for the understanding of some basic facts of finite group theory. These groups have their origin in elementary number theory; they, as well as related groups, will therefore be called "arithmetical groups". Their elements are either numbers as in the case of the additive group $\mathbb{Z}^+$ of the integers (cf. §1, ex. 3) or classes of numbers or of matrices with integral elements.

In subsection *a.* we state without reference some definitions and theorems from elementary number theory in a form in which they are needed in the subsequent development. All those given without proof can be found in every introductory book on number theory.

a. Facts of number theory. As usual we denote by $\mathbb{Z}$ the integral domain of the integers.

* Cf. O. Taussky, Über isomorphe Abbildungen von Gruppen, *Math. Annalen* **108** (1933), 615–620. We say that x, y generate the subgroup of $\mathfrak{G}$ all elements of which appear as products of powers of x and y (cf. Chap. II, §1, *d.*).

I. For every $a, b \in \mathbb{Z}$, $b > 0$, there is a unique pair $q, r \in \mathbb{Z}$, $0 \leqslant r < b$, so that $a = qb + r$. If $r = 0$ we say that $b \mid a$ (b divides a); then also $-b \mid a$.

II. For every $a, b \in \mathbb{Z}$ there is a unique number $d > 0$ with the following properties: $d \mid a$, $d \mid b$, and if $t \mid a$ and $t \mid b$ then $t \mid d$. This number d is called the *greatest common divisor* (g.c.d.) of a and b and denoted by the symbol (a, b). There are integers x, y such that $d = ax + by$ and d is the least positive value which the linear form $ax' + by'$ may assume for integral x', y'. Note that $(a+bc, b) = (a, b)$ for every $c \in \mathbb{Z}$.

If $(a, b) = 1$ the integers a, b are said to be *relatively prime.*

II′. For every $a, b \in \mathbb{Z}$ there is a unique number $v > 0$ which has the following properties: $a \mid v$, $b \mid v$ and if $a \mid u$ and $b \mid u$ then $v \mid u$. This number v is called the *least common multiple* (l.c.m.) of a and b and is denoted by the symbol $[a, b]$. It is related to the g.c.d. $d = (a, b)$ by the equation $dv = ab$.

III. If $m \mid ab$ and $(a, m) = 1$ then $m \mid b$.

IV. Let $a_1, \ldots, a_n \in \mathbb{Z}$. There is a unique positive integer $d = (a_1, \ldots, a_n)$ that satisfies the conditions: $d \mid a_1, \ldots, d \mid a_n$ and if $t \mid a_1, \ldots, t \mid a_n$ then $t \mid d$. The number d represents the least positive value which for integral $x_1, \ldots, x_n$ is assumed by the linear form $a_1x_1 + \ldots + a_nx_n$; it is called the *greatest common divisor* of the n integers $a_1, \ldots, a_n$. As an algebraic operation in $\mathbb{Z}$ the g.c.d. is *associative*:

$$(a_1, \ldots, a_m, b_1, \ldots, b_n) = ((a_1, \ldots, a_m), (b_1, \ldots, b_n)).$$

IV′. There is also a unique positive number $v = [a_1, \ldots, a_n]$ which satisfies the conditions: $a_1 \mid v, \ldots, a_n \mid v$ and if w is a multiple of every a_i ($i = 1, \ldots, n$) then $v \mid w$. This number v is called the *least common multiple* of the numbers $a_1, \ldots, a_n$. As an algebraic operation in $\mathbb{Z}$ the l.c.m. is *associative.* (cf. §1, ex. 5 and Chap. IV, §1.)

V. A positive integer $p > 1$ is called a *prime number* if it has exactly two different positive divisors, namely 1 and p. Thus from $p = ab$ it follows that either a or b equals ± 1. By III we have: If $p \mid ab$ and $p \nmid a$ then $p \mid b$.

VI. Every positive integer $n > 1$ either is a power of a prime number or

$$n = p_1^{r_1} \ldots p_k^{r_k} = \prod_{p \mid n} p^r$$

where $p_1, \ldots, p_k$ are different prime numbers, uniquely defined up to the order in which they occur as factors of n and $r_1, \ldots, r_k$ are uniquely defined positive integers.

VII. If $(a, b) = 1$ then for every fixed positive m in $\mathbb{Z}$ there are infinitely many λ in $\mathbb{Z}$ so that $(a+\lambda b, m) = 1$.

This is an immediate consequence of Dirichlet's famous theorem according to which the arithmetic progression $\{a+\lambda b \mid (a, b) = 1, \lambda \in \mathbb{Z}\}$ contains an infinity of prime numbers. Here is an elementary proof of the statement VII, due to N. Borba:

Let $(a, m) = 1$. It is readily seen that $(a+m^k b, m) = 1$, $(k = 1, 2, \ldots)$. In the case that $(a, m) > 1$ there are two possibilities:

(i) Every prime factor p of m is also a factor of a. Take $\lambda = q^k$ where q is a prime number, $q \nmid m$. Then $(a+\lambda b, m) = 1$. In fact as $p \mid m$ and $p \mid a$ we have $p \neq q$ and $p \nmid b$ whence $p \nmid q^k b$.

(ii) There are prime factors of m which do not divide a. Let n be the product of these prime factors and put $\lambda = n^k$. Let p be a prime factor of m. If $p \mid a$, then $p \nmid a+n^k b$. If $p \nmid a$, so that $p \mid n$, then $p \mid n^k b$ and thus also $p \nmid a+n^k b$.

b. Residue classes modulo m. Let m be an integer > 1. Two integers a, b are said to be *congruent modulo m*, in symbols

$$a \equiv b \pmod{m}$$

if $m \mid (b-a)$, i.e. $b-a = qm$, $q \in \mathbb{Z}$. Congruence mod m is readily seen to be an equivalence relation in $\mathbb{Z}$. The equivalence class containing the number a will be denoted by $\bar{a}$ so that $\bar{a} = \{x = a+sm \mid s \in \mathbb{Z}\}$. There are m disjoint classes, each integer being an element of exactly one class. Each class is defined by every one of its elements, e.g. by the (least non-negative) residue r obtained by dividing its members by m. Hence the classes are called residue classes. The system of the m residue classes will be denoted $\mathbf{R}_m$.

For every $x = a+sm$ $(s \in \mathbb{Z})$ and $y = b+tm$ $(t \in \mathbb{Z})$, i.e. $x \in \bar{a}$ and $y \in \bar{b}$, the sum $x+y$ lies in the class $\overline{a+b}$, and the product $xy \in \overline{ab}$. Hence we can define addition and multiplication of the classes:

$$\bar{a}+\bar{b} = \overline{a+b}, \quad \bar{a}\,\bar{b} = \overline{ab}.$$

With regard to these operations the system $\mathbf{R}_m$ of residue classes $\bar{r} (r = 0, 1, \ldots, m-1)$ forms a commutative ring $\mathbb{R}_m = (\mathbf{R}_m \mid +, \cdot)$; this implies that addition and multiplication of the classes are associative and commutative as well as distributive [cf. §1, ex. 3, (6)], that there is a zero class $\bar{0} = \{qm \mid q \in \mathbb{Z}\}$, and that for every class $\bar{a}$ there is a unique class $\overline{a'}$ such that $\bar{a}+\overline{a'} = \bar{0}$, namely the class $\overline{a'} = -\bar{a} = \{-x \mid x \in \bar{a}\}$. In particular the system $\mathbb{R}_m{}^+ = (\mathbf{R}_m \mid +)$ is an additive abelian group of order m.

The multiplicative system $(\mathbf{R}_m \mid \cdot)$ is a semigroup; it is in general not a group, for if m is not a prime number, $m = ab$ $(a \neq 1, b \neq 1)$ we have $\bar{a}\,\bar{b} = \overline{ab} = \bar{0}$ although $\bar{a} \neq \bar{0}, \bar{b} \neq \bar{0}$. We then shall say that $\bar{a}$ and $\bar{b}$ are zero divisors in $\mathbb{R}_m$.

A more judicious selection of classes from $\mathbb{R}_m$ can, however, lead to a multiplicative group within $\mathbb{R}_m$. Consider in particular those classes $\bar{a}$ in $\mathbb{R}_m$ where the representative a, and therefore every member $x = a+sm$ in $\bar{a}$, is relatively prime to m. These classes form an *abelian multiplicative group.* Indeed if $(a, m) = 1$ and $(b, m) = 1$ then also $(ab, m) = 1$. Thus the product $\bar{a}\cdot\bar{b} = \overline{ab}$ of two classes relatively prime to m is a class relatively prime to m. The unit element of $\mathbb{R}_m$, i.e. $\bar{1} = \{1+sm \mid s \in \mathbb{Z}\}$, obviously has this property. Finally for every $\bar{a}$, $(a, m) = 1$, there is a unique inverse $\bar{x} = \bar{a}^{-1}$, $(x, m) = 1$. Indeed the equation $\bar{a}\cdot\bar{x} = \bar{1}$ is equivalent to the condition that there exist x and s in $\mathbb{Z}$ such that $ax = 1+sm$, i.e. $ax-sm = 1$ or equivalently $(a, m) = 1$ which was assumed to be satisfied. Thus integers x and s which satisfy the condition do exist. As to the uniqueness of $\bar{x}$, suppose that $\bar{a}\cdot\bar{y} = \bar{1}$, i.e. $ay = 1+tm$ $(t \in \mathbb{Z})$. Then $a(y-x) = (t-s)m$. Since $(a, m) = 1$ we conclude (cf. *a.*, III) that $m \mid (y-x)$ whence $\bar{y} = \bar{x}$.

The multiplicative group of residue classes (mod m) which are relatively prime to m will be denoted by $\mathfrak{R}_m$.

If $m = p$ is a prime number, the condition $(a, m) = 1$ can be expressed in the form $p \nmid a$. Thus the ring $\mathbb{R}_p$ has no divisors of zero and therefore is a field: Its multiplicative group $\mathfrak{R}_p$ consists of the $p-1$ classes $\bar{a} \neq \bar{0}$.

The order of the group $\mathfrak{R}_m$ is customarily denoted by $\varphi(m)$ and φ is called Euler's (totient) function; it represents the number of integral residues r, $1 \leqslant r < m$ for which $(r, m) = 1$. We shall determine $\varphi(m)$ in terms of the prime factors of m.

If p is a prime number it is readily seen that $\varphi(p) = p-1$, $\varphi(p^2) = p^2-p$, more generally

(1) $\varphi(p^k) = p^k - p^{k-1} = p^k(1-p^{-1}).$

In order to settle the general case we shall now prove:

THEOREM 1. *Euler's function* φ *is multiplicative, that is*

(2) $\varphi(m \cdot n) = \varphi(m) \cdot \varphi(n)$ *if* $(m, n) = 1$.

Proof. By definition $\varphi(mn)$ represents the number of integers between 1 and mn which are relatively prime with mn, i.e. relatively prime with m and with n. All the integers from 1 to mn can be written in the form

$$rn+s,\ r = 0, 1, \ldots, m-1;\ s = 1, \ldots, n.$$

For fixed s there are m such integers which all have with n the same g.c.d.: $(rn+s, n) = (s, n)$. Thus for every r there are exactly $\varphi(n)$ such integers relatively prime with n.

Now consider the m integers $rn+s$ $(r = 0, \ldots, m-1)$ for which $(s, n) = 1$. On division by m they all yield distinct remainders (mod m). Indeed if two yield the same remainder u:

$$rn+s = qm+u, \quad r'n+s = q'm+u,$$

it follows that $(r-r')n = (q-q')m$; since $(m, n) = 1$ we conclude that $m \mid (r-r')$ whence $r' = r$. Thus the m integers $rn+s$ (s fixed) include exactly $\varphi(m)$ numbers relatively prime with m. This is so for each of the $\varphi(n)$ values of s for which $(s, n) = 1$. Thus (2) is proved.

If now $m = \prod_{p \mid m} p^k$ is the factorization of m in prime factors it follows that

(2′) $$\varphi(m) = \prod_{p \mid m} \varphi(p^k) = m \cdot \prod_{p \mid m} (1-p^{-1}).$$

We note an important property of Euler's function which is often applied in finite group theory: Let $d_1, d_2, \ldots, d_t = m$ be all the distinct divisors of m; assuming $\varphi(1) = 1$ one has the relation

(3) $$\varphi(d_1)+\varphi(d_2)+\ldots+\varphi(d_t) = m$$

which is frequently abbreviated into

(3′) $$\sum_{d \mid m} \varphi(d) = m.$$

To prove this let d be one of the divisors of m and $\mathbf{M}_d$ be the set of all those integers j, $1 \leqslant j \leqslant m$, for which $(j, m) = d$. Then $|\mathbf{M}_d| = \varphi(m/d)$. Indeed there are $\varphi(m/d)$ values of j for which $(j/d, m/d) = 1$. Further, if d and d' are two different divisors of m, then the two sets $\mathbf{M}_d$ and $\mathbf{M}_{d'}$ are disjoint. Since every j ($j = 1, \ldots, m$) is an element of $\mathbf{M}_d$ for a certain divisor d of m, all the $\mathbf{M}_d$ cover the set of the integers from 1 to m. Therefore $\sum_{d|m} \varphi(m/d) = m$. Since m/d runs through all divisors of m if d does, the last equation is actually the same as (3′).

c. We shall now generalize the preceding discussion. Let $\mathbb{R}$ be a possibly non-commutative ring with a (multiplicative) identity element e. An element u of $\mathbb{R}$ is called *a unit* of $\mathbb{R}$ if in $\mathbb{R}$ there is an inverse of u, that is an element $u' = u^{-1}$ such that $uu' = u'u = e$. The inverse u' then is unique; indeed, suppose that $uu'' = e$, then $u'uu'' = u'$, i.e. $eu'' = u'' = u'$.

All the units of $\mathbb{R}$ form a multiplicative group $\mathfrak{R}$. For, if u and v are units in $\mathbb{R}$ then so is the product uv; it is readily verified that the inverse of a unit is a unit and that $(uv)^{-1} = v^{-1}u^{-1}$. Obviously the identity element is a unit.

The multiplicative residue class group $\mathfrak{R}_m$ is precisely the group of the units of the residue class ring $\mathbb{R}_m$. In a field all elements except zero are units.

In a ring $\mathbb{R}$ with non-commutative multiplication one has *a priori* to distinguish between left units and right units. Let us say that an element u of $\mathbb{R}$ is a left unit if it has a left inverse u' in $\mathbb{R}$, i.e. $u'u = e$. It is a right unit if it has a right inverse u'', i.e. $uu'' = e$. The existence of a right unit does not imply the existence of a left unit. We shall be concerned only with rings in which we can verify that every right unit is a left unit and vice versa.

d. Matrix residue class groups (mod m). By $\mathbf{R}^{(n)}$ we denote the system of all $n \times n$ matrices

$$A = (a_{ij}) = \begin{pmatrix} a_{11} & a_{12} \ldots & a_{1n} \\ a_{21} & a_{22} \ldots & a_{2n} \\ \vdots & \vdots & \vdots \\ a_{n1} & a_{n2} & a_{nn} \end{pmatrix}, \quad B = (b_{ij}), P, Q, X, Y, \ldots$$

whose elements a_{ij}, b_{ij}, ... are ordinary integers; these matrices will be called "integral matrices". With respect to matrix addition and matrix multiplication $\mathbf{R}^{(n)}$ is a ring $\mathbb{R}^{(n)}$, non-commutative if $n \geqslant 2$. The ring $\mathbb{R}^{(1)}$ coincides with the integral domain $\mathbb{Z}$.

Let m be a fixed integer, $m \geqslant 2$. Two matrices A, B are said to be congruent modulo m, or $A \equiv B \pmod{m}$ if $m \mid (b_{ij} - a_{ij})$, for all $i, j = 1, \ldots, n$. This congruence is an equivalence relation in $\mathbb{R}^{(n)}$. The equivalence class containing the matrix A will be denoted by $\bar{A}$. Thus

$$\bar{A} = \{A + mP \mid P \in \mathbb{R}^{(n)}\}.$$

The class $\bar{A}$ may be considered as a matrix $(\overline{a_{ij}})$ whose elements $\overline{a_{ij}}$ are residue classes (mod m) in $\mathbb{Z}$, i.e. elements of the ring $\mathbb{R}_m$. Two classes, $\bar{A}$ and $\bar{B} = \{B + mQ \mid Q \in \mathbb{R}^{(n)}\}$, coincide if they have a common element, otherwise they are disjoint. Every integral matrix A is an element of some class: $A \in \bar{A}$.

For all matrices X in the class $\bar{A}$, $X = A + mP$, the g.c.d. $(\det X, m) = (\det A, m)$, independent of P. Indeed $\det X = \det (A + mP) = \det A + m \cdot \pi(m)$ where $\pi(m)$ is a polynomial in m having integral coefficients depending on A and P. Hence $\pi(m)$ is an integer.

The set of all classes (or class matrices) $\bar{A}$ is a ring $\mathbb{R}_m{}^{(n)}$ with respect to matrix addition and multiplication. Indeed let $X = A + mP \in \bar{A}$, $Y = B + mQ \in \bar{B}$. Then $X + Y = A + B + m(P + Q)$ which for all P and Q in $\mathbb{R}^{(n)}$ is a matrix in the class $\overline{A+B}$; hence we may define $\bar{A} + \bar{B} = \overline{A+B}$. Likewise $XY = AB + m(AQ + PB + mPQ) \in \overline{AB}$ which justifies the definition $\bar{A} \cdot \bar{B} = \overline{AB}$. The ring has as unit element the class $\bar{E} = \{E + mP \mid P \in \mathbb{R}^{(n)}\}$ where E denotes the $n \times n$-unit matrix, and as zero element the class $\bar{0} = \{mQ \mid Q \in \mathbb{R}^{(n)}\}$.

THEOREM 2. *All the left (right) units of $R_m{}^{(n)}$ are right (left) units and therefore units. An element $\bar{A}$ of $R_m{}^{(n)}$ is a unit if and only if the g.c.d.* $(\det A, m) = 1$.

Proof. If $\bar{A}$ is a left unit in $\mathbb{R}_m{}^{(n)}$ then there is a class $\bar{B}$ in $\mathbb{R}_m{}^{(n)}$ which satisfies the condition $\bar{B} \cdot \bar{A} = \bar{E}$. Thus for a representative A of $\bar{A}$ an integral matrix B can be found so that $B \cdot A = E + mS$, $S \in \mathbb{R}^{(n)}$, and the determinant $\det BA = 1 + m \cdot \sigma(m)$ where $\sigma(m)$ is a polynomial in m which has integral coefficients. Consequently $\sigma(m) \in \mathbb{Z}$. Therefore the g.c.d.

$$(\det BA, m) = 1.$$

Since det BA = det $B\cdot$det A we conclude also that (det A, m) = 1. Let det $A = d$. Since $\bar{d} \in \mathfrak{R}_m$ (cf. *b*.) there is an inverse $\overline{d'}$ of $\bar{d}$ and therefore an integer d' so that $d'd \equiv 1 \pmod{m}$. Clearly the argument is unchanged if $\bar{A}$ has a right inverse in $\mathfrak{R}_m$. The condition $(d, m) = 1$ is therefore necessary for the existence of a left inverse and of a right inverse of $\bar{A}$ in $\mathfrak{R}_m$.

We shall now show that it is also sufficient and that both inverses are equal. Recall the explicit expression for the inverse of a matrix A: Let $\hat{A}$ denote the transpose of the matrix of the algebraic complements* of the a_{ij} in A; then $A^{-1} = \hat{A}/d$. Since the elements of the matrix $\hat{A}$ are $(n-1)\times(n-1)$ subdeterminants of A it follows that $\hat{A} \in \mathbb{R}^{(n)}$ and therefore also

$$B = d'\cdot\hat{A} \in \mathbb{R}^{(n)}.$$

The class $\bar{B}$ then is the inverse of $\bar{A}$: Since $A\hat{A} = \hat{A}A = d\cdot E$ we have $B\cdot A = A\cdot B = d'd\cdot E \equiv E(\text{mod } m)$.

The group of units $\bar{A}$ in $\mathbb{R}_m{}^{(n)}$ will be denoted by $\mathfrak{R}_m{}^{(n)}$ (cf. *c*.). The order of the ring $\mathbb{R}_m{}^{(n)}$ is equal to $m^{(n^2)}$ An explicit expression for the order $|\mathfrak{R}_m{}^{(n)}|$ will be established in ex. 9.

e. We shall now determine the order of the group $\mathfrak{R}_m{}^{(2)}$. Let

$$A = \begin{pmatrix} a_1 & b_1 \\ a_2 & b_2 \end{pmatrix}$$

be a representative of one of its elements, $\bar{A}$; then

(4) $(\det A, m) = (a_1b_2 - a_2b_1, m) = 1.$

First we determine the number $\varphi_2(m)$ of possible choices a_1, a_2 in the range from 0 to $m-1$ for the first column of the matrix A. Clearly if $(a_1, a_2, m) > 1$ then no b_1, b_2 exist which satisfy the condition (4). Thus $\varphi_2(m)$ is the number of couples a_1, a_2 $(0 \leqslant a_i \leqslant m-1;\ i = 1, 2)$, i.e. first columns of the matrix A, for which

(4′) $(a_1, a_2, m) = 1.$

* The matrix Â is also called the adjugate of A.

In order to determine $\varphi_2(m)$ we proceed as in the case of Euler's function (cf. *b.*). In the case $m = p$, a prime number, only one of the p^2 couples a_1, a_2 $(0 \leqslant a_1, a_2 < p)$ does not satisfy the condition (4'); thus $\varphi_2(p) = p^2-1$. In the case $m = p^2$ the p^2 couples ip, jp $(i, j = 0, 1, \ldots, p-1)$ do not satisfy the condition (4'); thus $\varphi_2(p^2) = p^4-p^2$. Similarly we find

$$(5) \qquad \varphi_2(p^h) = p^{2h}-p^{2h-2} = p^{2h}(1-p^{-2}).$$

Now we shall prove that $\varphi_2(m)$ is multiplicative, that is

$$(5') \qquad \varphi_2(m_1m_2) = \varphi_2(m_1)\varphi_2(m_2) \quad \text{if} \quad (m_1, m_2) = 1.$$

Proof. Let us write the ${m_1}^2{m_2}^2$ couples a_1, a_2 in the form

$$a_1 = r_1m_2+s_1, \quad a_2 = r_2m_2+s_2,$$
$$(r_1, r_2 = 0, 1, \ldots, m_1-1, \quad s_1, s_2 = 0, 1, \ldots, m_2-1).$$

For fixed s_1, s_2 there are thus ${m_1}^2$ couples a_1, a_2 and for these

$$(a_1, a_2, m_2) = (s_1, s_2, m_2).$$

For every r_1, r_2 there are therefore $\varphi_2(m_2)$ couples a_1, a_2 such that $(a_1, a_2, m_2) = 1$.

Now let r_1, r_2 run through their range from 0 to m_1-1 and consider the set of a_1, a_2 for which $(s_1, s_2, m_2) = 1$. Let u_1, u_2 be the least non-negative remainders arising when a_1, a_2 respectively are divided by m_1:

$$a_1 = q_1m_1+u_1, \quad a_2 = q_2m_1+u_2.$$

If $a_1' = r_1'm_2+s_1$, $a_2' = r_2'm_2+s_2$ is another couple of this set which yields the same remainders (mod m_1):

$$a_1' = q_1'm_1+u_1, \quad a_2' = q_2'm_1+u_2,$$

we conclude that

$$a_1-a_1' = (r_1-r_1')\cdot m_2 = (q_1-q_1')\cdot m_1$$
$$a_2-a_2' = (r_2-r_2')\cdot m_2 = (q_2-q_2')\cdot m_1.$$

so that $m_1|(r_1-r_1')$, $m_1|(r_2-r_2')$. Hence $r_1' = r_1$, $r_2' = r_2$ and $a_1' = a_1$, $a_2' = a_2$. Thus the m_1 couples a_1, a_2 (with fixed s_1, s_2) are all distinct and include exactly $\varphi_2(m_1)$ couples relatively prime with m_1. This is so for each of the $\varphi_2(m_2)$ couples s_1, s_2 for which $(s_1, s_2, m_2) = 1$. Thus there are $\varphi_2(m_1)\varphi_2(m_2)$ couples a_1, a_2 which are relatively prime with m_1m_2 and the multiplicativity relation (5′) is proved.

An explicit expression for $\varphi_2(m)$ in terms of the prime factors of m now can be written down:

$$(5'') \qquad \varphi_2(m) = \prod_{p|m} \varphi_2(p^k) = m^2 \prod_{p|m} (1-p^{-2}) \quad \text{[cf. (2′)]}.$$

Once a first column of A has been chosen in one of the $\varphi_2(m)$ possible ways, how many second columns exist for A to satisfy the condition (4)?

Let $a_1 = \delta\cdot a_1'$, $a_2 = \delta\cdot a_2'$ so that $(a_1', a_2') = 1$ and $\det A = \delta(a_1'b_2-a_2'b_1)$. Then by (4) $(a_1'b_2-a_2'b_1, m) = 1$. Thus b_1, b_2 must be a solution of an equation

$$a_1'y_2-a_2'y_1 = s, \quad (s, m) = 1.$$

We shall show that for each of the $\varphi(m)$ possible values of s there are m solutions y_1, y_2. Indeed, every solution is obtained from one of them by adding to it a solution z_1, z_2 of the corresponding homogeneous equation: $a_1'z_2-a_2'z_1 = 0$ so that

$$z_1 = \lambda a_1', \quad z_2 = \lambda a_2', \quad \lambda = 0, 1, \ldots, m-1.$$

These m solutions are incongruent (mod m). In fact

$$\mu a_1' \equiv \lambda a_1', \quad \mu a_2' \equiv \lambda a_2' \pmod{m}, \quad |\mu-\lambda| < m$$

implies that

$$m|(\mu-\lambda)a_1', \quad m|(\mu-\lambda)a_2'.$$

Because $(a_1', a_2') = 1$ it follows that $\mu-\lambda = 0$. Hence for every s, $(s, m) = 1$, there are m different choices for b_1, b_2. Altogether there are $\bar{m}\cdot\varphi(m)$ second columns of A if a first column has been chosen. Therefore

$$(6) \qquad \left|\mathfrak{R}_m^{(2)}\right| = \psi_2(m) = m\cdot\varphi(m)\cdot\varphi_2(m).$$

Examples and exercises

1. Let $\mathbb{F}$ be a field. Show that the mappings $\mathbb{F} \to \mathbb{F}$:

$$x \to ax+\alpha, \quad a, \alpha \in F, \quad a \neq 0$$

form a group $\mathfrak{G}_{\mathbb{F}}$. In the case of a finite field of the order $|\mathbb{F}| = q$ determine the order $|\mathfrak{G}_{\mathbb{F}}|$.

2. (a) If $\mathbb{R}$ is a ring show that a unit of $\mathbb{R}$ cannot be a zero divisor.
(b) If $b \in \mathbb{R}_m$, $b \notin \mathfrak{R}_m$ show that b is a zero divisor in $\mathbb{R}_m$.

3. Find all multiplicative groups contained in $\mathbb{R}_{15}$. (Notice that there are groups in $\mathbb{R}_{15}$ whose elements are divisors of zero.) Generalize!

4. Let $\mathbb{R}$ be a ring and $\mathfrak{R}$ its unit group (cf. *c.*). Show that all the mappings $\mathbb{R} \to \mathbb{R}$:

$$x \to ax+\alpha, \quad a \in \mathfrak{R}, \quad \alpha \in \mathbb{R}$$

form a group. Determine its order for $\mathbb{R} = \mathbb{R}_m$.

5. In generalization of Theorem 1 show that if $(m, n) = d$ then

$$\varphi(mn) = \frac{d}{\varphi(d)}\,\varphi(m)\cdot\varphi(n).$$

6. Let Γ denote the system of the Gauss integers, that is the complex numbers

$$\alpha = a_1+a_2 i, \quad a_1, a_2 \in \mathbf{Z}, \quad i = \sqrt{(-1)}.$$

(a) Prove that, with respect to the ordinary addition and multiplication of the complex numbers, the system Γ is an integral domain.

(b) Determine the unit group of Γ.

7. Two Gauss integers $\alpha = a_1 + a_2 i$, $\beta = b_1 + b_2 i$ are said to be congruent (mod m) (m is an integer $\geqslant 2$), in symbols

$$\alpha \equiv \beta \pmod{m} \quad \text{if} \quad a_1 \equiv b_1, \quad a_2 \equiv b_2 \pmod{m}.$$

Congruence modulo m is an equivalence relation in Γ and

$$\bar{\alpha} = \{\alpha + m\xi \mid \xi \in \Gamma\}$$

is the congruence class (mod m) which contains the number α. Let $\mathbb{C}_m$ denote the system of all congruence classes $\bar{\alpha}$ ($\alpha \in \Gamma$) modulo m.

(a) Justify the definitions of addition and multiplication in $\mathbb{C}_m$, namely

$$\bar{\alpha} + \bar{\beta} = \overline{\alpha + \beta}, \quad \bar{\alpha} \cdot \bar{\beta} = \overline{\alpha\beta}$$

and show that with respect to these operations $\mathbb{C}_m$ is a ring.

(b) The class $\bar{\alpha}$ is a unit in $\mathbb{C}_m$ if there is a class $\bar{\beta}$ in $\mathbb{C}_m$ which satisfies the condition $\bar{\alpha}\,\bar{\beta} = \bar{1}$. Show that $\bar{\alpha}$ is a unit in $\mathbb{C}_m$ if and only if $(a_1{}^2 + a_2{}^2, m) = 1$.

(c) If $\bar{\alpha}$ is a unit in $\mathbb{C}_m$ then $(a_1, a_2, m) = 1$. Thus if $\mathfrak{C}_m$ represents the group of the units of $\mathbb{C}_m$ one has for the order of $\mathfrak{C}_m$ the inequality $|\mathfrak{C}_m| \leqslant \varphi_2(m)$.

Remark. The order $|\mathfrak{C}_m|$ may be considered as a generalized Euler function. An explicit expression for $|\mathfrak{C}_m|$ can be derived in terms of the prime factors of m in Γ. These do not in general coincide with the prime factors of m in $\mathbb{Z}$ as every prime number p in $\mathbb{Z}$ which equals the sum of two squares: $p = p_1{}^2 + p_2{}^2 = (p_1 + p_2 i)(p_1 - p_2 i)$ is evidently not a prime in Γ. For this side of the theory of the Gaussian integers reference may be made to Dirichlet–Dedekind's "Vorlesungen über Zahlentheorie" Supplement XI, §159. (4th edition, Braunschweig 1894.)

8. For small values of m the order of $\mathfrak{C}_m$ can be determined as follows: Set up a square table of the classes $\bar{\alpha}$ (mod m) represented by the m^2 couples a_1, a_2 ($0 \leqslant a_1, a_2 < m$), and a table of the corresponding values $a_1{}^2 + a_2{}^2$. The elements of $\mathfrak{C}_m$ are those $\bar{\alpha}$ for which $(a_1{}^2 + a_2{}^2, m) = 1$ [cf. ex. 7(b)].

If e.g. $m = 4$:

0,0	*1,0*	2,0	*3,0*	0	1	4	9
0,1	1,1	*2,1*	3,1	1	2	5	10
0,2	*1,2*	2,2	*3,2*	4	5	8	13
0,3	1,3	*2,3*	3,3	9	10	13	18

The couples representing elements of the group $\mathfrak{C}_4$ are italicized.

In this way the following results can be established:

$$|\mathfrak{C}_2| = 2, \quad |\mathfrak{C}_3| = 8, \quad |\mathfrak{C}_4| = 8, \quad |\mathfrak{C}_5| = 16,$$
$$|\mathfrak{C}_6| = 16, \quad |\mathfrak{C}_7| = 48, \quad |\mathfrak{C}_8| = 64, \quad |\mathfrak{C}_9| = 72.$$

Verify that the groups $\mathfrak{C}_3$ and $\mathfrak{C}_7$ are cyclic and that the rings $\mathbb{C}_3$ and $\mathbb{C}_7$ are fields with 3^2 and 7^2 elements respectively.

Remark. If $m = p$ is a prime number and $p \equiv 3 \pmod 4$, e.g. $p = 3, 7$, it can be shown that p is a prime number also in Γ. It is then found that $\mathbb{C}_p$ is a field and $\mathfrak{C}_p = \mathbb{C}_p \setminus \{0\}$ is its multiplication group which is cyclic (cf. Chap. II, §2, ex. 10). Therefore $|\mathfrak{C}_p| = p^2 - 1 = \varphi_2(p)$.

9. We shall extend the discussion of subsection *e.* with the aim to determine the order $\psi_n(m)$ of the group $\mathfrak{R}_m^{(n)}$ for $n \geqslant 2$. Clearly $\psi_1(m) = \varphi(m)$.

(a) The necessary and sufficient condition for the elements $a_i = a_{i1}$ of the first column of an integral matrix A to be representative of a unit class $\bar{A}$ in $\mathbb{R}_m^{(n)}$ is that the g.c.d. $(a_1, \ldots, a_n, m) = 1$.

(b) There are

$$\text{(7)} \qquad \varphi_n(m) = m^n \prod_{p|m} (1 - p^{-n})$$

possible first columns for a unit class $\bar{A}$ in $\mathbb{R}_m^{(n)}$.

$$\text{(c)} \qquad |\mathfrak{R}_m^{(n)}| = \psi_n(m) = m^{\frac{1}{2}n(n-1)} \varphi(m) \varphi_2(m) \ldots \varphi_n(m).$$

We shall indicate a proof for this formula.

Lemma. *Let the g.c.d.* $(a_1, \ldots, a_n) = \delta$. *For every natural number* m *there are* n *integers* $x_1, \ldots x_n$ *so that*

$$a_1x_1 + \ldots + a_nx_n = \delta \quad \textit{and} \quad (x_1, m) = 1.$$

Proof. Let first $n = 2$, $(a_1, a_2) = \delta$, $a_1 = a_1' \cdot \delta$, $a_2 = a_2' \cdot \delta$. Suppose that a pair of integers y_1, y_2 has been found so that $a_1y_1 + a_2y_2 = \delta$ and therefore

$$(8) \qquad a_1' \cdot y_1 + a_2' \cdot y_2 = 1.$$

Every solution of this equation appears in the form $y_1 + z_1$, $y_2 + z_2$ if $a_1'z_1 + a_2'z_2 = 0$, i.e. $z_1 = \lambda a_2'$, $z_2 = -\lambda a_1'$, $(\lambda = 0, \pm 1, \ldots)$. Since by (8) the g.c.d. $(y_1, a_2') = 1$, Theorem VII of subsection *a.* insures that an integer λ can be found so that if $x_1 = y_1 + \lambda a_2'$ we have $(x_1, m) = 1$.

If $n > 2$ let $a_\nu = \delta a_\nu'$ $(\nu = 1, \ldots, n)$; if $a_1y_1 + \ldots + a_ny_n = \delta$ we have

$$(8') \qquad a_1'y_1 + \ldots + a_n'y_n = 1.$$

Now let the g.c.d. $(a_2', \ldots, a_n') = \delta_1 = a_2'u_2 + \ldots + a_n'u_n$, $u_2, \ldots, u_n \in \mathbb{Z}$. Then $a_2'y_2 + \ldots + a_n'y_n = u \cdot \delta_1$, $u \in \mathbb{Z}$, and by equation (8′) we have $a_1'y_1 + u\delta_1 = 1$; again by Theorem VII there is an integral λ so that if $x_1 = y_1 + \lambda\delta_1$, the g.c.d. $(x_1, m) = 1$. Thus the lemma is proved.

We shall now show that there are

$$\psi_n(m) = m^{n-1}\psi_{n-1}(m) \cdot \varphi_n(m)$$

modulo m incongruent integral matrices

$$A = \begin{pmatrix} a_1 & a_{12} & \ldots & a_{1n} \\ \vdots & \vdots & & \vdots \\ a_n & a_{n2} & \ldots & a_{nn} \end{pmatrix}$$

for which $(\det A, m) = 1$, i.e. $\bar{A} \in \mathfrak{R}_m^{(n)}$.

Suppose that the first column of A has been chosen in one of the $\varphi_n(m)$ possible ways (cf. *b.*) and that $(a_1, \ldots, a_n) = \delta$, $(\delta, m) = 1$. Now let

$$X = \begin{pmatrix} x_1 & x_2 & \ldots & x_n \\ 0 & 1 & \ldots & 0 \\ \vdots & \vdots & & \vdots \\ 0 & 0 & \ldots & 1 \end{pmatrix}, \quad \det X = x_1,$$

assuming that $x_1, \ldots, x_n$ have been selected in $\mathbb{Z}$ according to the lemma: $a_1x_1 + \ldots + a_nx_n = \delta$, $(x_1, m) = 1$ so that the class $\bar{X} \in \mathfrak{R}_m^{(n)}$. We form the matrix product

$$XA = \begin{pmatrix} \delta & b_2 & \ldots & b_n \\ a_2 & a_{22} & \ldots & a_{2n} \\ \vdots & \vdots & & \vdots \\ a_n & a_{n2} & \ldots & a_{nn} \end{pmatrix}.$$

Further let

$$T = \begin{pmatrix} 1 & 0 & \ldots & 0 \\ -a_2' & 1 & \ldots & 0 \\ \vdots & \vdots & & \vdots \\ -a_n' & 0 & \ldots & 1 \end{pmatrix}, \quad \bar{T} \in \mathfrak{R}_m^{(n)}.$$

Then since $a_\nu'\delta = a_\nu$,

$$TXA = \begin{pmatrix} \delta & b_2 \ldots b_n \\ 0 & \\ \vdots & B_1 \\ 0 & \end{pmatrix}$$

with an $(n-1)\times(n-1)$ integral matrix B_1 and $TXA \in \mathfrak{R}_m^{(n)}$. Since $\det TXA = \det B_1 \cdot \delta$ we have $(\det B_1, m) = 1$. Thus, once one of the $\varphi_n(m)$ first columns has been chosen, one has $\psi_{n-1}(m)$ different possibilities for the matrix B_1 and m^{n-1} choices for the $(n-1)$-tuple $b_2, \ldots b_n$. This completes the proof.

§4 Geometrical groups

Groups having their origin in some branch of geometry may be called "geometrical groups". There exists a large variety of such groups, some of which are connected with more or less concealed geometrical facts. Those to be discussed in this section can be introduced without elaborate knowledge of geometric theories. The elements of linear algebra will be assumed.

Most of the groups in question have as elements non-singular linear homogeneous transformations of a finite dimensional space onto itself; therefore they can be represented by matrices. They are subgroups of the "full linear group" of all non-singular linear transformations, characterized by the restriction that their elements map a certain geometrical figure onto itself, moving or otherwise transforming it in such a way that in its new form it cannot be distinguished from the original one. To avoid triviality this condition requires that the figure possesses a kind of *symmetry*, a commonly accepted idea which eludes an immediate general definition. Its exact definition in every special case depends on the existence of a geometrical group whose elements are operations in space which leave the figure invariant, i.e. in appearance unchanged.* These geometrical groups are therefore often called *symmetry groups* of a figure.

In our discussion we shall concentrate on certain geometrical groups whose knowledge should help the reader to visualize the abstract concepts and operations of group theory. Appeal will be made to the reader's geometrical and physical intuition and not every statement will be strictly proved. Moreover routine calculations leading to algebraic representations of the groups will be only briefly indicated and details will be left to the reader. The contents of this section, although helpful, are not essential for the formal understanding of the later development.

a. Rotations and reflexions. As space of operation we choose the euclidean plane, oriented by a cartesian x_1, x_2 coordinate system with the origin O. Of all figures in the plane the most perfectly symmetric one is the circle. We shall be interested in the system of all linear homogeneous transformations of the plane onto itself which map the unit circle ${x_1}^2+{x_2}^2 = 1$ onto itself. This system is a group

* Cf. H. Weyl, *Symmetry*, Princeton University Press, 1952.

with respect to functional composition, i.e. the product of two such transformations, the identity, and the inverse of each belong to the system. The group will be denoted by $\mathfrak{O}_2$.

THEOREM 1. *The group $\mathfrak{O}_2$ consists of all rotations ρ of the plane about the fixed point O, the centre of the circle, and the reflexions with respect to an arbitrary straight line passing through O. All the rotations form a subgroup $\mathfrak{D}_2$ of $\mathfrak{O}_2$. The product of two reflexions is a rotation.*

Proof. Let A be the 2×2 real matrix of a transformation in $\mathfrak{O}_2$, i.e. a transformation which leaves the equation $x_1{}^2+x_2{}^2 = 1$ invariant. This condition means that the matrix A is orthogonal, that is, if A' denotes the transpose of A, we have

$$A'\cdot A = E = \begin{pmatrix}1 & 0\\ 0 & 1\end{pmatrix}.$$

There are the following two possibilities:

(i) $\det A = +1$. In this case it is readily established that

$$A = A^+ = \begin{pmatrix}\cos\alpha & -\sin\alpha\\ \sin\alpha & \cos\alpha\end{pmatrix} = A^+(\alpha)$$

so that the corresponding mapping ρ is a *rotation* with the angle α about O. The matrix A^+ is said to be *proper orthogonal.*

(ii) $\det A = -1$. The matrix A then is improper orthogonal:

$$A = A^- = \begin{pmatrix}\cos\alpha & \sin\alpha\\ \sin\alpha & -\cos\alpha\end{pmatrix} = A^-(\alpha)$$

and the corresponding transformation represents a reflexion of the plane with respect to the (pointwise fixed) straight line

$$L_\alpha\colon x_1 \sin(\alpha/2) - x_2\cos(\alpha/2) = 0$$

which passes through O and makes with the x_1-axis the angle $\alpha/2$.

It is readily established that the set of all matrices $A^+(\alpha)$ is a group with respect to matrix multiplication. This group is isomorphic to the group $\mathfrak{D}_2$. The set consisting of all the matrices

$$A^+(\alpha) \quad \text{and} \quad A^-(\beta) = \begin{pmatrix} \cos\beta & \sin\beta \\ \sin\beta & -\cos\beta \end{pmatrix}$$

is a group which is isomorphic to $\bar{\mathfrak{D}}_2$. The set of all the matrices $A^-(\alpha)$ is not a group, for a reflexion, if repeated (iterated) yields the identity mapping: $(A^-)^2 = E$, and the product of two reflexions $A^-(\alpha)A^-(\beta) = A^+(\alpha-\beta)$ is a rotation. On the other hand $A^+(\alpha)A^-(\beta) = A^-(\alpha+\beta)$ and $A^-(\beta)A^+(\alpha) = A^-(\beta-\alpha)$. Together these identities demonstrate the group properties and show that *the complete orthogonal group* $\bar{\mathfrak{D}}_2$ *is non-abelian*, in contrast to *the rotation group* $\mathfrak{D}_2$ which *is abelian*.

Between the set of all rotations and the set of all reflexions a one–one correspondence can be established in infinitely many ways; for instance, let σ_0 be that fixed reflexion which has as its matrix

$$A^-(0) = \begin{pmatrix} 1 & 0 \\ 0 & -1 \end{pmatrix}.$$

If ρ runs through all elements of $\mathfrak{D}_2$ then $\rho\sigma_0$ runs through all reflexions.

Finally it may be pointed out that a reflexion $A^-(\alpha)$ in the plane can be realized geometrically by a turn-over movement of the plane, that is, a space rotation through an angle of 180° having the line L_α as axis. If the plane is the x_1x_2-plane in a cartesian $x_1x_2x_3$-system, then this space rotation is represented by the proper orthogonal 3×3-matrix

$$\begin{pmatrix} \cos\alpha & \sin\alpha & 0 \\ \sin\alpha & -\cos\alpha & 0 \\ 0 & 0 & -1 \end{pmatrix}.$$

In order to imagine the process physically it may be preferable to restrict attention to the circle, considered as a two-faced disk, rather than trying to visualize the entire plane.

b. The dihedral groups. For the circular disk let us now substitute a disk Δ_n in the shape of a convex regular n-gon, i.e. a polygon with n equal edges, inscribed in the unit circle about O. The vertices of this "dihedron" Δ_n, in their natural order anticlockwise on the circle, will be denoted by the symbols $\hat{0}, \hat{1}, \hat{2}, \ldots, \widehat{n-1}$. To fix the position of Δ_n in the coordinate system let the vertex $\hat{0}$ have the coordinates $x_1 = 1, x_2 = 0$. The group of all rotations and reflexions by which Δ_n is brought into a position indistinguishable from its initial position, will be called the *dihedral group* and denoted by $\mathfrak{D}_n$. The elements of $\mathfrak{D}_n$ are called the covering transformations of Δ_n.

Let ρ denote the rotation of this disk about the origin O through the angle $2\pi/n$. It can be represented algebraically by the matrix $A = A^+(2\pi/n)$ and also by the cycle permutation $(\hat{0}\,\hat{1}\,\hat{2}\ldots\widehat{n-1})$. Like ρ, the powers (iterations) of ρ are also covering transformations of Δ_n and together they form the cyclic group

$$\langle\rho\rangle_n = \{\rho^0 = \epsilon, \rho, \rho^2, \ldots, \rho^{n-1}\}, \quad \rho^n = \epsilon \quad \text{(identity)}.$$

The cyclic subgroup $\langle A\rangle_n$ of $\mathfrak{O}_2$ as well as the cyclic subgroup $\langle(\hat{0}\,\hat{1}\,\hat{2}\ldots\widehat{n-1})\rangle_n < \mathfrak{S}_n$ are both isomorphic to $\langle\rho\rangle_n$.

The polygonal disk Δ_n has n symmetry axes. If n is even, then there are two types of axes: $n/2$ axes passing through opposite vertices and $n/2$ axes passing through the midpoints of opposite edges. If n is odd, then each edge has an opposite vertex and an axis passes through the midpoint of an edge and its opposite vertex.

Let us denote by σ the reflexion (symmetry) of Δ_n about the x_1-axis. To the reflexion σ then corresponds the reflexion matrix

$$B = A^-(0) = \begin{pmatrix} 1 & 0 \\ 0 & -1 \end{pmatrix}$$

and the other reflexions are given by the matrices $A^-(2j\pi/n) = A^jB$ $(j = 0, 1, \ldots, n-1)$, cf. subsection *a.*, (ii).

Accordingly

$$(A^jB)^2 = E$$

and

$$A^jB = (A^jB)^{-1} = B^{-1}A^{-j} = BA^{-j},$$

in particular

$$AB = BA^{-1} \quad \text{or} \quad (AB)^2 = E \quad \text{but} \quad AB \neq E,$$

and therefore

(1) $\rho\sigma = \sigma\rho^{-1}, \cdot \ \rho\sigma \neq \epsilon.$

By means of this rule it is readily verified that every product of powers of ρ and σ can be written in the form $\rho^j\sigma$ if σ has an odd exponent. It suffices to show that this is so for every product $\sigma\rho^k$. Indeed $\sigma\rho^k = \sigma\rho \cdot \rho^{k-1} = \rho^{-1}\sigma\rho^{k-1} = \rho^{-2}\sigma\rho^{k-2} = \ldots = \rho^{-k}\sigma$. Thus the n rotations ρ^j ($j = 0, 1, \ldots, n-1$) and the n reflexions $\rho^j\sigma$ form the group $\mathfrak{D}_n$ and $|\mathfrak{D}_n| = 2n$.

This result implies that the group $\mathfrak{D}_n$ is generated by the two elements ρ and σ of order n and 2 respectively which satisfy the condition (1), or $(\rho\sigma)^2 = \epsilon$. Extending the notation adopted earlier for the cyclic group of order n, namely $\langle a \rangle_n$, we write

$$\mathfrak{D}_n = \langle \rho, \sigma \,|\, \rho^n = \sigma^2 = (\rho\sigma)^2 = \epsilon \rangle.$$

Now we observe that in the preceding argument which led us to the representation of all elements of $\mathfrak{D}_n$ in the form ρ^j or $\rho^j\sigma$ the geometrical interpretation of ρ and σ as rotation and reflexion has not been used once the orders of these elements and the relation $(\rho\sigma)^2 = \epsilon$ has been established. It is therefore apparent that an abstract group $\bar{\mathfrak{D}}_n$ exists which is generated by two of its elements a and b of order n and 2 respectively which satisfy the condition $(ab)^2 = e$. This means that all elements of $\bar{\mathfrak{D}}_n$ are products of powers of a and b. Making use of the associative law and of the condition $(ab)^2 = e$ we can show as above that a^j and a^jb ($j = 0, 1, \ldots, n-1$) are all the elements of the group $\bar{\mathfrak{D}}_n$ generated by a and b. We write

$$\bar{\mathfrak{D}}_n = \langle a, b \,|\, a^n = b^2 = (ab)^2 = e \rangle$$

and call this a *presentation* of $\bar{\mathfrak{D}}_n$. It is readily seen that the mapping $a \to \rho$, $b \to \sigma$ induces an isomorphism $\bar{\mathfrak{D}}_n \simeq \mathfrak{D}_n$. The group $\bar{\mathfrak{D}}_n$ is called an *abstract dihedral group* and $\mathfrak{D}_n$ is a representation of $\bar{\mathfrak{D}}_n$ (cf. §1, *e.*).

Another representation of $\overline{\mathfrak{D}}_n$, namely by permutations, can be derived as follows: Together with the cyclic permutation $P = (\hat{0}\,\hat{1}\,\hat{2}\ldots\widehat{n-1})$ of the vertices induced by the rotation ρ consider the permutation

$$Q = \begin{cases} (\widehat{1\;2m-1})(\widehat{2\;2m-2})\ldots(\widehat{m-1\;m+1}) & \text{if } n = 2m \\ & \text{(leaving fixed } \hat{0} \text{ and } \hat{m}) \\ (\widehat{1\;2m})(\widehat{2\;2m-1})\ldots(\widehat{m-1\;\hat{m}}) & \text{if } n = 2m+1 \\ & \text{(leaving fixed } \hat{0} \text{ only).} \end{cases}$$

Then P and Q have the orders n and 2 respectively and satisfy the relation $(PQ)^2 = I$. Indeed if e.g. $n = 8$ we have

$$PQ = (\hat{0}\,\hat{1}\,\hat{2}\,\hat{3}\,\hat{4}\,\hat{5}\,\hat{6}\,\hat{7})(\hat{1}\,\hat{7})(\hat{2}\,\hat{6})(\hat{3}\,\hat{5}) = (\hat{0}\,\hat{1})(\hat{2}\,\hat{7})(\hat{3}\,\hat{6})(\hat{4}\,\hat{5})$$

and if $n = 9$:

$$PQ = (\hat{0}\,\hat{1}\,\hat{2}\,\hat{3}\,\hat{4}\,\hat{5}\,\hat{6}\,\hat{7}\,\hat{8})(\hat{1}\,\hat{8})(\hat{2}\,\hat{7})(\hat{3}\,\hat{6})(\hat{4}\,\hat{5}) = (\hat{0}\,\hat{1})(\hat{2}\,\hat{8})(\hat{3}\,\hat{7})(\hat{4}\,\hat{6}).$$

Thus the group $\langle P, Q\rangle$, generated by the two permutations P and Q is also a representation of the dihedral group $\overline{\mathfrak{D}}_n$.

c. *Rotations and reflexions in space.* As is the circle in the plane, so is the sphere in space the figure with perfect symmetry in all directions through its centre O. Every straight line through O is a symmetry axis and every plane through O is a mirror which reflects the sphere onto itself. We consider the system of all linear homogeneous transformations of the space onto itself which leave the unit sphere $x_1{}^2+x_2{}^2+x_3{}^2 = 1$ invariant. The system is evidently a group with composition as group multiplication; it is called the *orthogonal group*, denoted by $\mathfrak{O}_3$.

THEOREM 2. *Every element of the orthogonal group $\mathfrak{O}_3$ is either a rotation about an axis through O or a product of a reflexion with respect to a plane through O and a rotation about an axis through O. The set of all rotations is a subgroup $\mathfrak{D}_3$ of $\mathfrak{O}_3$.*

The subgroup $\mathfrak{D}_3$ is called the proper orthogonal group.

Proof. If A is the matrix of the coefficients of a linear homogeneous transformation in $\mathfrak{O}_3$ with respect to a cartesian coordinate system x_1, x_2, x_3 the invariance of the unit sphere leads to the condition $A'A = E$, where E represents the unit matrix:

$$E = \begin{pmatrix} 1 & 0 & 0 \\ 0 & 1 & 0 \\ 0 & 0 & 1 \end{pmatrix}.$$

Thus the matrix A is orthogonal: $A' = A^{-1}$. There are the following two possibilities:

(i) $\det A = +1$ (the matrix A is proper orthogonal). In order to prove that the corresponding element of $\mathfrak{D}_3$ is a rotation we show that it has an axis, i.e. there is a vector $a \neq 0$ which satisfies the condition $Aa = a$. Thus we have to establish that A has the number 1 as an eigenvalue, i.e. $\lambda = 1$ is a root of the characteristic polynomial

$$\det(\lambda E - A) = \lambda^3 - s\lambda^2 + s^*\lambda - 1$$

where s denotes the trace of A (i.e. the sum of the diagonal elements and equally the sum of the eigen values of A). Since $\lambda E - A = -\lambda(\lambda^{-1}E - A')A$ we can write

$$\begin{aligned} \det(\lambda E - A) &= -\lambda^3(\lambda^{-3} - s\lambda^{-2} + s^*\lambda^{-1} - 1) \\ &= \lambda^3 - s^*\lambda^2 + s\lambda - 1 \end{aligned}$$

so that $s^* = s$ and

$$\det(\lambda E - A) = \lambda^3 - s\lambda^2 + s\lambda - 1 = (\lambda - 1)(\lambda^2 - (s-1)\lambda + 1).$$

Thus $\lambda = 1$ is in fact an eigen value of A.

The motion in the plane $a'x = 0$ (perpendicular to the eigenvector a) is therefore a plane rotation about O. The angle α of the rotation is determined by the other two eigen values of A. These are the roots of the polynomial $\lambda^2 - (s-1)\lambda + 1$ and therefore two conjugate complex numbers $\lambda, \bar{\lambda}$ and $\lambda\bar{\lambda} = 1$. Hence $\lambda = e^{i\alpha}$ and $s - 1 = \lambda + \bar{\lambda} = 2\cos\alpha$. Thus α is determined up to a multiple of 2π. Let us choose now the axis of A as the x_3-axis of the coordinate system. Then

$$A = \begin{pmatrix} \cos\alpha & -\sin\alpha & 0 \\ \sin\alpha & \cos\alpha & 0 \\ 0 & 0 & 1 \end{pmatrix}$$

which, indeed, represents a rotation.

The product of two proper orthogonal matrices, the unit matrix and the inverse of a proper orthogonal matrix are proper orthogonal. Hence the rotations about O form a group $\mathfrak{D}_3 < \mathfrak{H}_3$.

(ii) $\det A = -1$. In this case we prove that the matrix A has the number -1 as eigen value. Indeed

$$\det(\lambda E - A) = \lambda^3 - s\lambda^2 + s^*\lambda + 1$$

and by the same argument as in (i) we find that $s^* = -s$ so that

$$\det(\lambda E - A) = \lambda^3 - s\lambda^2 - s\lambda + 1 = (\lambda+1)(\lambda^2 - (s+1)\lambda + 1).$$

Thus there is a vector $x \neq 0$ for which $Ax = -x$. Again the other two eigenvalues are conjugate complex, or equal real, numbers whose product is equal to 1 and so A induces a rotation in the plane through O perpendicular to the eigenvector x. The operation σ represented by the matrix A, therefore, is the product of a reflexion σ_1 in this plane and a rotation ρ which has the line carrying the vector x as its axis, that is

$$\sigma = \sigma_1 \rho = \rho \sigma_1.$$

Every element of the set $\mathfrak{H}_3 \setminus \mathfrak{D}_3$ can be represented in the form $A \cdot B_0$ where B_0 indicates a fixed reflexion in $\mathfrak{H}_3$ with respect to a plane through O, and A is a rotation. Indeed if $B \in \mathfrak{H}_3$, $B \notin \mathfrak{D}_3$, then $B \cdot B_0 = A \in \mathfrak{D}_3$ and since $B_0{}^2 = E$ it follows that $B = A \cdot B_0$.

This also shows that if a subgroup $\mathfrak{H}$ of $\mathfrak{H}_3$ contains one non-rotation σ, then there is a one–one correspondence between the rotations $\rho \in \mathfrak{H}$ and the non-rotations $\rho\sigma \in \mathfrak{H}$. If, in particular, the subgroup $\mathfrak{H}$ is finite, we conclude that the number of rotations equals the number of non-rotations in $\mathfrak{H}$.

d. The polyhedral groups. For every natural number $n \geqslant 2$ the circumference of a circle can be divided into n equal parts. Consequently to every n corresponds a regular n-gon. However, the number of regular polyhedra **P** is bounded. A regular polyhedron is a convex polyhedron whose faces are congruent regular polygons inscribed into a sphere. It was known in antiquity that there exist only the five so-called Platonic regular polyhedra **P**, namely:

I	II	III
Tetrahedron **T**	Octahedron **O**	Ikosahedron **J**
	Cube **C**	Dodekahedron **D**

To speak in general terms, we shall be interested in the group $\bar{\mathfrak{P}}$ of covering transformations of each of these polyhedra **P**. The group $\bar{\mathfrak{P}}$ consists of two classes of elements:

(i) Rotations ρ; these form a subgroup $\mathfrak{P} < \bar{\mathfrak{P}}$.

(ii) Rotation-reflexions $\rho\sigma_0$, $\rho \in \mathfrak{P}$, σ_0 is a fixed reflexion with respect to a plane through the centre O, $\sigma_0 \in \bar{\mathfrak{P}}$.

Since these operations carry vertices into vertices, edges into edges, and faces into faces, they can be represented as permutations of a finite number of objects. These groups $\mathfrak{P}$ and $\bar{\mathfrak{P}}$ are therefore finite and $|\bar{\mathfrak{P}}| = 2|\mathfrak{P}|$.

For a space figure to admit a group of covering rotations this figure must have symmetry axes, while to admit reflexions it must have symmetry planes. Both conditions are satisfied in the case of the five regular polyhedra. On simple cardboard models of these figures which the reader may wish to construct it appears that there are four distinct types of axes,* namely

vertex axes: through opposite vertices, occurring in **O, C, J, D**;
face axes: through midpoints of opposite faces, occurring in **O, C, J, D**;
t-axes: through a vertex and the midpoint of the opposite face, occurring in **T**;
edge axes: through midpoints of opposite edges, occurring in **T, O, C, J, D.**

The number of these axes occurring in **P** are denoted by n_v, n_f, n_t, n_e respectively. We use them to determine the order of $\mathfrak{P}$.

Let further m_v denote the number of edges (faces) issuing from (meeting at) a given vertex M of **P**. Then the cyclic group of rotations

* Cf. D. Hilbert and S. Cohn-Vossen, *Geometry and the Imagination*, New York 1952. (Translation of "Anschauliche Geometrie" Berlin 1932.) Chap. II, §14. Also H. S. M. Coxeter, *Introduction to Geometry*, New York 1969, Chap. 10, p. 151, and M. Herbig, *Drehgruppen*, Frankfurt a. M. 1974.

about the axis through M has order m_v. Next, there are $m_e = 2$ vertices lying on (faces meeting at) each of the edges. Thus, an edge axis carries a rotation group of order 2. Finally, if each face of **P** is an m_f-gon the rotation group about a face axis has order m_f. Here are the numbers of axes and their orders for each of the five regular polyhedra:

P	n_v	n_f	n_t	n_e	m_v	m_f	m_e	Form of face polygon
T	0	0	4	3	3	3	2	equilateral triangle
O	3	4	0	6	4	3	2	equilateral triangle
C	4	3	0	6	3	4	2	square
J	6	10	0	15	5	3	2	equilateral triangle
D	10	6	0	15	3	5	2	regular pentagon

Now we can determine the order of each of the groups $\mathfrak{P}$, that is, count the number of covering rotations of every regular polyhedron. First we observe that in a given **C** (**O**) we obtain the edges of an **O** (**C**) by joining the midpoints of adjacent faces in the given **C** (**O**). Hence the groups $\mathfrak{O}$ and $\mathfrak{C}$ are isomorphic. The same argument applies in the case of **J** and **D**; hence also $\mathfrak{D} \simeq \mathfrak{J}$. Thus we have to find only $|\mathfrak{T}|$, $|\mathfrak{C}|$ and $|\mathfrak{J}|$.

If **P** = **T** we note that each of the 3 edge axes carries a group of order 2; this yields $2+1+1$ elements. Each of the 4 t-axes has a group of order 3; thus, as the identity has already been counted, we get $4 \times 2 = 8$ new elements. Hence $|\mathfrak{T}| = 4+8 = 12$.

Similarly for the cube **C** we have 4 vertex axes, each with a group of order 3 which yields $3+3\times 2 = 9$ distinct elements. Further there are 3 face axes, each with a group of order 4, resulting in $3\times 3 = 9$ new elements. Finally we have 6 edge axes, each carrying a group of order 2, which gives us another 6 elements. Hence $|\mathfrak{C}| = 9+9+6 = 24$.

For the icosahedron **J** we have 6 vertex axes, each with a group of order 5, thus $5+5\times 4 = 25$ elements. Further there are 10 face axes, each with a group of order 3, thus $10\times 2 = 20$ elements. Finally we have 15 edge axes, each with a group of order 2 which yield another 15 elements. Hence $|\mathfrak{J}| = 25+20+15 = 60$.

e. We begin with the group $\mathfrak{O}$ of the octahedron. The octahedron may be inscribed in the unit sphere $x_1{}^2+x_2{}^2+x_3{}^2 = 1$ so that its 6 pairwise opposite vertices are given by the coordinate triplets

$$\pm 1, 0, 0; \quad 0, \pm 1, 0; \quad 0, 0, \pm 1$$

and that the symmetry axes passing through opposite vertices coincide with the coordinate axes. In view of this arrangement the elements of $\mathfrak{O}$ can be represented as follows: Let i, j, k be one of the six permutations* of the symbols 1, 2, 3 and let $\epsilon_l = \pm 1$, $(l = 1, 2, 3)$. If x_1, x_2, x_3 are the coordinates of one of the vertices of **O**, then the point with the coordinates

$$(2) \qquad y_1 = \epsilon_i x_i, \quad y_2 = \epsilon_j x_j, \quad y_3 = \epsilon_k x_k$$

is also a vertex; indeed, starting from an arbitrary fixed vertex x, each vertex of **O** may be obtained by a suitable choice of the permutation i, j, k and of the ϵ_l.

The matrix A of the linear homogeneous transformation (2) is evidently orthogonal. We list the $6 \cdot 2 \cdot 2 \cdot 2 = 48$ possible matrices A:

$$(2') \qquad \begin{pmatrix} \epsilon_1 & 0 & 0 \\ 0 & \epsilon_2 & 0 \\ 0 & 0 & \epsilon_3 \end{pmatrix}, \begin{pmatrix} 0 & \epsilon_2 & 0 \\ 0 & 0 & \epsilon_3 \\ \epsilon_1 & 0 & 0 \end{pmatrix}, \begin{pmatrix} 0 & 0 & \epsilon_3 \\ \epsilon_1 & 0 & 0 \\ 0 & \epsilon_2 & 0 \end{pmatrix}, \quad \det A = \epsilon_1 \epsilon_2 \epsilon_3;$$

$$(2'') \qquad \begin{pmatrix} \epsilon_1 & 0 & 0 \\ 0 & 0 & \epsilon_3 \\ 0 & \epsilon_2 & 0 \end{pmatrix}, \begin{pmatrix} 0 & 0 & \epsilon_3 \\ 0 & \epsilon_2 & 0 \\ \epsilon_1 & 0 & 0 \end{pmatrix}, \begin{pmatrix} 0 & \epsilon_2 & 0 \\ \epsilon_1 & 0 & 0 \\ 0 & 0 & \epsilon_3 \end{pmatrix}, \quad \det A = -\epsilon_1 \epsilon_2 \epsilon_3.$$

The subgroup $\mathfrak{O}$ of order 24 is characterized by the condition $\det A = +1$.

Another subgroup of the order 24 is defined by the condition

$$(3) \qquad \epsilon_1 \epsilon_2 \epsilon_3 = +1.$$

Indeed from (2) follows that $y_1 y_2 y_3 = \epsilon_1 \epsilon_2 \epsilon_3 x_1 x_2 x_3$. Hence if $x \to y$ and $y \to z$ are two of the transformations (2) subject to the condition (3), we have $x_1 x_2 x_3 = y_1 y_2 y_3 = z_1 z_2 z_3$; thus the ϵ-factors defining the mapping $x \to z$ have also the product $+1$. Let us denote this subgroup by $\mathfrak{H}$.

* The word "permutation" is used with two different meanings: (a) The mapping $\begin{pmatrix} 1 & 2 & 3 \\ i & j & k \end{pmatrix}$, also called "substitution", (b) the result of the mapping, i.e. the second row of the mapping symbol. At this instant we refer to the second meaning of the word.

We shall show that $\mathfrak{H} \simeq \mathfrak{T}$, the symmetry group of the tetrahedron. For this purpose we construct a cube so that the midpoints of its six faces are the six vertices of our octahedron **O**. The vertices of the cube are then the points with the coordinates

	x_1	x_2	x_3		x_1	x_2	x_3
P_1:	1	1	1	Q_1:	-1	-1	-1
P_2:	1	-1	-1	Q_2:	-1	1	1
P_3:	-1	1	-1	Q_3:	1	-1	1
P_4:	-1	-1	1	Q_4:	1	1	-1

$$x_1x_2x_3 = 1 \qquad\qquad x_1x_2x_3 = -1$$

so that the P_ν and Q_ν ($\nu = 1, 2, 3, 4$) are pairs of opposite vertices of the cube (cf. Fig. 2). The four points P_ν (Q_ν) are the vertices of a

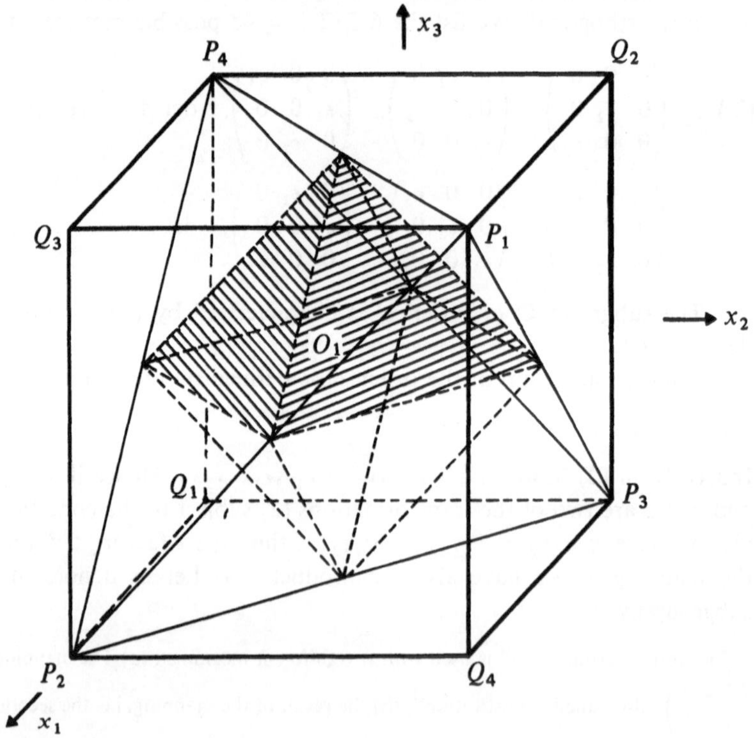

Fig. 2

tetrahedron $\mathbf{T}_P$ ($\mathbf{T}_Q$) whose edges are the six face diagonals of the cube which connect the points P_ν (Q_ν). We assume the two tetrahedrons as well as the octahedron **O** rigidly fixed in the cube. Every linear covering transformation of the tetrahedron $\mathbf{T}_P$ ($\mathbf{T}_Q$) is clearly a transformation in $\mathfrak{O}$ for which $\epsilon_1\epsilon_2\epsilon_3 = +1$; indeed the products of the coordinates of the vertices are not changed.

Conversely, every covering transformation of the octahedron **O** (and thus of the cube) either takes $\mathbf{T}_P \to \mathbf{T}_P$ and $\mathbf{T}_Q \to \mathbf{T}_Q$ *or* $\mathbf{T}_P \to \mathbf{T}_Q$ and $\mathbf{T}_Q \to \mathbf{T}_P$. The latter are those of the mappings (2) for which $\epsilon_1\epsilon_2\epsilon_3 = -1$. The former constitute the group $\mathfrak{H}$ which thus in fact is the tetrahedral group $\mathfrak{T}$.

For a discussion of the icosahedral group $\mathfrak{J}$ some knowledge of the elements of group theory is required. It will be taken up in Chap. II, §2, ex. 12.

Examples and exercises

1. (a) Every rotation in the plane is the product of two reflexions with respect to straight lines through the centre O.

(b) Every space rotation is the product of two reflexions with respect to planes through the centre.

2. (a) The dihedral group $\mathfrak{D}_2$ is isomorphic to the so-called four group $\mathfrak{V} = \{I, (12)(34), (13)(24), (14)(23)\}$.

(b) The group $\mathfrak{D}_3$ is isomorphic to the symmetric group $\mathfrak{S}_3$.

(c) The group $\mathfrak{D}_4$ is isomorphic to the subgroup of the symmetric group $\mathfrak{S}_4$ (acting on the indeterminates x_0, x_1, x_2, x_3) whose permutations leave the polynomial $x_0x_2+x_1x_3$ invariant.

(d) All groups $\mathfrak{D}_n$ ($n \geqslant 3$) are non-abelian.

3. (a) Show that the two reflexions σ, τ corresponding to the two matrices $A^-(0)$, $A^-(2\pi/n)$ are generators of $\mathfrak{D}_n$. Find a system of relations involving σ and τ.

(b) Do every two reflexions in $\mathfrak{D}_n$ form a set of generators for $\mathfrak{D}_n$?

4. If $n = 4m$ then $\mathfrak{D}_{2m}$ is a subgroup of $\mathfrak{D}_n$ containing one half of the elements of $\mathfrak{D}_n$.

5. (a) Show that the group $\mathfrak{O}_3$ is non-commutative.

(b) Show that in $\bar{\mathfrak{O}}_3$ not every reflexion commutes with every rotation. Find conditions for commutativity.

6. *Matrix representation of reflexions in $\mathfrak{D}_3$.* By a, x, y, y^* we denote vectors in the real 3-dimensional space, represented by coordinate columns:

$$x = \begin{pmatrix} x_1 \\ x_2 \\ x_3 \end{pmatrix}, \text{ etc.}$$

The inner product of two vectors a, x is the row-column product $a'x = a_1x_1 + a_2x_2 + a_3x_3$ and the column-row product ax' denotes the matrix of rank one (or zero if one of the vectors is the zero vector) with the elements a_ix_j.

Let the mirror plane Π of the reflexion σ be given by the equation $a'x = 0$. This equation expresses the fact that the vector $a \neq 0$ is perpendicular to all vectors x in this plane. If y is a point in space (endpoint of its position vector which is also denoted by y), its image y^* by the reflexion with respect to Π lies on a line through y, perpendicular to Π:

$$y^* = y + \lambda a.$$

The factor λ has to be determined so that y and y^* are equidistant from Π, that is, $y + y^*$ must be a point of the plane Π. Thus $a'(y + y^*) = 0$ and $a'(2y + \lambda a) = 0$ whence

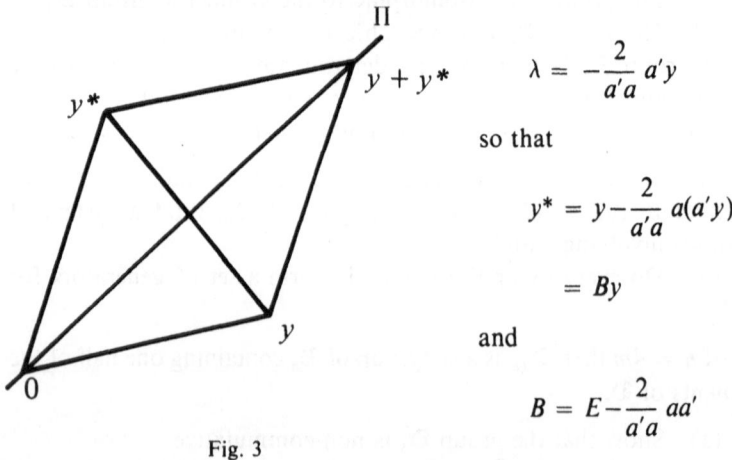

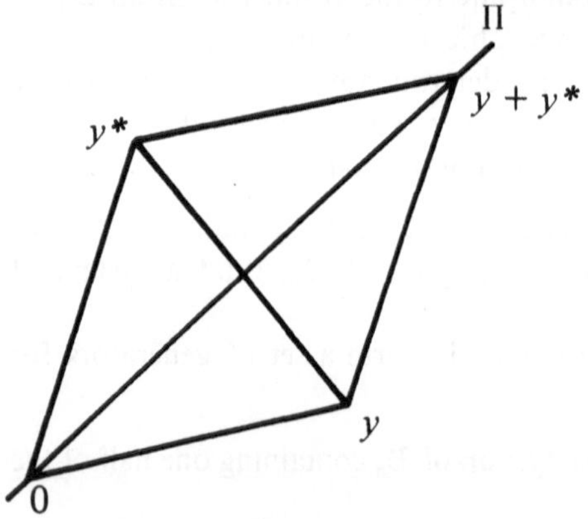

Fig. 3

$$\lambda = -\frac{2}{a'a}\, a'y$$

so that

$$y^* = y - \frac{2}{a'a}\, a(a'y)$$

$$= By$$

and

$$B = E - \frac{2}{a'a}\, aa'$$

is the matrix of the reflexion σ with respect to the plane Π.

7. Show that the tetrahedral group $\tilde{\mathfrak{T}}$ is isomorphic to the symmetric group $\mathfrak{S}_4$ and that the rotation group $\mathfrak{T}$ is isomorphic to the alternating group $\mathfrak{A}_4$.

8. From the list of matrices (2′), (2″) determine those which correspond to

(a) the rotations of **O** about the axes through the midpoints of opposite edges of **O**;

(b) the rotations of **O** about the axes through the midpoints of opposite faces of **O**;

(c) the reflexions of **O** with respect to the plane containing the x-axis and the midpoint of two opposite edges of **O**.

9. Verify the statement at the end of subsection *a.* by showing that the straight line L_α is the axis of the rotation represented by the proper orthogonal matrix

$$\begin{pmatrix} \cos\alpha & \sin\alpha & 0 \\ \sin\alpha & -\cos\alpha & 0 \\ 0 & 0 & -1 \end{pmatrix}$$

10. Let a, b, c, d be constant complex numbers, $ad-bc \neq 0$; let x be a complex variable representing a point in the complex number plane. The mapping

$$x \to y = \frac{ax+b}{cx+d}$$

sends circles and straight lines in the complex number plane onto circles or straight lines. The set of all such mappings forms a group $\mathfrak{M}$ with respect to functional composition.

Remark. It can be shown that the full group of all "circle preserving" transformations is obtained by adjoining to $\mathfrak{M}$ the mapping $x \to \bar{x}$ if $\bar{x}$ denotes the complex conjugate of x, that is, by adding to $\mathfrak{M}$ the set of mappings

$$x \to y = \frac{a\bar{x}+b}{c\bar{x}+d} \quad (ad-bc \neq 0).$$

Chapter II

SUBSETS, SUBGROUPS, HOMOMORPHISMS

§1 The algebra of subsets in a group

When should we say that "we know a group $\mathfrak{G}$"? A simple answer: If for every pair of elements $a, b \in \mathfrak{G}$ we know the product ab. For finite groups of very low order, say $|G| < 15$, this may be a satisfactory answer. For groups of larger order or for infinite groups the group table is usually unwieldy and shows only few—if any—interesting features of the group under consideration.

This suggests that in order to arrive at facts which help us to understand the restrictions imposed on the set of the elements of $\mathfrak{G}$ by the group axioms, we should concentrate not on single elements, but on certain subsets of $\mathfrak{G}$. Several suitable choices of such subset systems will be discussed in due course. First we shall consider a system of subsets of $\mathfrak{G}$ without further restrictions.

Apart from the purely set-theoretical operations $\mathbf{A} \cap \mathbf{B}$ (intersection) and $\mathbf{A} \cup \mathbf{B}$ (union), $\mathbf{A}, \mathbf{B} \subseteq \mathfrak{G}$, for the purposes of group theory we introduce operations which are based on the multiplication law in $\mathfrak{G}$.

a. For non-empty subsets $\mathbf{A}$ and $\mathbf{B}$ of $\mathfrak{G}$ the product $\mathbf{AB}$ consists of all elements of $\mathfrak{G}$ which can be represented in the form ab for some $a \in \mathbf{A}$ and some $b \in \mathbf{B}$. Briefly

$$\mathbf{AB} = \{ab \mid a \in \mathbf{A}, b \in \mathbf{B}\}.$$

In particular $\mathbf{A}$ and $\mathbf{B}$ may coincide. Also $\mathbf{A}$ (or $\mathbf{B}$) may consist of one single element.

It is easy to verify that the product of three or more subsets of $\mathfrak{G}$ is associative: $\mathbf{A}(\mathbf{BC}) = (\mathbf{AB})\mathbf{C}$. There are distributive laws with the union operation, but not in general with the intersection (cf. exx. 6, 7).

If $\mathfrak{A}$ is a subgroup of $\mathfrak{G}$, $\mathfrak{A} \leqslant \mathfrak{G}$, and $a \in \mathfrak{A}$, then the product

$ax \in \mathfrak{A}$ for every $x \in \mathfrak{A}$; hence $a\mathfrak{A} \subseteq \mathfrak{A}$. But for every $y \in \mathfrak{A}$ there is an $x \in \mathfrak{A}$ such that $ax = y$; thus $a\mathfrak{A} = \mathfrak{A}$:

THEOREM 1. *If* $\mathfrak{A} \leqslant \mathfrak{G}$ *then* $\mathfrak{A}\mathfrak{A} = \mathfrak{A}$.

b. We define

$$\mathbf{A}^{-1} = \{a^{-1} \mid a \in \mathbf{A}\}, \quad \mathbf{A} \subseteq \mathfrak{G}.$$

It is readily seen that $\mathbf{A}$ and $\mathbf{A}^{-1}$ may be disjoint, but if $\mathfrak{A} \leqslant \mathfrak{G}$ then $\mathfrak{A}^{-1} = \mathfrak{A}$. Indeed, for every $a \in \mathfrak{A}$ an element $x \in \mathfrak{A}$ can be found such that $x^{-1} = a$, namely $x = a^{-1}$. The condition $\mathbf{A}^{-1} = \mathbf{A}$ is not sufficient for $\mathbf{A}$ to be a subgroup of $\mathfrak{G}$. If $\mathbf{B}$ is a subset of $\mathfrak{G} \setminus \{e\}$, then $\mathbf{A} = \mathbf{B} \cup \mathbf{B}^{-1}$ satisfies the condition and is not a subgroup of $\mathfrak{G}$. We can state, however, the following theorem:

THEOREM 2. *The condition* $\mathbf{A}^{-1}\mathbf{A} \subseteq \mathbf{A}$ *is necessary and sufficient for* $\mathbf{A}$ *to be a subgroup of* $\mathfrak{G}$.

Indeed, if $\mathfrak{A} \leqslant \mathfrak{G}$, then $\mathfrak{A}^{-1}\mathfrak{A} = \mathfrak{A}$. Conversely if $\mathbf{A}^{-1}\mathbf{A} \subseteq \mathbf{A}$ we conclude that for every $a \in \mathbf{A}$ the product $a^{-1}a = e \in \mathbf{A}$ and therefore $a^{-1}e = a^{-1} \in \mathbf{A}$. Similarly $a \in \mathbf{A}^{-1}$ and if $b \in \mathbf{A}$ then also $ab \in \mathbf{A}$. Therefore $\mathbf{A} = \mathfrak{A} \leqslant \mathfrak{G}$.

COROLLARY. *Let* $\mathfrak{A} \leqslant \mathfrak{G}$, $\mathfrak{B} \leqslant \mathfrak{G}$. *The product* $\mathfrak{A}\mathfrak{B}$ *is a subgroup of* $\mathfrak{G}$ *if and only if* $\mathfrak{A}\mathfrak{B} = \mathfrak{B}\mathfrak{A}$.

Indeed if $\mathfrak{A}\mathfrak{B} = \mathfrak{B}\mathfrak{A}$, then $(\mathfrak{A}\mathfrak{B})^{-1}\mathfrak{A}\mathfrak{B} = (\mathfrak{B}^{-1}\mathfrak{A}^{-1})\mathfrak{A}\mathfrak{B} = \mathfrak{B}^{-1}(\mathfrak{A}^{-1}\mathfrak{A})\mathfrak{B} \subseteq \mathfrak{B}^{-1}(\mathfrak{A}\mathfrak{B}) = \mathfrak{B}^{-1}(\mathfrak{B}\mathfrak{A}) \subseteq \mathfrak{B}\mathfrak{A}$, hence $\mathfrak{A}\mathfrak{B} \leqslant \mathfrak{G}$. Conversely if $\mathfrak{A}\mathfrak{B} \leqslant \mathfrak{G}$, then $\mathfrak{A}\mathfrak{B} = (\mathfrak{A}\mathfrak{B})^{-1} = \mathfrak{B}^{-1}\mathfrak{A}^{-1} = \mathfrak{B}\mathfrak{A}$.

c. Let $\varphi : x \to y = \varphi(x)$ be a one–one mapping of $\mathfrak{G}$ onto itself. We define $\varphi(\mathbf{A}) = \{\varphi(a) \mid a \in \mathbf{A}\}$. The mapping φ is invertible: $\varphi^{-1}(\varphi(x)) = x = \varphi(\varphi^{-1}(x))$ for all $x \in \mathfrak{G}$. Therefore also

$$(1) \qquad \varphi^{-1}(\varphi(\mathbf{A})) = \mathbf{A} = \varphi(\varphi^{-1}(\mathbf{A})).$$

Now let us drop the restrictions "one–one and onto" for the mapping φ. The image of $\mathfrak{G}$ or of a subset $\mathbf{A}$ of $\mathfrak{G}$ under φ is a subset $\mathbf{C} = \varphi(\mathbf{A}) \subseteq \mathfrak{G}$. The symbol $\varphi^{-1}(a)$ $(a \in \mathbf{A})$ then denotes in general not a single element (cf. Chap. I, §2 *a.*), but rather $\varphi^{-1}(a) = \mathbf{B}$, the set of all $b \in \mathfrak{G}$ for which $\varphi(b) = a$. Thus $\varphi(\mathbf{B}) = a$, so that $\varphi(\varphi^{-1}(a)) = a$, but $\varphi^{-1}(\varphi(a))$ is not always equal to a. The first part of (1) is no longer valid in the more general situation.

d. Let $a \in \mathfrak{G}$. The cyclic subgroup $\langle a \rangle$ generated by the element a is clearly the smallest subgroup of $\mathfrak{G}$ which contains a, i.e. the intersection of all subgroups of $\mathfrak{G}$ which contain this element. Indeed $\langle a \rangle$ is contained in every subgroup which contains a. Likewise, given two elements a, b of $\mathfrak{G}$, let us denote by $\langle a, b \rangle$ the smallest subgroup of $\mathfrak{G}$ which contains a and b (cf. Chap. I, §4, *b.*). The subgroup $\langle a, b \rangle$ contains all the powers of a and all the powers of b and all possible products of these powers, i.e. all elements of the form

$$(2) \qquad a^{i_1}b^{j_1}a^{i_2}b^{j_2}\ldots a^{i_s}b^{j_s} \quad (i_\sigma, j_\sigma \in \mathbf{Z}, \quad \sigma = 1, \ldots, s).$$

Conversely it is readily seen that all such products form a subgroup $\mathfrak{H} \leqslant \mathfrak{G}$ and that $\mathfrak{H}$ is the subgroup generated by a and b. We convince ourselves that $\mathfrak{H} \leqslant \langle a, b \rangle$ and $\langle a, b \rangle \leqslant \mathfrak{H}$. Hence $\mathfrak{H} = \langle a, b \rangle$.

More generally let $\mathbf{A} = \{a, b, c, \ldots\} \subseteq \mathfrak{G}$. We define $\tilde{\mathbf{A}}$ as the intersection of all subgroups of $\mathfrak{G}$ which contain the subset $\mathbf{A}$, i.e. the smallest subgroup of $\mathfrak{G}$ which contains all the elements of $\mathbf{A}$. Further consider the subset $\langle \mathbf{A} \rangle$ of $\mathfrak{G}$ whose elements are all possible finite products of elements of $\mathbf{A} \cup \mathbf{A}^{-1}$, for example

$$(3) \qquad aabcb^{-1}cacc, \ldots.$$

Obviously $\langle \mathbf{A} \rangle$ is a group; we call it the *subgroup of* $\mathfrak{G}$ *generated by* $\mathbf{A}$. We notice that $\langle \mathbf{A} \rangle \leqslant \tilde{\mathbf{A}}$. Since $\mathbf{A} \subseteq \langle \mathbf{A} \rangle$ we have $\tilde{\mathbf{A}} \leqslant \langle \mathbf{A} \rangle$. Hence

$$(4) \qquad \langle \mathbf{A} \rangle = \tilde{\mathbf{A}}.$$

This concept can be extended to two or more generating subsets $\mathbf{A}, \mathbf{B}, \ldots$. We put

$$\langle \mathbf{A}, \mathbf{B} \rangle = \langle \mathbf{A} \cup \mathbf{B} \rangle, \quad \langle \mathbf{A}, \mathbf{B}, \mathbf{C} \rangle = \langle \mathbf{A} \cup \mathbf{B} \cup \mathbf{C} \rangle, \quad \text{etc.}$$

Then

$$\mathbf{AB} \subseteq \langle \mathbf{A}, \mathbf{B} \rangle.$$

Let us denote by $\mathbf{S}$ the set of all subgroups of the group $\mathfrak{G}$. Each of the two operations $\cap$ and $\langle\,,\rangle$ is a composition law on $\mathbf{S}$.

This means that if $\mathfrak{H}, \mathfrak{K} \leqslant \mathfrak{G}$, i.e. $\mathfrak{H}, \mathfrak{K} \in \mathbf{S}$, then $\mathfrak{H} \cap \mathfrak{K} \in \mathbf{S}$ and $\langle \mathfrak{H}, \mathfrak{K} \rangle \in \mathbf{S}$. Hence $(\mathbf{S} \mid \cap, \langle\, ,\rangle)$ is an algebraic system.

On the other hand neither the union $\mathfrak{H} \cup \mathfrak{K}$ nor the product $\mathfrak{H}\mathfrak{K}$ is necessarily a subgroup of $\mathfrak{G}$. In the case of the union this is evident from the following theorem

THEOREM 3. *If $\mathfrak{H}, \mathfrak{K} \leqslant \mathfrak{G}$ then $\mathfrak{H} \cup \mathfrak{K}$ is a subgroup of $\mathfrak{G}$ if and only if either $\mathfrak{H} \leqslant \mathfrak{K}$ or $\mathfrak{K} \leqslant \mathfrak{H}$.*

Proof. The condition is clearly sufficient for $\mathfrak{H} \cup \mathfrak{K}$ to be a subgroup of $\mathfrak{G}$, for

$$(5) \qquad \mathfrak{H} \cup \mathfrak{K} = \mathfrak{K} \Leftrightarrow \mathfrak{H} \leqslant \mathfrak{K}$$

$$(6) \qquad \mathfrak{H} \cup \mathfrak{K} = \mathfrak{H} \Leftrightarrow \mathfrak{K} \leqslant \mathfrak{H}.$$

Conversely let us assume that $\mathfrak{H} \cup \mathfrak{K} = \mathfrak{L} < \mathfrak{G}$ and neither (5) nor (6) is valid. Let $\mathfrak{H} \cap \mathfrak{K} = \mathfrak{D} < \mathfrak{H}$ and $\mathfrak{D} < \mathfrak{K}$. Then there is an element $a \in \mathfrak{H}$, $a \notin \mathfrak{D}$ and an element $b \in \mathfrak{K}$, $b \notin \mathfrak{D}$. Clearly the product $ab \in \mathfrak{H} \cup \mathfrak{K}$ and therefore either $ab \in \mathfrak{H}$ or $ab \in \mathfrak{K}$. Suppose that $ab \in \mathfrak{H}$; then $b \in a^{-1}\mathfrak{H} = \mathfrak{H}$ and therefore $b \in \mathfrak{D}$, a contradiction. Likewise in the case $ab \in \mathfrak{K}$ it follows that $a \in \mathfrak{D}$. Hence (5) or (6) is unavoidable.

COROLLARY. *No group is the union of two of its proper subgroups.*

e. We conclude this section with two theorems on pairs of subsets of a group. Although these theorems will not be referred to later, they seem to be of independent interest.

THEOREM 4. *Let $\mathfrak{G}$ be a finite group and $\mathbf{A}$, $\mathbf{B}$ subsets of $\mathfrak{G}$ for which $\mathbf{A} \cup \mathbf{B} = \mathfrak{G}$ and $\mathbf{A} \cap \mathbf{B} = \varnothing$. Let φ denote a one–one mapping of $\mathfrak{G}$ onto itself. Then $\mathbf{A} \cdot \varphi(\mathbf{B}) = \mathbf{B} \cdot \varphi(\mathbf{A})$.*

Proof. We apply the indirect method. Assume that the relation is not valid, that is, for some $a_1 \in \mathbf{A}$, $b_1 \in \mathbf{B}$ the product $a_1\varphi(b_1) = c \notin \mathbf{B}\,\varphi(\mathbf{A})$ and therefore $b\varphi(a) \neq c$ for every choice of $a \in \mathbf{A}$, $b \in \mathbf{B}$. Hence for every $a \in \mathbf{A}$ the element $c\varphi(a)^{-1} \notin \mathbf{B}$. Thus $c\varphi(a)^{-1} = a' \in \mathbf{A}$. If a runs through all the elements of $\mathbf{A}$, then $c\varphi(a)^{-1}$ assumes $|\mathbf{A}|$ distinct values $a' \in \mathbf{A}$; thus $c\varphi(a)^{-1}$ also runs through the whole of $\mathbf{A}$ and $a' = c\varphi(a)^{-1} = \psi(a)$ represents a one–one mapping of $\mathbf{A}$ onto itself. Hence every element a' of $\mathbf{A}$ has a unique pre-image and thus ψ is invertible.

In particular let $a' = a_1$. The element a corresponding to the element a_1 is defined by $c\varphi(a)^{-1} = a_1\varphi(b_1)\varphi(a)^{-1} = a_1$. Thus $a_1\varphi(b_1) = a_1\varphi(a)$ whence $\varphi(a) = \varphi(b_1)$ and $a = b_1$ which is impossible. Consequently $\mathbf{A}\varphi(\mathbf{B}) \subseteq \mathbf{B}\varphi(\mathbf{A})$. Repeating the argument with **A** and **B** interchanged we obtain $\mathbf{B}\varphi(\mathbf{A}) \subseteq \mathbf{A}\varphi(\mathbf{B})$. Hence both sides are equal and the theorem is proved.

Let $\varphi(x) = x$ $(x \in \mathfrak{G})$; it follows that for two disjoint subsets $\mathbf{A}, \mathbf{B}$ of $\mathfrak{G}$, $\mathbf{A} \cup \mathbf{B} = \mathfrak{G}$, one has

(7) $\quad \mathbf{AB} = \mathbf{BA}.$

Likewise if $\varphi(x) = x^{-1}$ the theorem yields the relation

(8) $\quad \mathbf{AB}^{-1} = \mathbf{BA}^{-1}.$ (cf. exx. 3–4).

THEOREM 5. (H. B. Mann, 1952). *Let $\mathfrak{G}$ be a finite group and $\mathbf{A}, \mathbf{B} \subset \mathfrak{G}$. If $|\mathbf{A}| + |\mathbf{B}| > |\mathfrak{G}|$, then $\mathbf{AB} = \mathfrak{G}$.*

Proof. Indirect: Suppose that $\mathbf{AB} \subset \mathfrak{G}$ (excluding equality) so that there is an element $c \in \mathfrak{G}$, $c \notin \mathbf{AB}$. The set $c\mathbf{B}^{-1} = \{cb^{-1} | b \in \mathbf{B}\}$ has no element in common with **A**, for $cb^{-1} = a \in \mathbf{A}$ implies that $c = ab \in \mathbf{AB}$. Hence $|\mathbf{A}| + |c\mathbf{B}^{-1}| \leqslant |\mathfrak{G}|$. Since $|c\mathbf{B}^{-1}| = |\mathbf{B}^{-1}| = |\mathbf{B}|$ we conclude that $|\mathbf{A}| + |\mathbf{B}| \leqslant |\mathfrak{G}|$, a contradiction to the assumption of the theorem.

For the far-reaching consequences of this theorem we refer to the monograph by H. B. Mann, *Addition Theorems: The Addition Theorems of Group Theory and Number Theory*, Interscience Publishers, New York 1965.

Examples and exercises

1. Prove that the intersection of a pair $\mathfrak{A}, \mathfrak{B}$ of subgroups of a group $\mathfrak{G}$ is again a subgroup of $\mathfrak{G}$. Generalize this to the case of a finite number and an infinity of subgroups.

2. Determine two subgroups $\mathfrak{A}, \mathfrak{B}$ of the symmetric group $\mathfrak{S}_3$ for which the product is not a subgroup of $\mathfrak{S}_3$. Notice that $\mathfrak{AB} \neq \mathfrak{BA}$. Do the same for $\mathfrak{S}_4$.

3. If $\mathfrak{G}$ is a finite group and **A** and **B** are subsets of $\mathfrak{G}$ for which $\mathbf{A} \cup \mathbf{B} = \mathfrak{G}$ and $\mathbf{A} \cap \mathbf{B} = \emptyset$ (cf. Theorem 4) show that for every fixed element c in $\mathfrak{G}$ one has $\mathbf{A}c\mathbf{B} = \mathbf{B}c\mathbf{A}$.

4. If $\mathbf{A}$ and $\mathbf{B}$ are both non-empty and otherwise as in ex. 3, show that

(a) $\mathfrak{G} \setminus \mathbf{AB}^{-1}$ is a subgroup $\mathfrak{H}$ of $\mathfrak{G}$;

(b) if $e \in \mathbf{A}$, then $\mathfrak{H} \leqslant \mathbf{A}$ and $\mathfrak{H}\mathbf{A} = \mathbf{A}$.

5. Let a, b, c be three elements of a group $\mathfrak{G}$. If $a^2 = b$, $b^2 = c$, $c^2 = a$ show that $\langle a, b, c\rangle$ is an abelian subgroup of $\mathfrak{G}$.

6. For any three subsets $\mathbf{A}, \mathbf{B}, \mathbf{C}$ of a group $\mathfrak{G}$ verify that

$$(\mathbf{A} \cup \mathbf{B})\mathbf{C} = \mathbf{AC} \cup \mathbf{BC}, \quad \mathbf{C}(\mathbf{A} \cup \mathbf{B}) = \mathbf{CA} \cup \mathbf{CB}$$

$$(\mathbf{A} \cap \mathbf{B})\mathbf{C} \subseteq \mathbf{AC} \cap \mathbf{BC}, \quad \mathbf{C}(\mathbf{A} \cap \mathbf{B}) \subseteq \mathbf{CA} \cap \mathbf{CB}.$$

7. If in ex. 6 the subset $\mathbf{C}$ consists of a single element c only, one has exact distributivity for union and intersection with the group multiplication.

8. Prove the converse of theorem 1: *A* subset $\mathbf{A}$ of a finite group $\mathfrak{G}$ is a subgroup of $\mathfrak{G}$ if (and only if) $\mathbf{AA} = \mathbf{A}^2 = \mathbf{A}$.

9. Let $\mathbf{A}$ be a subset of a finite group $\mathfrak{G}$. Consider the sequence

(9) $\quad \mathbf{A}, \mathbf{A}^2, \mathbf{A}^3, \ldots.$

As it cannot contain more than a finite number of different terms there must be an index r so that $\mathbf{A}^r$ is the first term which equals one of its predecessors, say $\mathbf{A}^m$, $m < r$. If $\mathbf{A}$ is a single element we have $m = 1$; in general $m > 1$. Let $r = m+p$; then the first $m+p-1$ terms of the sequence (9) are all different and the terms

$$\mathbf{A}^m, \mathbf{A}^{m+1}, \ldots, \mathbf{A}^{m+p-1}$$

represent the first period of the sequence. From the mth term on the sequence is periodic: If $n, n' \geqslant m$ the two relations $\mathbf{A}^{n'} = \mathbf{A}^n$ and $n' \equiv n \pmod{p}$ are equivalent.

Now assume that $p \mid n$ and $m \leqslant n < m+p$; then

$$(\mathbf{A}^n)^2 = \mathbf{A}^{2n} = \mathbf{A}^n$$

and by ex. 8 $\mathbf{A}^n$ is a group.

(a) Show that among the powers of **A** there is only one group, namely $\mathbf{A}^n$.

(b) Show that if n' is an exponent for which $(\{e\} \cup \mathbf{A})^{n'}$ is a group then $(\{e\} \cup \mathbf{A})^{n'} = \langle \mathbf{A} \rangle$.

10. If $\mathfrak{A}, \mathfrak{B} \leqslant \mathfrak{G}$ show that $\langle \mathfrak{A}, \mathfrak{B} \rangle = \langle \mathfrak{A}\mathfrak{B} \rangle$.

11. *The Frattini subgroup.* Definition: Let **A** be an arbitrary subset of the group $\mathfrak{G}$. An element x of $\mathfrak{G}$ is called a *non-generator* of $\mathfrak{G}$ if from $\langle \mathbf{A}, x \rangle = \mathfrak{G}$ it follows that $\langle \mathbf{A} \rangle = \mathfrak{G}$. Prove the following theorem:

THEOREM. *The set $\Phi = \Phi(\mathfrak{G})$ consisting of all non-generators of a group $\mathfrak{G}$ is a subgroup of $\mathfrak{G}$.*

The subgroup Φ has been introduced by G. Frattini in 1885. It plays a role in some more recent developments of group theory.

According to the definition, the Frattini subgroup Φ consists of all those elements of a group $\mathfrak{G}$ which can be dropped from a system of generators without affecting its being a system of generators. Another definition of $\Phi(\mathfrak{G})$ will be discussed in Chap. III, §1, ex. 18. Cf. also Chap. II, §2, ex. 15 and §3, ex. 10.

§2 A subgroup and its cosets. Lagrange's theorem

Every group $\mathfrak{G}$ has been seen to have at least two subgroups, the trivial subgroup $\mathfrak{e}$ and the improper subgroup $\mathfrak{G}$. For some of the groups examined in Chap. I the existence of proper subgroups could readily be established. In the present section we shall begin to discuss general properties of a subgroup in relation to the given group $\mathfrak{G}$.

a. Let $\mathfrak{H}$ be a subgroup of $\mathfrak{G} : \mathfrak{H} \leqslant \mathfrak{G}$. If $\mathfrak{H} < \mathfrak{G}$ one can find an element $a \in \mathfrak{G}$, $a \notin \mathfrak{H}$, and form the "left coset $a\mathfrak{H}$ of $\mathfrak{H}$ in $\mathfrak{G}$" (or "modulo $\mathfrak{H}$ in $\mathfrak{G}$"):

$$a\mathfrak{H} = \{ax \mid x \in \mathfrak{H}\}.$$

If $a' = ax$ is an element of $a\mathfrak{H}$ it is readily seen that $a'\mathfrak{H} = a(x\mathfrak{H}) = a\mathfrak{H}$. Thus every element of $a\mathfrak{H}$ can be taken as a representative of this coset: A coset is defined by any one of its elements.

If $a \notin \mathfrak{H}$ then none of the $ax \in a\mathfrak{H}$ is an element of $\mathfrak{H}$, for if

$ax = y \in \mathfrak{H}$ it follows that $a = yx^{-1} \in \mathfrak{H}$; hence $\mathfrak{H}$ and $a\mathfrak{H}$ are disjoint: $\mathfrak{H} \cap (a\mathfrak{H}) = \varnothing$ if $a \notin \mathfrak{H}$.

It may happen that $\mathfrak{G} = \mathfrak{H} \cup a\mathfrak{H}$. If, however, there is an element $b \in \mathfrak{G}$, but $b \notin \mathfrak{H}$ and $b \notin a\mathfrak{H}$ we can form the left coset $b\mathfrak{H}$. Again $\mathfrak{H} \cap b\mathfrak{H} = \varnothing$ and also $a\mathfrak{H} \cap b\mathfrak{H} = \varnothing$. For if $a\mathfrak{H}$ and $b\mathfrak{H}$ have a common element ax so that $ax = by$ for some $y \in \mathfrak{H}$ we conclude that $b^{-1}a = yx^{-1} \in \mathfrak{H}$ and $a = byx^{-1} \in b\mathfrak{H}$ whence $a\mathfrak{H} = b\mathfrak{H}$. By the same argument it is verified that two arbitrary left cosets having a common element coincide, or: *Two distinct cosets $a\mathfrak{H}$ and $b\mathfrak{H}$ of a subgroup $\mathfrak{H} < \mathfrak{G}$ are disjoint.*

This process of selecting an element in $\mathfrak{G}$ and forming another coset may (or may not) be continued. It leads to the decomposition of $\mathfrak{G}$ into mutually disjoint left cosets of $\mathfrak{H}$ in $\mathfrak{G}$, namely

$$\mathfrak{G} = \mathfrak{H} \cup a\mathfrak{H} \cup b\mathfrak{H} \cup \dots$$

The system $e, a, b, \dots$ of representatives of the different left cosets is called a *left transversal* of $\mathfrak{H}$ in $\mathfrak{G}$. If $\mathfrak{H} = \mathfrak{e}$ this is simply an arrangement of all the elements of $\mathfrak{G}$.

If i is the number of distinct cosets of a subgroup $\mathfrak{H}$ in $\mathfrak{G}$ we call i the *index of $\mathfrak{H}$ in $\mathfrak{G}$*; this index is often denoted by the symbol $(\mathfrak{G} : \mathfrak{H})$. It may represent an infinite cardinal number. If it is finite, we can find i representatives $e, a_2, \dots, a_i$ in $\mathfrak{G}$ such that

$$\text{(1)} \qquad \mathfrak{G} = \mathfrak{H} \cup a_2\mathfrak{H} \cup a_3\mathfrak{H} \cup \dots \cup a_i\mathfrak{H}, \quad a_2 \notin \mathfrak{H},$$
$$a_3 \notin \mathfrak{H} \cup a_2\mathfrak{H}, \dots a_i \notin \mathfrak{H} \cup a_2\mathfrak{H} \cup \dots \cup a_{i-1}\mathfrak{H}.$$

The index of a subgroup of a finite group is always finite.

If $\mathfrak{G}$ is infinite and the number of cosets of a subgroup $\mathfrak{H}$ in $\mathfrak{G}$ is not finite, then $\mathfrak{H}$ is called a *subgroup of infinite index*. Every finite subgroup of an infinite group is a subgroup of infinite index.

Instead of the left cosets $a\mathfrak{H}$ we may consider the right cosets $\mathfrak{H}b$. All statements made for the left cosets remain valid *mutatis mutandis* for the right cosets.

If $\mathfrak{H}$ is a finite subgroup of the (finite or infinite) group $\mathfrak{G}$ we find that $a\mathfrak{H}$ has the same number of elements as $\mathfrak{H}$ regardless of the element a in $\mathfrak{G}$; in symbols: $|a\mathfrak{H}| = |\mathfrak{H}| = |\mathfrak{H}b|$. Indeed, to distinct elements x, y of $\mathfrak{H}$ correspond distinct elements ax, ay in $a\mathfrak{H}$ and vice versa. Thus, if $|\mathfrak{G}| = g$ and $|\mathfrak{H}| = m$ we conclude from (1) that $g = mi$ and $(\mathfrak{G} : \mathfrak{H}) = g/m = i$. This then is the content of Lagrange's theorem, one of the basic facts of group theory:

THEOREM 1. (Lagrange, 1770). ***The order m and the index i of a subgroup of a finite group $\mathfrak{G}$ are divisors of the order g of $\mathfrak{G}$. In fact $g = mi$.***

COROLLARY 1. *Every element a of a finite group $\mathfrak{G}$ satisfies the condition $a^g = e$. If a has order m, then $m|g$.*

Proof. If $\mathfrak{G}$ is finite and $a \in \mathfrak{G}$ then the cyclic subgroup $\langle a \rangle$ is of finite order, say m (i.e. the number of distinct powers of a in $\mathfrak{G}$). Hence $g = mi$ if i is the index of $\langle a \rangle = \langle a \rangle_m$ in $\mathfrak{G}$. Since $a^m = e$ it follows that $a^g = (a^m)^i = e$.

COROLLARY 2. *A group of prime order is cyclic; it is generated by every one of its elements except by the unit element itself.*

Indeed if $g = p$, a prime number, the possible orders of the group elements are either 1 or p. Only the unit element e has order 1 and thus each one of the $p-1$ other elements has order p.

b. Index theorems. Between the left cosets and the right cosets of a subgroup $\mathfrak{H} \leqslant \mathfrak{G}$ there exists a one–one correspondence: If $e, a, b, \ldots$ is a left transversal, then $e, a^{-1}, b^{-1}, \ldots$ is a right transversal. Indeed for every left coset $a\mathfrak{H}$ the set $(a\mathfrak{H})^{-1} = \mathfrak{H}^{-1}a^{-1} = \mathfrak{H}a^{-1}$ is a right coset and conversely. If $a\mathfrak{H} \neq b\mathfrak{H}$ it follows that $\mathfrak{H}a^{-1} \neq \mathfrak{H}b^{-1}$. Thus whether the group $\mathfrak{G}$ is finite or infinite, the index $i = (\mathfrak{G} : \mathfrak{H})$ of a subgroup $\mathfrak{H}$ of $\mathfrak{G}$ may be defined equally as the number of left cosets or the number of right cosets of $\mathfrak{H}$ in $\mathfrak{G}$ (cf. ex. 3).

THEOREM 2. *Let* $\mathfrak{K} < \mathfrak{H} < \mathfrak{G}$ *with finite indices* $(\mathfrak{G} : \mathfrak{H}) = i$, $(\mathfrak{G} : \mathfrak{K}) = j$; *then* $i|j$ *and*

$$\frac{j}{i} = \frac{(\mathfrak{G} : \mathfrak{K})}{(\mathfrak{G} : \mathfrak{H})} = (\mathfrak{H} : \mathfrak{K}).$$

Proof. Since $\mathfrak{K}$ has a finite number j of left cosets in $\mathfrak{G}$ it cannot have more than j cosets in $\mathfrak{H}$. Thus $l = (\mathfrak{H} : \mathfrak{K})$ is finite. Let $\mathfrak{H} = \mathfrak{K} \cup b_2\mathfrak{K} \cup \ldots \cup b_l\mathfrak{K}$, $b_2 \notin \mathfrak{K}$, $b_3 \notin \mathfrak{K} \cup b_2\mathfrak{K}, \ldots$. Substituting this into the decomposition (1) of $\mathfrak{G}$ (mod $\mathfrak{H}$) we obtain the decomposition of $\mathfrak{G}$ (mod $\mathfrak{K}$) where the left cosets appear in the form $a_\rho b_\lambda \mathfrak{K}$, $a_\rho \in \mathfrak{G}$, $b_\lambda \in \mathfrak{H}$ ($\rho = 1, \ldots, i$; $\lambda = 1, \ldots, l$). The a_ρ lie in distinct cosets of $\mathfrak{H}$ in $\mathfrak{G}$, the b_λ in distinct cosets of $\mathfrak{K}$ in $\mathfrak{H}$. The $il = j$ cosets $a_\rho b_\lambda \mathfrak{K}$ are all distinct. Suppose that $a_\rho b_\lambda \mathfrak{K} = a_\sigma b_\mu \mathfrak{K}$. Multiplying this equation by $\mathfrak{H}$ from the right side, we obtain $a_\rho \mathfrak{H} = a_\sigma \mathfrak{H}$; thus $\rho = \sigma$ and $b_\lambda \mathfrak{K} = b_\mu \mathfrak{K}$ whence $\lambda = \mu$. Consequently there are il distinct cosets of $\mathfrak{K}$ in $\mathfrak{G}$.

By a similar argument the following statement can be verified:

THEOREM 2′. *Let* $\mathfrak{K} < \mathfrak{H} < \mathfrak{G}$ *with finite indices* $(\mathfrak{G} : \mathfrak{H})$ *and* $(\mathfrak{H} : \mathfrak{K})$; *then* $(\mathfrak{G} : \mathfrak{K}) = (\mathfrak{G} : \mathfrak{H})(\mathfrak{H} : \mathfrak{K})$.

c. Further we consider the case of the intersection $\mathfrak{D} = \mathfrak{H} \cap \mathfrak{K}$ of two subgroups $\mathfrak{H}$, $\mathfrak{K}$ of finite index in $\mathfrak{G}$. The intersection $\mathfrak{D}$ is a subgroup of $\mathfrak{H}$ and of $\mathfrak{K}$. Let

$$(2) \qquad \mathfrak{H} = \mathfrak{D} \cup b_2\mathfrak{D} \cup b_3\mathfrak{D} \cup \ldots, \quad b_2 \notin \mathfrak{D},$$
$$b_3 \notin \mathfrak{D} \cup b_2\mathfrak{D}, \ldots, \quad b_\lambda \in \mathfrak{H}.$$

Then

$$(3) \qquad \mathfrak{G} = \mathfrak{K} \cup b_2\mathfrak{K} \cup b_3\mathfrak{K} \cup \ldots \cup b_l\mathfrak{K} \cup a_{l+1}\mathfrak{K} \cup \ldots \cup a_j\mathfrak{K},$$
$$a_\nu \notin \mathfrak{H},$$

and we observe

1. Two cosets $b_\lambda\mathfrak{K}$, $b_\mu\mathfrak{K}$ ($\lambda, \mu = 1, \ldots, l$; $b_1 = e$, $\mu \neq \lambda$) are distinct, for $b_\mu\mathfrak{K} = b_\lambda\mathfrak{K}$ implies $b_\lambda^{-1}b_\mu \in \mathfrak{K}$. On the other hand $b_\lambda, b_\mu \in \mathfrak{H}$; hence $b_\lambda^{-1}b_\mu \in \mathfrak{D}$, that is $b_\mu\mathfrak{D} = b_\lambda\mathfrak{D}$ which contradicts the assumption.

2. Every $b\mathfrak{K}$ ($b \in \mathfrak{H}$) coincides with one of the $b_\lambda\mathfrak{K}$ in (3). Indeed for every $b \in \mathfrak{H}$ there exists a b_λ such that $b \in b_\lambda\mathfrak{D}$ since by supposition the $b_1, b_2, \ldots, b_l$ represent a left transversal of $\mathfrak{D}$ in $\mathfrak{H}$. Hence $b\mathfrak{D} = b_\lambda\mathfrak{D}$ which implies $b^{-1}b_\lambda \in \mathfrak{D}$ and since $\mathfrak{D} \leqslant \mathfrak{K}$ also $b\mathfrak{K} = b_\lambda\mathfrak{K}$.

Thus there cannot be more cosets $b_\lambda\mathfrak{D}$ of $\mathfrak{D}$ in $\mathfrak{H}$ [cf. (2)] than there are cosets $b_\lambda\mathfrak{K}$ of $\mathfrak{K}$ in $\mathfrak{G}$ [cf. (3)] which means that the index $(\mathfrak{H} : \mathfrak{D})$ is finite. Hence by Theorem 2′

$$(4) \qquad (\mathfrak{G} : \mathfrak{D}) = (\mathfrak{G} : \mathfrak{H})(\mathfrak{H} : \mathfrak{D})$$

is finite.

As a further consequence of the preceding discussion we note that

(5) $(\mathfrak{G}:\mathfrak{K}) \geqslant (\mathfrak{H}:\mathfrak{D})$

Thus by (4)

(6) $(\mathfrak{G}:\mathfrak{D}) \leqslant (\mathfrak{G}:\mathfrak{H})(\mathfrak{G}:\mathfrak{K}).$

In (4) and (5) $\mathfrak{H}$ and $\mathfrak{K}$ are interchangeable. The principal result is the following theorem:

THEOREM 3. (Poincaré). *The intersection of two subgroups of finite index in a group $\mathfrak{G}$ is a subgroup of finite index in $\mathfrak{G}$.*

Remark. If $\mathfrak{H}$ and $\mathfrak{K}$ have finite index in $\mathfrak{G}$ and if $\mathfrak{H} \cap \mathfrak{K}$ is finite, the group $\mathfrak{G}$ is necessarily finite. If $\mathfrak{H} \cap \mathfrak{K} = \mathfrak{e}$ the equation (4) is another expression of Lagrange's theorem.

In connection with the preceding theorem we shall derive another important index relation. Multiply both sides of the equation (2) by $\mathfrak{K}$ from the right. Using the fact that $\mathfrak{D}\mathfrak{K} = \mathfrak{K}$ we have

$$\mathfrak{H}\mathfrak{K} = \mathfrak{K} \cup b_2\mathfrak{K} \cup b_3\mathfrak{K} \cup \ldots$$

irrespective whether $\mathfrak{H}\mathfrak{K} \leqslant \mathfrak{G}$ or only $\mathfrak{H}\mathfrak{K} \subseteq \mathfrak{G}$. As shown before, the $b_2\mathfrak{K}$ are all mutually disjoint. Hence the following result:

THEOREM 3′. *There is a one–one correspondence between the cosets of $\mathfrak{H} \cap \mathfrak{K}$ in $\mathfrak{H}$ and the cosets of $\mathfrak{K}$ in $\mathfrak{H}\mathfrak{K}$. Thus if either of the indices $(\mathfrak{H}\mathfrak{K}:\mathfrak{K})$ * or $(\mathfrak{H}:\mathfrak{H} \cap \mathfrak{K})$ is finite, then so is the other and*

$$(\mathfrak{H}\mathfrak{K}:\mathfrak{K}) = (\mathfrak{H}:\mathfrak{H} \cap \mathfrak{K}).$$

d. Finite cyclic groups. We are now in the position to develop the general theory of the simplest type of a finite group, the cyclic group

$$\langle a \rangle_m = \{a^0 = e, a, a^2, \ldots, a^{m-1} \,|\, a^m = e, a^\mu \neq e, 0 < \mu < m\}.$$

Some of its properties have been established in earlier sections (cf. *a.*, and Chap. I, §1, *f.* and ex. 3 and 4).

* Even if $\mathfrak{H}\mathfrak{K}$ is not a group, it is a union of complete cosets of $\mathfrak{K}$ and the symbol $(\mathfrak{H}\mathfrak{K}:\mathfrak{K})$ indicates the number of left cosets of $\mathfrak{K}$ contained in $\mathfrak{H}\mathfrak{K}$.

Every element $x \in \langle a \rangle_m$ has as its order a certain divisor d of m and for every such d there is an element of order d, namely $a^{m/d}$. (This is not in general the case for a non-cyclic group of order m; for example, $\mathfrak{S}_4$ contains no element of order 6.)

An element $x = a^\mu$ which has order m, as for instance the element a itself, is said to be *primitive* in $\langle a \rangle_m$. In order to determine all the primitive elements of $\langle a \rangle_m$ let μ be one of the $\varphi(m)$ distinct positive integers less than m which are relatively prime to m. We shall show that the corresponding element a^μ has order m, that is

$$\langle a^\mu \rangle_m = \langle a \rangle_m.$$

To prove this fact let us assume that a^μ has order n. This implies that $\mu n \equiv 0 \pmod m$, i.e. $m \mid \mu n$. As $(\mu, m) = 1$ we conclude that $m \mid n$. On the other hand by Lagrange's theorem $n \mid m$. Hence the order n of a^μ must equal m.

It is readily seen that a^ν has order $k = m/\delta$ if $(\nu, m) = \delta$. Indeed $(a^\nu)^k = (a^{\nu/\delta})^m = e$ and $a^{\nu x} = e$ implies that $\nu x \equiv 0 \pmod m$ for which congruence $x = k$ is the least positive solution.

Now let us assume that in the group $\langle a \rangle_m$ there are $\psi(d)$ elements of order d. From the preceding result we know that there are $\varphi(d)$ elements of order d in $\langle a^{m/d} \rangle_d$. Hence $\psi(d) \geqslant \varphi(d)$. But every one of the m elements of $\langle a \rangle_m$ has a certain order d where $d \mid m$, and therefore

$$\sum_{d/m} \psi(d) = m.$$

With regard to Chap. I, §3, (3′)

$$\sum_{d/m} (\psi(d) - \varphi(d)) = 0.$$

Since the terms of this sum are non-negative they must equal zero; thus $\psi(d) = \varphi(d)$; hence all elements of order d in $\langle a \rangle_m$ are elements of the cyclic subgroup $\langle a^{m/d} \rangle_d$.

Finally we show that every subgroup $\mathfrak{H}$ of the cyclic group $\langle a \rangle_m$ is itself cyclic and therefore coincides with one of the subgroups $\langle a^{m/d} \rangle_d$, $d \mid m$. Indeed let $a^\lambda \in \mathfrak{H}$ and $a^\mu \in \mathfrak{H}$ where a^μ is not a power of a^λ and let $\rho = (\lambda, \mu) = \alpha\lambda + \beta\mu$ for certain integers α, β. We conclude that $a^\rho \in \mathfrak{H}$ and since $\lambda = \lambda'\rho$, $\mu = \mu'\rho$ for integers λ', μ', it follows that a^λ and a^μ are powers of a^ρ. If further $a^\nu \in \mathfrak{H}$ and a^ν is not a power of a^ρ, let $\sigma = (\nu, \rho)$; then $a^\sigma \in \mathfrak{H}$ and a^λ, a^μ, a^ν are powers of a^σ. Now either $\mathfrak{H} = \langle a^\sigma \rangle$ and the statement is established, or we can continue using the same argument. After a finite number of repetitions we obtain $\mathfrak{H} = \langle a^\tau \rangle_d$ for a certain divisor d of m. This group contains all those elements of $\langle a \rangle_m$ whose order divides d. It is therefore unique and coincides with $\langle a^{m/d} \rangle_d$.

We restate the principal facts concerning finite cyclic groups:

THEOREM 4. *The cyclic group $\langle a \rangle_m$ of order $m > 1$ has $\varphi(m)$ primitive elements. An element a^μ is primitive if and only if $(\mu, m) = 1$ $(0 < \mu < m)$. Every subgroup of $\langle a \rangle_m$ is cyclic and for every divisor d of m there is exactly one cyclic subgroup of order d in $\langle a \rangle_m$, namely $\langle a^{m/d} \rangle_d$.*

Concerning infinite cyclic groups cf. ex. 5.

Examples and exercises

1. Let $\mathfrak{G}$ be a group and $\mathfrak{H}$ a subgroup of index 2 in $\mathfrak{G}$. Show that for every $a \in \mathfrak{G}$ the square $a^2 \in \mathfrak{H}$.

2. Is it true that if $\mathfrak{H} < \mathfrak{G}$ and the index $(\mathfrak{G} : \mathfrak{H}) = 3$, then for every $a \in \mathfrak{G}$ the element $a^3 \in \mathfrak{H}$?

3. Provide the details in the proofs of the following statements:

(a) All cosets (left and right) of a finite subgroup of a group contain the same number of elements.

(b) If a subgroup $\mathfrak{H} < \mathfrak{G}$ has a finite number i of left cosets (incl. $\mathfrak{H}$ itself) in $\mathfrak{G}$, then it also has i right cosets in $\mathfrak{G}$.

4. Carry through the proof of theorem 2′.

5. Let $\langle a \rangle = \langle a \rangle_\infty$ denote the infinite cyclic group generated by the element a, i.e. $\langle a \rangle = \ldots a^{-2}, a^{-1}, a^0 = e, a, a^2, \ldots$.

(a) Show that $\langle a \rangle$ contains exactly two primitive (generating) elements.

(b) Prove that every subgroup $\mathfrak{H}$ of $\langle a \rangle$ is cyclic and infinite, except $\mathfrak{H} = \mathfrak{e}$; thus $\mathfrak{H} = \langle a^r \rangle$ if r denotes the least positive exponent for which $a^r \in \mathfrak{H}$.

(c) Determine the cosets and the index of $\langle a^r \rangle$ in $\langle a \rangle$.

6. In the symmetric group $\mathfrak{S}_4$ consider the subgroup $\mathfrak{H}$ of all the permutations of 1, 2, 3, 4 which leave the symbol 4 fixed so that $\mathfrak{H} \simeq \mathfrak{S}_3$.

(a) Show that not every left transversal of $\mathfrak{H}$ in $\mathfrak{S}_4$ is a right transversal.

(b) Find a system of permutations in $\mathfrak{S}_4$ which is simultaneously a left as well as a right transversal of $\mathfrak{H}$ in $\mathfrak{S}_4$.

Remark. A left (right) transversal of a subgroup $\mathfrak{H}$ in $\mathfrak{G}$ ($\mathfrak{H} > \mathfrak{e}$) is not unique. If $e, a, b, \ldots$ is a left transversal of $\mathfrak{H}$ in $\mathfrak{G}$, then so is $e', a', b', \ldots, e' \in \mathfrak{H}, a' \in a\mathfrak{H}, b' \in b\mathfrak{H}, \ldots$ with no other restriction.

For a generalization of ex. 6(b) cf. §6, *c*.

7. Apply Corollary 1 of Lagrange's theorem to the multiplicative residue class group $\mathfrak{R}_m$ (cf. Chap. I, §3, *b*.). Hence

$$a^{\varphi(m)} \equiv 1 \pmod m, \quad a \in \mathbb{Z}, \quad (a, m) = 1$$

which is Euler's theorem of elementary number theory, Fermat's theorem if $m = p$, p a prime, and $\varphi(p) = p-1$.

8. Let $\mathfrak{G}$ be a group with elements $a, b, \ldots$ and $\mathfrak{H} < \mathfrak{G}$. Suppose that $(\mathfrak{G}:\mathfrak{H}) = 2$. If $a \notin \mathfrak{H}$ and $b \notin \mathfrak{H}$ then it follows that $ab \in \mathfrak{H}$.

9. Let $x, y \in \mathfrak{G}$ and $\mathfrak{H} \leqslant \mathfrak{G}$. Let the relation $x \sim y$ be defined by $x^{-1}y \in \mathfrak{H}$. Show that this is an equivalence relation in $\mathfrak{G}$ and that the left cosets (mod $\mathfrak{H}$) are the corresponding equivalence classes.

10. THEOREM. *The multiplicative group* $\mathfrak{F}$ *of a finite field* $\mathbb{F}$ *is cyclic.*

Proof. Let $|\mathbb{F}| = q$. (According to a theorem of algebra, not to be used in this proof, $q = p^m$ where p is a prime number and m a positive integer.) Then $|\mathfrak{F}| = q-1$ and therefore by Lagrange's theorem $\alpha^{q-1} = 1$ for every $\alpha \in \mathfrak{F}$, i.e. $\alpha \in \mathbb{F}$, $\alpha \neq 0$. Moreover the order t of α in $\mathfrak{F}$ is a divisor of $q-1$.

Let $\psi(t)$ denote the number of elements of order t in $\mathfrak{F}$ so that

$$\sum_{t|q-1} \psi(t) = q-1.$$

If α is an element of order t, it is a root of the equation $\xi^t - 1 = 0$ of degree t and $1, \alpha, \alpha^2, \ldots, \alpha^{t-1}$ constitute a set of t distinct roots of this equation. According to a theorem of elementary algebra an algebraic equation of degree t with coefficients from a field cannot have more than t roots in this field. In the present case the t roots are the elements of the cyclic group $\langle\alpha\rangle_t$ which contains $\varphi(t)$ elements of order t; hence $\psi(t) = \varphi(t)$. If an element of order t does not exist in $\mathfrak{F}$ we have $\psi(t) = 0$. Thus $\psi(t) \leqslant \varphi(t)$. With regard to Chap. I, §3, (3′) we conclude that $\psi(t) = \varphi(t)$ for every divisor t of $q-1$ and in particular

$$\psi(q-1) = \varphi(q-1).$$

Thus $\mathfrak{F}$ contains $\varphi(q-1)$ elements of order $q-1$ and each of the $\varphi(q-1)$ primitive elements is a generator of the group $\mathfrak{F}$.

COROLLARY. *Let p be a prime number. The multiplicative residue class group $\mathfrak{R}_p$ is a cyclic group of order $p-1$.*

If $\bar{a}$ is one of the $\varphi(p-1)$ generators of $\mathfrak{R}_p$ then the number a is called a *primitive congruence root* (mod p).

Remark. Let p be an odd prime. It can be shown that the group $\mathfrak{R}_m$ is cyclic if and only if $m = 2, 4, p^k, 2p^k$ $(k = 1, 2, \ldots)$.

11. Let m and n be relatively prime natural numbers and $c \in \mathfrak{G}$ an element of order mn. Then there is one and only one pair of commuting elements a and b, a of order m and b of order n, for which $c = ab$.

12. *The icosahedral group* $\mathfrak{J}$. A cardboard model of the regular icosahedron **J** is readily produced from a piece of paper in the shape shown in Fig. 4. For the following discussion we assume that the reader has before him such a model. From it we recognize that **J** has three types of symmetry axes:

6 axes L_0 through opposite vertices of **J**,
15 axes L_1 through the midpoints of opposite edges,
10 axes L_2 through the midpoints of opposite faces

(cf. Chap. I, §4, *d.*). We denote by $\mathfrak{L}_0, \mathfrak{L}_1, \mathfrak{L}_2$ the subgroup of all rotations of **J** about a fixed axis L_0, L_1, L_2 respectively.

Because every vertex of **J** is common to 5 edges (faces), $|\mathfrak{L}_0| = 5$. Every edge has two vertices as endpoints (and belongs to the boundary

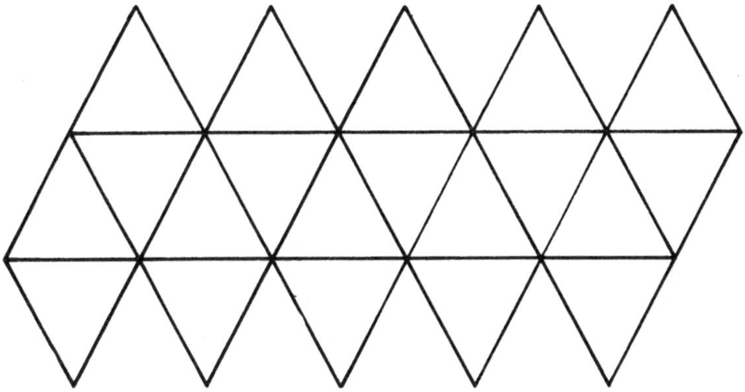

Figure 4

of two faces); hence $|\mathfrak{L}_1| = 2$. Finally, since every face is an equilateral triangle, we have $|\mathfrak{L}_2| = 3$. No two of the 31 subgroups $\mathfrak{L}_0, \mathfrak{L}_1, \mathfrak{L}_2$ can have an element different from the identity in common. Thus the 60 elements of the rotation group $\mathfrak{J}$ are

$$\begin{aligned} &1 \text{ identity}\\ 30-\ 6 &= 24 \text{ rotations of order 5}\\ 30-15 &= 15 \text{ rotations of order 2}\\ 30-10 &= 20 \text{ rotations of order 3.}\end{aligned}$$

Further we learn from our model that the icosahedron **J** can be placed into a cartesian $x_1x_2x_3$-system in such a way that the coordinate axes are three of the 15 axes of the second type (L_1). Perpendicular to each of the three coordinate axes there is a pair of opposite edges; indeed if the pair perpendicular to the x_1-axis lies in the x_1x_2-plane, then the pair perpendicular to the x_2-axis lies in the x_2x_3-plane, and the pair perpendicular to the x_3-axis lies in the x_3x_1-plane. Thus in each of the coordinate planes lie four of the 12 vertices of **J**. Fig. 5 is to illustrate this situation.

The constellation, consisting of three perpendicular axes of the type L_1 together with the opposite pairs of edges, is rigidly connected with the icosahedron **J** and every covering rotation of the constellation is a covering rotation of **J**, but not conversely. The constellation uniquely defines a cube **C** so that the midpoints of the faces of the cube coincide with the midpoints of the edges of **J** intersecting the

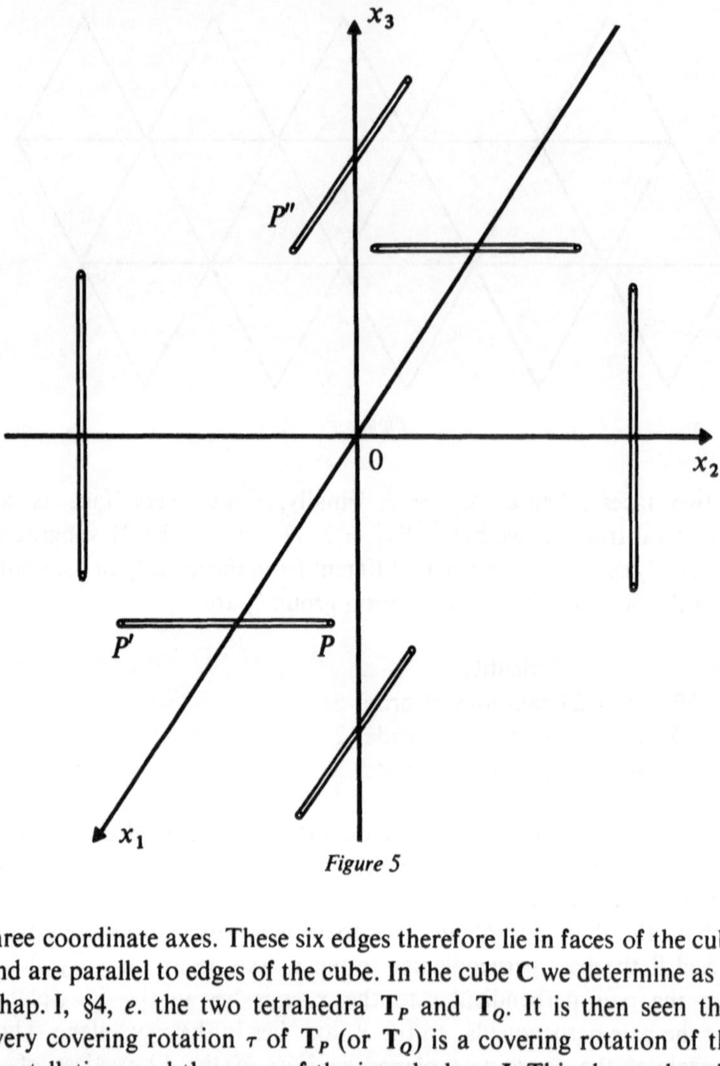

Figure 5

three coordinate axes. These six edges therefore lie in faces of the cube and are parallel to edges of the cube. In the cube **C** we determine as in Chap. I, §4, *e*. the two tetrahedra $\mathbf{T}_P$ and $\mathbf{T}_Q$. It is then seen that every covering rotation τ of $\mathbf{T}_P$ (or $\mathbf{T}_Q$) is a covering rotation of the constellation, and thus one of the icosahedron **J**. This shows that the group $\mathfrak{T}$ of the covering rotations of the tetrahedron $\mathbf{T}_P$ is a subgroup of $\mathfrak{J}$. We have $|\mathfrak{T}| = 12$, $(\mathfrak{J} : \mathfrak{T}) = 5$.

The group $\mathfrak{T}$ cannot be cyclic; by Chap. I, §4, ex. 7: $\mathfrak{T} \simeq \mathfrak{A}_4$. Hence none of its elements has the order 12 and for all 12 elements τ of $\mathfrak{T}$ we have

(7) $\tau^6 = \epsilon$ (identity).

We shall now derive the decomposition of $\mathfrak{J}$ into cosets of $\mathfrak{T}$. Since no element of $\mathfrak{T}$ can have the order 5 we conclude that $\mathfrak{T} \cap \mathfrak{L}_0 = \epsilon$ for the six groups $\mathfrak{L}_0$. Let us designate by P and P' those vertices of **J** which lie on the edge intersecting the positive x-axis (P in the first quadrant of the x_1x_2-plane, cf. Fig. 6) and denote by L_0 the straight line $\overline{OP}$. Further let α be the rotation about L_0 through the angle $2\pi/5$ so that $\mathfrak{L}_{0,} = \langle\alpha\rangle_5$. We shall show that

$$\mathfrak{J} = \mathfrak{T} \cup \alpha\mathfrak{T} \cup \alpha^2\mathfrak{T} \cup \alpha^3\mathfrak{T} \cup \alpha^4\mathfrak{T}.$$

The 60 elements $\alpha^i\tau$, $i = 0, 1, 2, 3, 4$, $\tau \in \mathfrak{T}$, are distinct. Indeed let $\tau, \tau' \in \mathfrak{T}$ and suppose that $\alpha^i\tau = \alpha^j\tau'$: Then $\alpha^{i-j} = \tau'\tau^{-1} = \tau'' \in \mathfrak{T}$ and therefore by (7) $\alpha^{6(i-j)} = \epsilon$. This implies that $6(i-j) \equiv 0 \pmod 5$ or $i \equiv j \pmod 5$ whence $\alpha^i = \alpha^j$ and $\tau' = \tau$.

Now we shall derive a matrix representation for $\mathfrak{J}$. A representation of $\mathfrak{T}$ by 3×3-matrices has been obtained in Chap. I, §4, *e*. Thus it will be sufficient to establish a proper orthogonal matrix A representing the rotation α. From the definition of α it follows that A must have the vector $\overrightarrow{OP}$ as eigen vector. Moreover, if P'' is the vertex of **J** situated in the positive quadrant of the x_1x_3-plane, then the image of P'' under the mapping represented by the matrix A will be the vertex P'.

So let us determine the coordinates of the three vertices P, P', P'' (cf. Fig. 6). If s is the length of an edge of **J** and r the distance of an edge from the origin (centre) O, the coordinates of P are

$$x_1 = r, \quad x_2 = \tfrac{1}{2}s, \quad x_3 = 0.$$

From the figure the relation between r and s is given by the quadratic equation

$$(r - \tfrac{1}{2}s)^2 + r^2 = \tfrac{3}{4}s^2.$$

If we choose $s = 2$ we have the relation

$$(8) \qquad r^2 - r - 1 = 0.$$

Hence

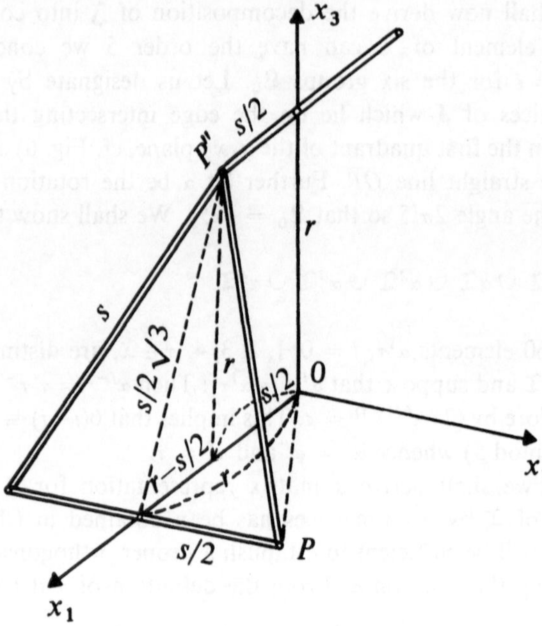

Figure 6

$$P: (r, 1, 0), \quad P': (r, -1, 0), \quad P'': (1, 0, r)$$

where r denotes the positive root of the equation (8).

Now we determine the properly orthogonal matrix

$$A = \begin{pmatrix} a_{11} & a_{12} & a_{13} \\ a_{21} & a_{22} & a_{23} \\ a_{31} & a_{32} & a_{33} \end{pmatrix}$$

so that it corresponds to the rotation α. To have $\overrightarrow{OP}$ as eigen vector with the eigen value 1 the matrix A must satisfy the conditions

$$a_{11}r + a_{12} = r, \quad a_{21}r + a_{22} = 1, \quad a_{31}r + a_{32} = 0.$$

Using the orthogonality conditions for the columns (or the fact that $A' = A^{-1}$ has the same axis as A) we obtain

$$a_{11}r + a_{21} = r, \quad a_{12}r + a_{22} = 1, \quad a_{13}r + a_{23} = 0.$$

The condition that A rotates the vector $\overline{OP''}$ into $\overline{OP'}$ leads to the three equations

$$a_{11}+a_{13}r = r, \quad a_{21}+a_{23}r = -1, \quad a_{31}+a_{33}r = 0,$$

whence again by orthogonality

$$a_{11}r-a_{21} = 1, \quad a_{12}r-a_{22} = 0, \quad a_{13}r-a_{23} = r.$$

From these relations, together with the quadratic equation (8), we find that

$$\text{(9)} \qquad A = \tfrac{1}{2}\begin{pmatrix} r & r-1 & 1 \\ r-1 & 1 & -r \\ -1 & r & r-1 \end{pmatrix}.$$

Finally we make the connection with a permutation group:

THEOREM. *The icosahedral group $\mathfrak{J}$ is isomorphic to the alternating group $\mathfrak{A}_5$.*

Proof. From the 15 axes of type L_1 one can choose 5 triplets $\Sigma_0, \Sigma_1, \Sigma_2, \Sigma_3, \Sigma_4$, each of which has three mutually perpendicular axes; one of these triplets, e.g. Σ_0, served us as reference system with the axes x_1, x_2, x_3. The rotations τ in the tetrahedral group $\mathfrak{T}$ have the following property: They carry $\Sigma_0 \to \Sigma_0$, thereby interchanging the axes x_1, x_2, x_3 as well as the other triplets $\Sigma_1, \Sigma_2, \Sigma_3, \Sigma_4$. The rotation α, however, about the axis $\overline{OP}$ leaves none of the Σ_i fixed. If it carries Σ_0 into Σ_1 then all $\alpha\tau$ $(\tau \in \mathfrak{T})$ carry Σ_0 into Σ_1. Let us choose the numeration of the Σ_i so that

$$\alpha^i\Sigma_0 = \Sigma_i \quad (i = 0, 1, 2, 3, 4).$$

In view of the model of the icosahedron it should be clear that one can do this without loss of generality. So it is suggested to associate with the rotation α the cycle permutation $C = (0\ 1\ 2\ 3\ 4)$. Moreover $\mathfrak{J}$ contains the group $\mathfrak{T}$ which is isomorphic to the alternating group $\mathfrak{A}_4$. The group $\mathfrak{T}$ leaves Σ_0 fixed. The corresponding permutation group therefore acts on the symbols 1, 2, 3, 4. We conclude that the group $\mathfrak{G}$ of permutations of degree 5 contains the set

$$\mathfrak{A}_4 \cup C\mathfrak{A}_4 \cup C^2\mathfrak{A}_4 \cup C^3\mathfrak{A}_4 \cup C^4\mathfrak{A}_4.$$

It is sufficient to show that this union consists of 60 different elements. Indeed, suppose that for some permutations P, Q in $\mathfrak{A}_4$ we had $C^i P = C^j Q$, i.e. $C^{i-j} \in \mathfrak{A}_4$; then $C^{i-j} = I$ and $Q = P$. Thus $\mathfrak{G} = \mathfrak{A}_5$.

13. Show that the group $\mathfrak{J}$ has the presentation

$$\langle \alpha, \rho \,|\, \alpha^5 = \epsilon, \, \rho^2 = \epsilon, \, (\alpha \cdot \rho)^3 = \epsilon \rangle.$$

14. *The group* $\mathfrak{J}$. The example of the tetrahedral groups, $\mathfrak{T} \simeq \mathfrak{A}_4$ and $\tilde{\mathfrak{T}} \simeq \mathfrak{S}_4$, and the fact that $\mathfrak{J} \simeq \mathfrak{A}_5$ might suggest that $\tilde{\mathfrak{J}}$ is isomorphic to the symmetric group $\mathfrak{S}_5$. This, however, is not the case. Indeed, the icosahedron **J** has 15 symmetry planes, one through every pair of opposite edges; therefore $\tilde{\mathfrak{J}}$ contains 15 reflections, that is at least 15 operations of order 2. Adding to these the 15 rotations of order 2, contained in $\mathfrak{J}$, raises the number to 30. But the symmetric group $\mathfrak{S}_5$ contains only 10 transpositions and 12 products of two disjoint transpositions, i.e. 22 operations of order 2. Thus $\tilde{\mathfrak{J}} \neq \mathfrak{S}_5$. (cf. §4, ex. 17.)

From Chap. I, §4, *d.* we know that if σ is one of the covering reflexions (or reflexion-rotations) of **J**, then $\tilde{\mathfrak{J}} = \mathfrak{J} \cup \sigma \mathfrak{J}$. Accordingly the orthogonal representation of $\tilde{\mathfrak{J}}$ is obtained by adjoining to the representation of $\mathfrak{J}$ a single reflexion matrix, e.g. the matrix B_1 which corresponds to the reflexion σ_1 with respect to the $x_2 x_3$-plane of the system Σ_0, i.e.

$$B_1 = \begin{pmatrix} -1 & 0 & 0 \\ 0 & 1 & 0 \\ 0 & 0 & 1 \end{pmatrix}.$$

A representation of $\tilde{\mathfrak{J}}$ by a group of permutations of degree 5 is obviously impossible. The group $\mathfrak{G}$ representing $\mathfrak{J}$ isomorphically was the alternating group on the five systems $\Sigma_0, \ldots, \Sigma_4$, or briefly, on the symbols $0, \ldots, 4$: Every operation in $\mathfrak{J}$ applied to Σ_i reproduces Σ_i (including its orientation) or changes Σ_i into a system Σ_j with the same orientation. But every reflexion σ applied to Σ_i changes the orientation of Σ_i. The same is therefore true for every operation in $\sigma_1 \mathfrak{J}$. Thus, for the representation of $\tilde{\mathfrak{J}}$ we need 10 symbols.

Denoting by Σ_i' the system Σ_i with negative orientation and letting

$$0, 1, 2, 3, 4 \quad \text{stand for} \quad \Sigma_0, \Sigma_1, \Sigma_2, \Sigma_3, \Sigma_4$$

$$0', 1', 2', 3', 4' \quad \text{stand for} \quad \Sigma_0', \Sigma_1', \Sigma_2', \Sigma_3', \Sigma_4'$$

we have the rotation α, (i.e. the matrix A of ex. 12) represented by the permutation (0 1 2 3 4) (0′ 1′ 2′ 3′ 4′), and similarly the whole subgroup $\mathfrak{J}$ represented by the "duplicated" alternating group $\mathfrak{A}_4^{(2)}$ which is, of course, isomorphic to $\mathfrak{A}_4$.

For the representation of the coset $\sigma_1\mathfrak{J}$ we shall construct the permutation of degree 10 which corresponds to the reflexion σ_1 (or to the matrix B_1). For this purpose we notice that

the columns of A^0 represent the axis vectors of Σ_0,
the columns of A represent the axis vectors of Σ_1,
the columns of A^2 represent the axis vectors of Σ_2, etc.

Now $A^0 = E$ and

$$A = \tfrac{1}{2}\begin{pmatrix} r & r-1 & 1 \\ r-1 & 1 & -r \\ -1 & r & r-1 \end{pmatrix}, \quad A^2 = \tfrac{1}{2}\begin{pmatrix} 1 & r & r-1 \\ r & 1-r & -1 \\ 1-r & 1 & -r \end{pmatrix},$$

$$A^3 = \tfrac{1}{2}\begin{pmatrix} 1 & r & 1-r \\ r & 1-r & 1 \\ r-1 & -1 & -r \end{pmatrix}, \quad A^4 = \tfrac{1}{2}\begin{pmatrix} r & r-1 & -1 \\ r-1 & 1 & r \\ 1 & -r & r-1 \end{pmatrix}$$

$$= A' = A^{-1}.$$

So we find $\sigma_1\Sigma_0 = \Sigma_0'$, $\sigma_1\Sigma_1 = \Sigma_2'$, $\sigma_1\Sigma_3 = \Sigma_4'$. Accordingly the reflexion σ_1 is represented by the permutation

(0 0′) (1 2′) (2 1′) (3 4′) (4 3′),

a product of five transpositions, therefore an odd permutation of degree 10.

15. Determine the Frattini subgroup (cf. §1, ex. 11) of a finite cyclic group.

16. Let $\mathfrak{G}$ be a finite abelian group and let $\mathfrak{H}_n$ and $\mathfrak{L}_n$ be the two subgroups of $\mathfrak{G}$ which have been introduced in Chap. I, §2, ex. 14. Show that

$$|\mathfrak{H}_n||\mathfrak{L}_n| = |\mathfrak{G}|.$$

§3 Homomorphisms, normal subgroups and factor groups

a. In Chapter I, §1, *e.* we defined an isomorphism from a group $\mathfrak{G}$ onto a group $\mathfrak{G}'$ to be an invertible mapping $f : \mathfrak{G} \to \mathfrak{G}'$ which satisfies the functional equation

$$(1) \qquad f(xy) = f(x)f(y), \quad x, y \in \mathfrak{G}.$$

This concept of isomorphism will now be generalized by eliminating the restriction that f be one–one and onto. A mapping $f : \mathfrak{G} \to \mathfrak{G}'$ from a group $\mathfrak{G}$ *into* a group $\mathfrak{G}'$ will be called a *homomorphism* if f satisfies the functional equation (1). If the homomorphism f is one–one we shall say that it is a *monomorphism.* If it is onto it is called an *epimorphism.* A homomorphism which is simultaneously an epimorphism and a monomorphism is an isomorphism.

Remark. In general a homomorphism is a multi-one mapping. This means that to several elements in $\mathfrak{G}$ may correspond one and the same element of $\mathfrak{G}'$. Since $f(ex) = f(e)f(x) = f(x)$ we conclude that $f(e) = e'$, the unit element of $\mathfrak{G}'$. Also $f(x \cdot x^{-1}) = f(x)f(x^{-1}) = f(e) = e'$, so that $f(x^{-1}) = f(x)^{-1}$.

THEOREM 1. *Let $f : \mathfrak{G} \to \mathfrak{G}'$ be a homomorphism of $\mathfrak{G}$ into $\mathfrak{G}'$. Then the image set $f(\mathfrak{G})$ is a subgroup of $\mathfrak{G}'$.*

Indeed if $x', y' \in f(\mathfrak{G})$ then there exists at least one element x and one element y in $\mathfrak{G}$ such that $x' = f(x)$ and $y' = f(y)$. But then $x'y' = f(x)f(y) = f(xy) \in f(\mathfrak{G})$; thus there is in $\mathfrak{G}$ an element, namely xy, whose image is $x'y'$. Since $f(e) = e'$ and $f(x^{-1}) = f(x)^{-1}$, the theorem is established.

It may be pointed out that in the present text most homomorphisms are epimorphisms. From Theorem 1 it follows that a monomorphism $f : \mathfrak{G} \to \mathfrak{G}'$ is an isomorphism of $\mathfrak{G}$ onto a subgroup of $\mathfrak{G}'$.

b. To further investigate the relation between a group $\mathfrak{G}$ and its homomorphic image $f(\mathfrak{G}) \leqslant \mathfrak{G}'$ we begin by considering the set of all those elements x in $\mathfrak{G}$ which by the homomorphism $f : \mathfrak{G} \to \mathfrak{G}'$ are mapped into the unit element $e' \in \mathfrak{G}'$, that is the set $f^{-1}(e')$ in $\mathfrak{G}$. If $f(x) = e'$ and $f(y) = e'$ it follows that $f(xy) = f(x)f(y) = e'$, that is if $x, y \in f^{-1}(e')$ then $xy \in f^{-1}(e')$. Moreover, from the preceding remark we also know that $e \in f^{-1}(e')$ and $x^{-1} \in f^{-1}(e')$. Thus we have the following result:

THEOREM 2. *If $f : \mathfrak{G} \to \mathfrak{G}'$ is a homomorphism then the set $f^{-1}(e')$ is a subgroup of $\mathfrak{G}$.*

This subgroup is called the *kernel* of the homomorphism f; it is denoted by ker f.

Can every subgroup $\mathfrak{H} \leqslant \mathfrak{G}$ appear as the kernel of a homomorphism? To answer this question let $f : \mathfrak{G} \to \mathfrak{G}'$ and $\mathfrak{H} = \ker f$. Consider the coset $a\mathfrak{H} = \{ax \mid x \in \mathfrak{H}\}$ for any fixed $a \in \mathfrak{G}$. Then $f(ax) = f(a)f(x) = f(a)e' = f(a)$ or briefly $f(a\mathfrak{H}) = f(a)$. Likewise $f(\mathfrak{H}a) = f(a)$. Thus $f(a\mathfrak{H}) = f(\mathfrak{H}a) = f(a)$.

Evidently $f(a) = e'$ implies that $a \in f^{-1}(e') = \mathfrak{H}$. Hence the equation $f(a) = f(b)$, that is $f(b)^{-1}f(a) = f(b^{-1}a) = e'$, $[f(ab^{-1}) = e']$ implies that $b^{-1}a \in \mathfrak{H}$ $(ab^{-1} \in \mathfrak{H})$ or $a\mathfrak{H} = b\mathfrak{H}$ $(\mathfrak{H}a = \mathfrak{H}b)$. Thus, elements a, b of $\mathfrak{G}$ for which the values of f are equal lie in the same left (right) coset of $\mathfrak{H}$ in $\mathfrak{G}$. Hence for all $a \in \mathfrak{G}$ and $x \in \mathfrak{H}$ we have $f(a) = f(ax) = f(xa)$; therefore $e' = f(a)^{-1}f(xa) = f(a^{-1}xa)$ which implies that $a^{-1}xa \in \mathfrak{H}$ or $\mathfrak{H}a \subseteq a\mathfrak{H}$. Likewise $a\mathfrak{H} \subseteq \mathfrak{H}a$; hence

(2) $\quad \mathfrak{H}a = a\mathfrak{H} \quad$ for all $\quad a \in \mathfrak{G}$.

A subgroup $\mathfrak{H} < \mathfrak{G}$ which satisfies the condition (2) is called a *normal subgroup* of $\mathfrak{G}$. If $\mathfrak{H}$ is normal in $\mathfrak{G}$ we shall write $\mathfrak{H} \trianglelefteq \mathfrak{G}$, and $\mathfrak{H} \triangleleft \mathfrak{G}$ if $\mathfrak{H}$ is normal and $\mathfrak{H} \neq \mathfrak{G}$. Clearly if $|\mathfrak{G}| > 1$ then $\mathfrak{e} \triangleleft \mathfrak{G}$. Also $\mathfrak{G} \trianglelefteq \mathfrak{G}$. In an abelian group $\mathfrak{G}$ every subgroup is normal. The fact that not every subgroup of a non-abelian group $\mathfrak{G}$ is normal is established by the example of §2, ex. 6(a).

All this is to show that the answer to the last question is No, or more accurately:

THEOREM 3. *The kernel of every homomorphism $\mathfrak{G} \to \mathfrak{G}'$ is a normal subgroup of $\mathfrak{G}$. The kernel of an isomorphism $\mathfrak{G} \to \mathfrak{G}'$, and thus also* *the kernel of a monomorphism, is the trivial subgroup consisting of the unit element e in* $\mathfrak{G}$.

The condition (2) by which a normal subgroup $\mathfrak{H}$ of $\mathfrak{G}$ has been defined can be expressed in the following equivalent ways:

1. For a normal subgroup $\mathfrak{H}$ of $\mathfrak{G}$ every left coset in $\mathfrak{G}$ is also a right coset;
2. Under group multiplication a normal subgroup $\mathfrak{H}$ commutes with every element $a \in \mathfrak{G}$. This does not imply that $ax = xa$ for every $a \in \mathfrak{G}$, $x \in \mathfrak{H}$; it rather means that for every $a \in \mathfrak{G}$, $x \in \mathfrak{H}$ there is an element $y \in \mathfrak{H}$ such that $ax = ya$. Thus $axa^{-1} \in \mathfrak{H}$.

c. The next natural question is: *Can every normal subgroup* $\mathfrak{H}$ *of* $\mathfrak{G}$ *appear as the kernel of a homomorphism of* $\mathfrak{G}$ *onto some group* $\mathfrak{G}'$? Let $\mathfrak{H} \trianglelefteq \mathfrak{G}$ and consider the cosets $a\mathfrak{H}$, $a \in \mathfrak{G}$. The product of two of these cosets $a\mathfrak{H}$, $b\mathfrak{H}$ is again a coset of $\mathfrak{H}$ as can be seen by using associativity and the condition (2), for

$$(3) \qquad a\mathfrak{H}b\mathfrak{H} = a(\mathfrak{H}b)\mathfrak{H} = a(b\mathfrak{H})\mathfrak{H} = ab\mathfrak{H}\mathfrak{H} = (ab)\mathfrak{H}.$$

In this coset multiplication the subgroup $\mathfrak{H}$ itself plays the role of a unit element: $\mathfrak{H}\cdot a\mathfrak{H} = a\mathfrak{H}\cdot\mathfrak{H} = a\mathfrak{H}$ and the inverse of the coset $a\mathfrak{H}$ is $a^{-1}\mathfrak{H}$ because $a^{-1}\mathfrak{H}\cdot a\mathfrak{H} = a^{-1}a\mathfrak{H}\mathfrak{H} = e\mathfrak{H} = \mathfrak{H}$. Thus we have the following important result:

THEOREM 4. *The cosets of a normal subgroup* $\mathfrak{H}$ *of* $\mathfrak{G}$ *form a group with respect to set multiplication in* $\mathfrak{G}$.

This group is called the *factor group* or the *quotient group* of $\mathfrak{G}$ over $\mathfrak{H}$ (or of $\mathfrak{H}$ in $\mathfrak{G}$). It is denoted by $\mathfrak{G}/\mathfrak{H}$. If $\mathfrak{H}$ has finite index in $\mathfrak{G}$ then $\mathfrak{G}/\mathfrak{H}$ is finite and has order

$$|\mathfrak{G}/\mathfrak{H}| = (\mathfrak{G}:\mathfrak{H}).$$

From the multiplication law (3) for the cosets (which are the elements of $\mathfrak{G}/\mathfrak{H}$) it is evident that the mapping $f_{\mathfrak{H}}: \mathfrak{G} \to \mathfrak{G}/\mathfrak{H}$ defined by

$$a \to f_{\mathfrak{H}}(a) = a\mathfrak{H} \quad (a \in \mathfrak{G})$$

constitutes an epimorphism which is called the *natural* or *canonical homomorphism* of $\mathfrak{G}$ with respect to the normal subgroup $\mathfrak{H}$. Its kernel is the given normal subgroup $\mathfrak{H}$ of $\mathfrak{G}$. Thus the second question has been answered Yes.

d. THEOREM 5. *Let* $f : \mathfrak{G} \to \mathfrak{G}'$ *be an epimorphism and let* $\ker f = \mathfrak{H}$. *Then the mapping* $g : \mathfrak{G}/\mathfrak{H} \to \mathfrak{G}'$ *defined by* $a\mathfrak{H} \to g(a\mathfrak{H}) = a' = f(a)$ *is an isomorphism.*

Proof. The mapping g is invertible because the homomorphism f maps elements of distinct cosets of $\mathfrak{H}$ onto different elements of $\mathfrak{G}'$. The mapping g is an isomorphism; indeed

$$g(ab\mathfrak{H}) = f(ab) = f(a)\cdot f(b) = a'\cdot b' = g(a\mathfrak{H})\cdot g(b\mathfrak{H}).$$

The given epimorphism $f : \mathfrak{G} \to \mathfrak{G}'$ can then be represented in the form

$$f(x) = g(f_{\mathfrak{H}}(x))$$

or briefly $f = gf_{\mathfrak{H}}$. This is called the *canonical product* representing the homomorphism f.

THEOREM 6. *Let* $f : \mathfrak{G} \to \mathfrak{G}'$ *be an epimorphism.*

(a) *If* $\mathfrak{H}$ *is a subgroup of* $\mathfrak{G}$, *then* $f(\mathfrak{H}) = \mathfrak{H}'$ *is a subgroup of* $\mathfrak{G}'$.

(b) *If* $\mathfrak{H}$ *is a normal subgroup of* $\mathfrak{G}$, *then* $f(\mathfrak{H}) = \mathfrak{H}'$ *is a normal subgroup of* $\mathfrak{G}'$.

Proof. (a) This follows immediately from Theorem 1.

(b) Let $x \in \mathfrak{H}$, thus $txt^{-1} \in \mathfrak{H}$ for all $t \in \mathfrak{G}$. For an arbitrary $s' \in \mathfrak{G}'$ there is an element t in $\mathfrak{G}$ such that $f(t) = s'$. Hence if $x' \in \mathfrak{H}'$ we have

$$s'x's'^{-1} = f(t)f(x)f(t^{-1}) = f(txt^{-1}) \in \mathfrak{H}'$$

and therefore $\mathfrak{H}' \trianglelefteq \mathfrak{G}'$.

This theorem is valid in particular if $f : \mathfrak{G} \to \mathfrak{G}'$ is an isomorphism (cf. Chap. I, §1, *e.*).

We shall return to these considerations in §5.

Examples and exercises

1. A subgroup $\mathfrak{H}$ of index 2 in a group $\mathfrak{G}$ is always normal.

Thus the alternating group $\mathfrak{A}_n$ is normal in the symmetric group $\mathfrak{S}_n$; so is the proper orthogonal group $\mathfrak{D}_n$ in the full orthogonal group $\mathfrak{H}_n$. Cf. Chap. I, §4, *a*.

2. In the group $\mathfrak{G}_{\mathbb{F}}$ whose elements are represented by the symbols $A = (a, \alpha)$ where a, α are elements of a field $\mathbb{F}$, $a \neq 0$, with the multiplication law $AB = (ab, a\beta + \alpha)$ (cf. Chap. I, §3, ex. 4) the subgroup

$$\mathfrak{H} = \{(1, \tau) \mid \tau \in \mathbb{F}\}$$

is normal. Indeed for some fixed $A \in \mathfrak{G}_{\mathbb{F}}$

$$A\mathfrak{H} = \{(a, a\tau + \alpha) \mid \tau \in \mathbb{F}\}, \quad \mathfrak{H}A = \{(a, \tau + \alpha) \mid \tau \in \mathbb{F}\}$$

and both these cosets are identical: If τ runs through all the elements of the field $\mathbb{F}$, then so does $a\tau + \alpha$ as well as $\tau + \alpha$.

The cosets $\mathfrak{H}, \mathfrak{H}A, \mathfrak{H}B, \ldots$ are the elements of the factor group $\mathfrak{G}_{\mathbb{F}}/\mathfrak{H}$. Let $\mathfrak{H}B = \{(b, \sigma + \beta) \mid \sigma \in \mathbb{F}\}$. Then

$$\begin{aligned} \mathfrak{H}A \cdot \mathfrak{H}B = \mathfrak{H} \cdot AB &= \{(ab, a(\beta + \sigma) + \alpha + \tau) \mid \sigma, \tau \in \mathbb{F}\} \\ &= \{(ab, \rho) \mid \rho \in \mathbb{F}\}. \end{aligned}$$

Thus the factor group $\mathfrak{G}_{\mathbb{F}}/\mathfrak{H}$ is abelian and if we denote by $\mathfrak{F}$ the multiplication group of the field $\mathbb{F}$, then there is an isomorphism $\psi : \mathfrak{G}_{\mathbb{F}}/\mathfrak{H} \to \mathfrak{F}$ defined by $\psi(\mathfrak{H}A) = a$.

3. A subgroup $\mathfrak{H} < \mathfrak{G}$ is normal in the group $\mathfrak{G}$ if its left cosets coincide with its right cosets; this implies that if

$$\mathfrak{G} = \mathfrak{H} \cup a_2\mathfrak{H} \cup a_3\mathfrak{H} \cup \ldots = \mathfrak{H} \cup \mathfrak{H}b_2 \cup \mathfrak{H}b_3 \cup \ldots$$

then for every a_μ there is a b_ν such that $a_\mu\mathfrak{H} = \mathfrak{H}b_\nu$; thus $a_\mu \in \mathfrak{H}b_\nu$, hence $\mathfrak{H}b_\nu = \mathfrak{H}a_\mu$, i.e. $a_\mu\mathfrak{H} = \mathfrak{H}a_\mu$.

4. (a) The intersection of two normal subgroups of $\mathfrak{G}$ is a normal subgroup of $\mathfrak{G}$.

(b) If $\mathfrak{L} \leqslant \mathfrak{G}$ and $\mathfrak{H} \trianglelefteq \mathfrak{G}$ then $\mathfrak{H} \cap \mathfrak{L} \trianglelefteq \mathfrak{L}$.

5. As in Chap. I, §3 we denote by $\mathbb{R}_n^+$ the additive group of the residue classes (mod n). Let $n = qm$.

(a) Show that there is an epimorphism $f : \mathbb{R}_n^+ \to \mathbb{R}_m^+$.

(b) Determine $\ker f$. This means: Find a known group which is isomorphic to the subgroup $\ker f \trianglelefteq \mathbb{R}_n^+$.

(c) Determine the factor group $\mathbb{R}_n^+/\ker f$.

6. Let $\mathfrak{L}_n = GL(n, \mathbb{F})$ denote the group of all non-singular $n \times n$ matrices A with elements in a field $\mathbb{F}$. The multiplication in $\mathfrak{L}_n$ is ordinary matrix multiplication.

(a) Show that the mapping $\delta : A \to \det A$ is an epimorphism of $\mathfrak{L}_n \to \mathfrak{F}$, the multiplicative group of the field $\mathbb{F}$.

(b) Describe $\ker \delta = \mathfrak{K}_n$ and the factor group $\mathfrak{L}_n/\mathfrak{K}_n$.

7. If $\mathfrak{H} < \mathfrak{G}$ and $\mathfrak{K} \trianglelefteq \mathfrak{G}$ show that $\mathfrak{H}\mathfrak{K} \leqslant \mathfrak{G}$. If also $\mathfrak{H} \trianglelefteq \mathfrak{G}$ show that $\mathfrak{H}\mathfrak{K} \trianglelefteq \mathfrak{G}$.

8. Show that every subgroup of the quaternion group $\mathfrak{Q}$ is normal in $\mathfrak{Q}$. (Cf. Chap. I, §2, ex. 9.)

9. Show that a homomorphic image of a cyclic group is a cyclic group and that every cyclic group is a homomorphic image of the infinite cyclic group.

10. Show that the Frattini subgroup $\Phi(\mathfrak{G})$ of a group $\mathfrak{G}$ is a normal subgroup of $\mathfrak{G}$. (Cf. §1, ex. 11.)

11. Let $\mathfrak{G}$ be a finite group, $\mathfrak{H} \trianglelefteq \mathfrak{G}$, $x \in \mathfrak{G}$, $|\langle x \rangle| = m$. If the coset $x\mathfrak{H}$, as an element of $\mathfrak{G}/\mathfrak{H}$, has the order n, then $n \mid m$.

§4. Transformation. Conjugate elements. Invariant subsets

a. Let $\mathfrak{G}$ be a group and $t \in \mathfrak{G}$. We call *transformation* or *conjugation* by t the mapping $\tau : \mathfrak{G} \to \mathfrak{G}$ which carries an element a of $\mathfrak{G}$ onto the element $\tau(a) = tat^{-1}$, often denoted by a^t and referred to as the *conjugate* of a by t. This mapping is evidently invertible; for every fixed t in $\mathfrak{G} : a = t^{-1}\tau(a)t$.

Similarly we define for a subset $\mathbf{A} \subseteq \mathfrak{G}$ the conjugate set $\tau(\mathbf{A}) = t\mathbf{A}t^{-1} = \{tat^{-1} \mid a \in \mathbf{A}\} = \mathbf{A}^t$.

We derive a few simple properties of conjugation. Let $\mathbf{B} \subseteq \mathfrak{G}$. According to the relations stated in §1, exx. 6 and 7 we have

$$(1) \qquad t(\mathbf{A} \cup \mathbf{B})t^{-1} = t\mathbf{A}t^{-1} \cup t\mathbf{B}t^{-1},$$

$$(2) \qquad t(\mathbf{A} \cap \mathbf{B})t^{-1} = t\mathbf{A}t^{-1} \cap t\mathbf{B}t^{-1},$$

and since

(3) $t(ab)t^{-1} = tat^{-1} \cdot tbt^{-1}$

also

(3′) $t(\mathbf{AB})t^{-1} = t\mathbf{A}t^{-1} \cdot t\mathbf{B}t^{-1}$.

THEOREM 1. *If $\mathfrak{H} \leqslant \mathfrak{G}$ and $t \in \mathfrak{G}$, then $\mathfrak{H}^t = t\mathfrak{H}t^{-1} \leqslant \mathfrak{G}$ and $\mathfrak{H}^t \simeq \mathfrak{H}$.*

Proof. Let $a, b \in \mathfrak{H}$ so that $a^t, b^t \in \mathfrak{H}^t$. Every element of $\mathfrak{H}^t$ appears in the form a^t with a certain $a \in \mathfrak{H}$. By (3) $a^t b^t = (ab)^t \in \mathfrak{H}^t$. Also $e^t = e \in \mathfrak{H}^t$ and $(a^t)^{-1} = ta^{-1}t^{-1} \in \mathfrak{H}^t$. Again by (3), the mapping $a \to \tau(a) = a^t$ being invertible, is an isomorphism $\tau: \mathfrak{H} \to \mathfrak{H}^t$.

The subgroup $\mathfrak{H}^t$ is named a *conjugate subgroup* of $\mathfrak{H}$ in $\mathfrak{G}$.

COROLLARY. *Let $\mathbf{A} \subseteq \mathfrak{G}$; then $t\langle \mathbf{A} \rangle t^{-1} = \langle t\mathbf{A}t^{-1} \rangle$.*

b. A subset $\mathbf{A}$ of $\mathfrak{G}$ is said to be *invariant in* $\mathfrak{G}$ if $\mathbf{A}^t \subseteq \mathbf{A}$ for all $t \in \mathfrak{G}$, i.e. $a^t \in \mathbf{A}$ for all $a \in \mathbf{A}$ and all $t \in \mathfrak{G}$. Since $\mathbf{A}^{t^{-1}} = t^{-1}\mathbf{A}t \subseteq \mathbf{A}$ and therefore $\mathbf{A} \subseteq t\mathbf{A}t^{-1}$ we conclude that $\mathbf{A}$ *is invariant in* $\mathfrak{G}$ *if and only if* $\mathbf{A}^t = \mathbf{A}$ *for every* t *in* $\mathfrak{G}$. Assuming that $\mathbf{A}$ is invariant and that its elements are arranged in a certain order, conjugation effects in general a change in this order, a permutation of the elements of $\mathbf{A}$.

A subgroup $\mathfrak{H}$ of $\mathfrak{G}$ is invariant in $\mathfrak{G}$ if and only if it is a normal subgroup of $\mathfrak{G}$; indeed the two conditions

$$t\mathfrak{H} = \mathfrak{H}t \quad \text{and} \quad t\mathfrak{H}t^{-1} = \mathfrak{H}$$

are equivalent. We can therefore continue our list of characterizations of a normal subgroup (cf. §3, *b.*) $\mathfrak{H}$ by the following two statements:

3. $\mathfrak{H}$ is invariant in $\mathfrak{G}$.
4. Each subgroup $\mathfrak{H}^t$ conjugate to $\mathfrak{H}$ coincides with $\mathfrak{H}$, i.e. $\mathfrak{H}$ is a *self-conjugate subgroup* (cf. also §6, ex. 4).

The unit element e is invariant in every group $\mathfrak{G}$. Every element a is invariant in $\mathfrak{G}$ if and only if $\mathfrak{G}$ is abelian. The set of all invariant elements z in $\mathfrak{G}$ is called the *centre* (center, central, German: Zentrum) of $\mathfrak{G}$ and denoted by

$$\mathbf{Z} = \mathbf{Z}(\mathfrak{G}) = \{z \in \mathfrak{G} \mid z^t = z \quad \text{for all} \quad t \in \mathfrak{G}\}.$$

If the centre $\mathbf{Z}(\mathfrak{G})$ consists only of the unit element e of $\mathfrak{G}$ then $\mathbf{Z}(\mathfrak{G})$ is said to be "trivial"; we also say that $\mathfrak{G}$ has "no centre". We have $\mathbf{Z} = \mathbf{G}$ if and only if $\mathfrak{G}$ is abelian.

THEOREM 2. *The centre of $\mathfrak{G}$ is a normal subgroup of $\mathfrak{G}$.*

Proof. If $z_1, z_2 \in \mathbf{Z}$, i.e. $z_1{}^t = z_1$, $z_2{}^t = z_2$ for all t in $\mathfrak{G}$, then also

$$(z_1 z_2)^t = z_1{}^t z_2{}^t = z_1 z_2 \in \mathbf{Z}.$$

Obviously $e \in \mathbf{Z}$ and $(z^{-1})^t = tz^{-1}t^{-1} = (tzt^{-1})^{-1} = z^{-1} \in \mathbf{Z}$. Since every element of $\mathbf{Z}$ is invariant, the whole set $\mathbf{Z}$ is invariant. Hence its elements form a normal subgroup of $\mathfrak{G}$ which from now on will be denoted by $\mathfrak{Z} = \mathfrak{Z}(\mathfrak{G})$.

COROLLARY. *The factor group $\mathfrak{G}/\mathfrak{Z}(\mathfrak{G})$ cannot be cyclic of order greater than one.*

Proof. Let us assume that $\mathfrak{G}/\mathfrak{Z}$ is cyclic, $\mathfrak{G}/\mathfrak{Z} = \langle a\mathfrak{Z}\rangle$ for some element a of $\mathfrak{G}$. The elements of $\mathfrak{G}/\mathfrak{Z}$ are then the cosets

$$(a\mathfrak{Z})^\mu = a^\mu \mathfrak{Z} \quad (\mu = 0, \pm 1, \pm 2, \ldots).$$

Thus every element x of $\mathfrak{G}$ appears in the form $x = a^\mu z$, $z \in \mathfrak{Z}$. Let $y = a^\nu z'$, $z' \in \mathfrak{Z}$; then $xy = a^\mu z a^\nu z' = a^{\mu+\nu} z z' = a^\nu z' a^\mu z = yx$. Hence $\mathfrak{G}$ is abelian and $\mathfrak{Z} = \mathfrak{G}$ so that $|\mathfrak{G}/\mathfrak{Z}| = 1$.

c. The classes. We now ask for the smallest invariant subsets of a group $\mathfrak{G}$. Such a set contains at least one element a of $\mathfrak{G}$ and at the same time all the conjugates a^t, $t \in \mathfrak{G}$, of a and no other element. It is called a *class of conjugate elements* or a *conjugacy class* or briefly a *class* in $\mathfrak{G}$. The conjugacy class containing the element a is denoted

$$\mathscr{C}(a) = \{a^t \mid t \in \mathfrak{G}\}.$$

All conjugates of a^s (fixed s in $\mathfrak{G}$) are also conjugates of a because

$$(a^s)^t = t(sas^{-1})t^{-1} = (ts)a(ts)^{-1} = a^{ts}.$$

Hence $\mathscr{C}(a^s) \subseteq \mathscr{C}(a)$. But $(a^s)^{s^{-1}} = a$; therefore $\mathscr{C}(a) \subseteq \mathscr{C}(a^s)$. Consequently

$$s\mathscr{C}(a)s^{-1} = \mathscr{C}(a^s) = \mathscr{C}(a) \quad \text{for all} \quad s \in \mathfrak{G}.$$

Conjugacy is an equivalence relation on $\mathfrak{G}$; hence every element a of $\mathfrak{G}$ is an element of a certain class, namely $\mathscr{C}(a)$, and two classes coincide if and only if they have an element in common.

The union of two or more classes $\mathscr{C}(a), \mathscr{C}(b), \ldots$ is clearly an invariant subset of $\mathfrak{G}$. In fact we have the following theorem:

THEOREM 3. *A subset* $\mathbf{A}$ *of* $\mathfrak{G}$ *is invariant if and only if it is a union of conjugacy classes, namely*

$$\mathbf{A} = \bigcup_{a \in \mathbf{A}} \mathscr{C}(a). \tag{4}$$

Proof. Suppose that $\mathbf{A}^t = \mathbf{A}$ for all $t \in \mathfrak{G}$. If $a \in \mathbf{A}$ then also $a^t \in \mathbf{A}$ whatever t in $\mathfrak{G}$ and therefore $\mathscr{C}(a) \subseteq \mathbf{A}$; hence $\bigcup_{a\in\mathbf{A}} \mathscr{C}(a) \subseteq \mathbf{A}$. Since if $a \in \mathbf{A}$ also $a \in \mathscr{C}(a)$ it follows that $\mathbf{A} \subseteq \bigcup_{a\in\mathbf{A}} \mathscr{C}(a)$. Thus the invariance of $\mathbf{A}$ implies (4). The converse, namely the invariance of $\bigcup_{a\in\mathbf{A}} \mathscr{C}(a)$, is obvious. Thus the theorem is proved.

In particular we have:

COROLLARY 1. *A subgroup* $\mathfrak{H}$ *of* $\mathfrak{G}$ *is normal in* $\mathfrak{G}$ *if and only if it is a union of conjugacy classes.*

In the case of the centre $\mathfrak{Z} = \mathfrak{Z}(\mathfrak{G})$ each element itself constitutes a class in $\mathfrak{G}$; therefore the central elements are often called "isolated elements" of $\mathfrak{G}$.

COROLLARY 2. *Let* $\mathfrak{H} \leqslant \mathfrak{G}$. *The intersection* $\mathfrak{D} = \bigcap_{t\in\mathfrak{G}} \mathfrak{H}^t$ *of all subgroups* $\mathfrak{H}^t$ *conjugate to* $\mathfrak{H}$ *in* $\mathfrak{G}$ *is a normal subgroup of* $\mathfrak{G}$.

Indeed $\mathfrak{D}$ is a subgroup of $\mathfrak{G}$ and with every element a of $\mathfrak{D}$ every conjugate $a^t \in \mathfrak{D}$. Hence $\mathfrak{D}$ consists of complete classes.

We note that if $\mathfrak{K} \lhd \mathfrak{G}$ and $\mathfrak{K} \leqslant \mathfrak{H}$, then $t\mathfrak{K}t^{-1} = \mathfrak{K}$; thus $\mathfrak{K} \leqslant t\mathfrak{H}t^{-1}$ for all t in $\mathfrak{G}$. Hence $\mathfrak{K} \leqslant \mathfrak{D}$.

The notion of the class $\mathscr{C}(a)$ of conjugates of a in $\mathfrak{G}$ can be extended. For a subset $\mathbf{A}$ of $\mathfrak{G}$ we may define

$$\mathscr{C}(\mathbf{A}) = \{\mathbf{A}^t \mid t \in \mathfrak{G}\},$$

i.e. the class of all subsets $\mathbf{A}^t$ conjugate to $\mathbf{A}$ in $\mathfrak{G}$. Its elements are distinct subsets of $\mathfrak{G}$ which need not be disjoint. If a is an element of $\mathbf{A}$, then all the conjugates a^t are distributed over the elements of

$\mathscr{C}(\mathbf{A})$, the subsets $\mathbf{A}^t$. Hence the set $\bar{\mathscr{C}}(\mathbf{A})$ of all elements of $\mathfrak{G}$ contained in the subsets $\mathbf{A}^t$ is a union of conjugacy classes of single elements of $\mathfrak{G}$ and therefore invariant in $\mathfrak{G}$.

d. The normalizer. It is often important to know the number of conjugates of an element a, or of a subgroup $\mathfrak{H}$, or of a subset $\mathbf{A}$ of $\mathfrak{G}$. With this problem in mind we introduce the concept of the *normalizer* $\mathfrak{N}_{\mathfrak{G}}(\mathbf{A}) = \mathfrak{N}(\mathbf{A})$ *of* $\mathbf{A}$ *in** $\mathfrak{G}$:

$$\mathfrak{N}(\mathbf{A}) = \{x \in \mathfrak{G} \,|\, x\mathbf{A}x^{-1} = \mathbf{A}\}.$$

This is readily seen to be the largest subset of $\mathfrak{G}$ whose elements transform the set $\mathbf{A}$ into itself. We note that $\mathbf{A}$ is not necessarily a subset of $\mathfrak{N}(\mathbf{A})$ (cf. ex. 15). *The normalizer* $\mathfrak{N}(\mathbf{A})$ *is* in fact *a subgroup of* $\mathfrak{G}$, for, if $\mathbf{A}^x = \mathbf{A}$ and $\mathbf{A}^y = \mathbf{A}$, then $\mathbf{A}^{xy} = \mathbf{A}^x = \mathbf{A}$ while obviously $\mathbf{A}^e = \mathbf{A}$ and $\mathbf{A}^{x^{-1}} = \mathbf{A}$.

If $b \notin \mathfrak{N}(\mathbf{A})$ then $\mathbf{A}^b \neq \mathbf{A}$ and $\mathfrak{N}(\mathbf{A}) = \mathfrak{G}$ if and only if $\mathbf{A}$ is invariant in $\mathfrak{G}$.

THEOREM 4. *Conjugate subsets* $\mathbf{A}$ *and* $\mathbf{A}^t$ *of* $\mathfrak{G}$ *have conjugate normalizers*:

$$\mathfrak{N}(t\mathbf{A}t^{-1}) = t\mathfrak{N}(\mathbf{A})t^{-1}.$$

Proof. By definition

$$\mathfrak{N}(t\mathbf{A}t^{-1}) = \{u \in \mathfrak{G} \,|\, u\mathbf{A}^t u^{-1} = \mathbf{A}^t\}.$$

This implies that $t^{-1}ut\mathbf{A}t^{-1}u^{-1}t = \mathbf{A}$. Therefore $t^{-1}ut \in \mathfrak{N}(\mathbf{A})$ and $u \in t\mathfrak{N}(\mathbf{A})t^{-1}$. Hence $\mathfrak{N}(t\mathbf{A}t^{-1}) \leqslant t\mathfrak{N}(\mathbf{A})t^{-1}$.

Conversely $t\mathfrak{N}(\mathbf{A})t^{-1} = \{v \in \mathfrak{G} \,|\, t^{-1}vt \in \mathfrak{N}(\mathbf{A})\}$. Thus if $t^{-1}vt \in \mathfrak{N}(\mathbf{A})$ we have $(t^{-1}vt)\mathbf{A}(t^{-1}vt)^{-1} = \mathbf{A}$ whence $v(t\mathbf{A}t^{-1})v^{-1} = t\mathbf{A}t^{-1}$, i.e. $v \in \mathfrak{N}(\mathbf{A}^t)$ and therefore $t\mathfrak{N}(\mathbf{A})t^{-1} \leqslant \mathfrak{N}(\mathbf{A}^t)$. From the two opposite inclusions follows equality.

Here now is the principal theorem:

THEOREM 5. *Let* $\mathbf{A}$ *be a subset of the group* $\mathfrak{G}$ *such that the index* $(\mathfrak{G} : \mathfrak{N}(\mathbf{A})) = i$ *is finite. If* $|\mathscr{C}(\mathbf{A})| = \gamma_{\mathbf{A}}$, *i.e.* $\gamma_{\mathbf{A}}$ *is the number of different sets* $\mathbf{A}^t$ $(t \in \mathfrak{G})$, *then* $\gamma_{\mathbf{A}} = i$.

* The subscript $\mathfrak{G}$ in $\mathfrak{N}_{\mathfrak{G}}(\mathbf{A})$ will usually be omitted.

Proof. Let $b \in \mathfrak{G}$; then $b \cdot \mathfrak{N}(\mathbf{A})$ is one of the i left cosets of $\mathfrak{N}(\mathbf{A})$ in $\mathfrak{G}$. If $x \in \mathfrak{N}(\mathbf{A})$, the element $bx \in b\mathfrak{N}(\mathbf{A})$ and transforms $\mathbf{A}$ into the conjugate subset

$$\mathbf{A}^{bx} = (bx)\mathbf{A}(bx)^{-1} = b(x\mathbf{A}x^{-1})b^{-1} = b\mathbf{A}b^{-1} = \mathbf{A}^{b},$$

which is the same for all $x \in \mathfrak{N}(\mathbf{A})$. Hence the number of subsets conjugate to $\mathbf{A}$ is bounded above by the number of cosets of $\mathfrak{N}(\mathbf{A})$ in $\mathfrak{G}$, that is, $\gamma_{\mathbf{A}} \leqslant i$.

On the other hand, if for two elements b, c in $\mathfrak{G}$ we have $\mathbf{A}^{b} = \mathbf{A}^{c}$ it follows that $\mathbf{A}^{c^{-1}b} = \mathbf{A}$ and $c^{-1}b \in \mathfrak{N}(\mathbf{A})$; hence $b \in c\mathfrak{N}(\mathbf{A})$ and $c\mathfrak{N}(\mathbf{A}) = b\mathfrak{N}(\mathbf{A})$. Equivalently we can say: If $b\mathfrak{N}(\mathbf{A}) \neq c\mathfrak{N}(\mathbf{A})$ it follows that $\mathbf{A}^{b} \neq \mathbf{A}^{c}$. Therefore the number of cosets of $\mathfrak{N}(\mathbf{A})$ in $\mathfrak{G}$ cannot be greater than the number of sets $\mathbf{A}^{t}$ conjugate to $\mathbf{A}$ in $\mathfrak{G}$. Hence $i \leqslant \gamma_{\mathbf{A}}$.

Thus we have shown that $i = \gamma_{\mathbf{A}}$.

COROLLARY 1. *If* $\mathbf{A}$ *consists of a single element* a *of* $\mathfrak{G}$ *one has (by definition)* $\mathfrak{N}(a) = \{x \mid xax^{-1} = a, x \in \mathfrak{G}\}$, *and*

$$\gamma_a = |\mathscr{C}(a)| = (\mathfrak{G} : \mathfrak{N}(a)).$$

COROLLARY 2. *If* $\mathfrak{G}$ *is a finite group then for every* $\mathbf{A} \subseteq \mathfrak{G}$ *the number* $\gamma_{\mathbf{A}}$ *of conjugates* $\mathbf{A}^{t}$ $(t \in \mathfrak{G})$ *is a divisor of the order* $|\mathfrak{G}| = g$.

In fact $g = \gamma_{\mathbf{A}} \cdot |\mathfrak{N}(\mathbf{A})|$.

e. Generalization of Cayley's theorem. Let $\mathfrak{G}$ be a finite group of order g. Cayley's theorem states that there is a permutation group $\mathfrak{G}^*$ of degree g, i.e. a subgroup of the symmetric group $\mathfrak{S}_g$, which is isomorphic to the given group $\mathfrak{G}$ (cf. Chap. I, §2, *f.*). If g is large, operation with permutations of degree g may be tedious. Therefore it is natural to ask the question: Is it possible to find for $\mathfrak{G}$ an isomorphic permutation group of a degree smaller than g? To answer this question we shall resume the approach which was used in the proof of Cayley's theorem.

For the start we consider the following slightly more general situation: Let $\mathfrak{G}$ be a group and $\mathfrak{H} < \mathfrak{G}$ such that the index $i = (\mathfrak{G} : \mathfrak{H})$ is finite. With each of the elements $a, b, \ldots$ of $\mathfrak{G}$ we associate a permutation of the i cosets $x\mathfrak{H}$ of $\mathfrak{H}$ in $\mathfrak{G}$ by the mapping

$$\varphi_{\mathfrak{H}} : a \to A = \begin{pmatrix} xH \\ axH \end{pmatrix}, \quad b \to B = \begin{pmatrix} xH \\ bxH \end{pmatrix}, \ldots \quad (x \in \mathfrak{G}).$$

Then in fact $\varphi_{\mathfrak{H}} : \mathfrak{G} \to \mathfrak{S}_i$ and moreover

$$\varphi_{\mathfrak{H}}(ab) = \begin{pmatrix} x\mathfrak{H} \\ abx\mathfrak{H} \end{pmatrix} = \begin{pmatrix} bx\mathfrak{H} \\ abx\mathfrak{H} \end{pmatrix}\begin{pmatrix} x\mathfrak{H} \\ bx\mathfrak{H} \end{pmatrix} = AB = \varphi_{\mathfrak{H}}(a)\varphi_{\mathfrak{H}}(b).$$

Thus the mapping $\varphi_{\mathfrak{H}}$ of $\mathfrak{G}$ into $\mathfrak{S}_i$ is a homomorphism of $\mathfrak{G}$, and it remains to deal with the question: When is this homomorphism a monomorphism, i.e. an isomorphism onto a certain subgroup of $\mathfrak{S}_i$?

First we shall establish: When is $\varphi_{\mathfrak{H}}$ not a monomorphism? This is evidently the case if and only if $\ker \varphi_{\mathfrak{H}}$ contains an element $c \neq e$, that is, if and only if there is in $\mathfrak{G}$ an element $c \neq e$ for which

$$\varphi_{\mathfrak{H}}(c) = I, \quad I = \begin{pmatrix} x\mathfrak{H} \\ x\mathfrak{H} \end{pmatrix}.$$

If c is such an element then for all $x \in \mathfrak{G}$ we have $cx\mathfrak{H} = x\mathfrak{H}$, that is, $x^{-1}cx\mathfrak{H} = \mathfrak{H}$ or equivalently $x^{-1}cx \in \mathfrak{H}$ and we conclude that

$$c \in x\mathfrak{H}x^{-1} \quad \text{for all} \quad x \in \mathfrak{G}.$$

Thus the homomorphism $\varphi_{\mathfrak{H}}$ fails to be a monomorphism if and only if the intersection $\mathfrak{D} = \bigcap_{x \in \mathfrak{G}} \mathfrak{H}^x$ contains an element $c \neq e$. Since the elements of $\mathfrak{D}$, and only these, are mapped under $\varphi_{\mathfrak{H}}$ onto the identity permutation I we conclude that

$$\mathfrak{D} = \ker \varphi_{\mathfrak{H}}.$$

Thus we obtain the following theorem which answers the original question:

THEOREM 6. *The homomorphism $\varphi_{\mathfrak{H}}$ of $\mathfrak{G}$ into the symmetric group $\mathfrak{S}_i$, $i = (\mathfrak{G} : \mathfrak{H})$, is a monomorphism if and only if the intersection $\mathfrak{D}$ of all the subgroups conjugate to $\mathfrak{H}$ in $\mathfrak{G}$ consists only of the unit element e.*

An obvious necessary (but not sufficient) condition for $\varphi_{\mathfrak{H}}$ to be a monomorphism is that the group $\mathfrak{G}$ be finite.

COROLLARY. *If $\mathfrak{G}$ is an infinite group and $\mathfrak{H} < \mathfrak{G}$, a subgroup whose index $(\mathfrak{G} : \mathfrak{H})$ is finite, then the intersection $\mathfrak{D} = \cap \mathfrak{H}^x$ cannot consist of the unit element alone* (cf. ex. 12).

Examples and exercises

1. *Conjugation in permutation groups.*

(a) Let

$$P = \begin{pmatrix} \nu \\ p_\nu \end{pmatrix}, \quad T = \begin{pmatrix} \nu \\ t_\nu \end{pmatrix} \quad (\nu = 1, \ldots, n)$$

be two permutations of degree n $(n > 2)$. We wish to determine the permutation TPT^{-1}. Let

$$(5) \qquad P = C_1 C_2 \ldots C_r$$

be the product decomposition of P into disjoint cycles C_ρ $(\rho = 1, \ldots, r)$. Then

$$TPT^{-1} = TC_1T^{-1}TC_2T^{-1} \ldots TC_rT^{-1}.$$

Hence it is sufficient to find TCT^{-1} for a single cycle, e.g. $C = (1\ 2 \ldots k)$:

$$TCT^{-1} = \begin{pmatrix} \nu \\ t_\nu \end{pmatrix} (1\ 2 \ldots k) \begin{pmatrix} t_\nu \\ \nu \end{pmatrix} = (t_1\ t_2 \ldots t_k).$$

Thus, TPT^{-1} is obtained from P by replacing ν by t_ν, i.e.

$$(5') \qquad TPT^{-1} = \begin{pmatrix} t_1 & t_2 & \ldots & t_n \\ t_{p_1} & t_{p_2} & \ldots & t_{p_n} \end{pmatrix}.$$

It is readily seen that conjugation (transformation) of a permutation P with a permutation T preserves the *cycle type* of P, that is, P and TPT^{-1} split into the same number of disjoint cycles and corresponding cycles C and TCT^{-1} have the same length.

(b) Conversely show that if two permutations of degree n have equal cycle types, then they are conjugate in $\mathfrak{S}_n$.

(c) Suppose that in the representation (5) of P as a product of r disjoint cycles C_ρ there occur k_1 cycles of order 1, i.e. k_1 fixed symbols, k_2 cycles of order 2 (transpositions), in general k_σ cycles of order $\sigma \leqslant n$. Then

$$(6) \qquad k_1 + 2k_2 + 3k_3 + \ldots + sk_s = n \ (0 < s \leqslant n, \quad k_\sigma \geqslant 0, \quad k_n \leqslant 1).$$

This condition is the same for every $TPT^{-1} \in \mathscr{C}(P)$, $T \in \mathfrak{S}_n$; the partition (6) of n uniquely defines the class $\mathscr{C}(P)$ and therefore also the number $|\mathscr{C}(P)|$. We shall determine it now.

For this purpose we form with the $n!$ different elements T of $\mathfrak{S}_n$ the $n!$ different expressions TPT^{-1} as in (5′), not all of which represent distinct permutations. Suppose that $C_1, \ldots, C_{k_1}$ are the different 1-cycles in P; they can be written in $k_1!$ different orders. The k_2 disjoint 2-cycles in P can be written in $k_2!$ different orders, and each of them in 2 ways $[(\alpha\beta) = (\beta\alpha)]$. The k_3 disjoint 3-cycles can be written in $k_3!$ different orders and each of them in 3 ways $[(\alpha\beta\gamma) = (\beta\gamma\alpha) = (\gamma\alpha\beta)]$ .... The k_s disjoint s-cycles can be written in $k_s!$ different orders and each of them in s ways. Hence among the $n!$ expressions there are

$$k_1!1^{k_1}\, k_2!2^{k_2} \ldots k_s!s^{k_s}$$

different representations of one and the same conjugate of P. Thus the number of distinct conjugates of P in $\mathfrak{S}_n$ equals

$$(7) \qquad |\mathscr{C}(P)| = \frac{n!}{k_1!1^{k_1} \cdot k_2!2^{k_2} \ldots k_s!s^{k_s}}.$$

2. (a) Show that the permutations I, (12)(34), (13)(24), (14)(23) form a normal subgroup $\mathfrak{V}$ in $\mathfrak{A}_4$ as well as in $\mathfrak{S}_4$.

(b) Verify that $\mathfrak{S}_4/\mathfrak{V} \simeq \mathfrak{S}_3$ and $\mathfrak{A}_4/\mathfrak{V} \simeq \mathfrak{A}_3$. (This should not suggest that there are homomorphisms $\mathfrak{S}_n \to \mathfrak{S}_{n-1}$, $\mathfrak{A}_n \to \mathfrak{A}_{n-1}$ if $n > 4$.)

(c) Every proper subgroup (of order 2) of $\mathfrak{V}$ is normal in $\mathfrak{V}$ because $\mathfrak{V}$ is abelian, but not normal in $\mathfrak{A}_4$ or in $\mathfrak{S}_4$. This shows that from $\mathfrak{L} \lhd \mathfrak{H} \lhd \mathfrak{G}$ one cannot conclude that $\mathfrak{L} \lhd \mathfrak{G}$.

3. Show that the symmetric group $\mathfrak{S}_n$ $(n > 2)$ has "no centre", that is: It has the "trivial centre" consisting of the identity element e only.

4. Find the conjugacy classes in $\mathfrak{S}_4$ and in $\mathfrak{A}_4$.

5. In the group $\mathfrak{G}_{\mathbb{F}}$ (cf. §3, ex. 2) the conjugacy classes are:
the unit element $E = (1, 0)$,
the set $\mathfrak{H} \setminus E = \{(1, \tau) \mid \tau \in \mathbb{F}, \tau \neq 0\}$,
the cosets $A\mathfrak{H} = \{(a, \rho) \mid \rho \in \mathbb{F}, a \text{ fixed in } \mathbb{F}, a \neq 0, a \neq 1\}$.

6. (a) Every subgroup of the centre $\mathfrak{Z}(\mathfrak{G})$ is normal in $\mathfrak{G}$.
(b) If $\mathfrak{H} \lhd \mathfrak{G}$ and $b \in \mathfrak{H}$ show that $\mathscr{C}_{\mathfrak{G}}(b) \subset \mathfrak{H}$.

7. If $\mathfrak{N}(a)$ is the normalizer of an element $a \in \mathfrak{G}$ show that
(a) $\mathfrak{Z} = \mathfrak{Z}(\mathfrak{G}) \trianglelefteq \mathfrak{N}(a)$ and $\mathfrak{Z} \trianglelefteq \mathfrak{Z}(\mathfrak{N}(a))$.
(b) $\mathfrak{Z} = \bigcap_{x \in \mathfrak{G}} \mathfrak{N}(x) = \bigcap_{x \in \mathfrak{G}} \mathfrak{Z}(\mathfrak{N}(x))$.
(c) Discuss the special case that $\mathfrak{Z}(\mathfrak{G}) = \mathfrak{N}(a)$.

8. If **A** is an invariant subset of the group $\mathfrak{G}$, then $\langle \mathbf{A} \rangle \trianglelefteq \mathfrak{G}$.

9. Let $|\mathfrak{G}| = g$, $|\langle a \rangle| = \alpha$ the order of a in $\mathfrak{G}$, $\gamma_a = |\mathscr{C}(a)|$, $|\mathfrak{Z}(\mathfrak{G})| = \zeta$. Show that γ_a is a divisor of g/α and of g/ζ.

10. Determine all finite groups which have
(a) exactly two distinct conjugacy classes;
(b) exactly three distinct conjugacy classes.

11. Show that the method of subsection *e.* cannot be applied to obtain an isomorphic permutation representation of degree $< |\mathfrak{G}|$ for the group $\mathfrak{G}$ if
(a) $\mathfrak{G}$ is cyclic of prime order;
(b) $\mathfrak{G}$ is the quaternion group $\mathfrak{Q}$ of order 8 (cf. §3, ex. 8).

12. Let $\mathfrak{G}$ be an infinite group and $\mathfrak{H}$ a subgroup of $\mathfrak{G}$ with finite index $(\mathfrak{G} : \mathfrak{H}) = i > 2$. Show that $\mathfrak{H}$ contains a subgroup $\mathfrak{K}$ of finite index $j = (\mathfrak{G} : \mathfrak{K})$ in $\mathfrak{G}$ so that $i \mid j$ and $j \mid i!$.

13. (a) Let $\mathfrak{G}$ be a finite group and $\mathfrak{H} < \mathfrak{G}$, $(\mathfrak{G} : \mathfrak{H}) > 1$. Show that the system of all conjugate subgroups $\mathfrak{H}^t$ $(t \in \mathfrak{G})$ can never cover the whole group $\mathfrak{G}$.

(b) Let $\mathfrak{G}$ be a non-commutative group. Construct a smallest subset $\mathbf{A} \subset \mathfrak{G}$ such that the system of all conjugate subsets $\mathbf{A}^t$ $(t \in \mathfrak{G})$ covers the whole group $\mathfrak{G}$.

14. Let x_1, x_2, x_3, x_4 be the coordinates of four points on a straight line. The expression

$$\lambda = (x_1, x_2 \mid x_3, x_4) = \frac{x_1 - x_3}{x_1 - x_4} \Big/ \frac{x_2 - x_3}{x_2 - x_4}$$

is called the *cross ratio* of the four points. If x_1, x_2, x_3, x_4 are subjected to all the 24 permutations of $\mathfrak{S}_4$, it is found that the cross ratio assumes the six values: λ, $1/\lambda$, $1-\lambda$, $1-1/\lambda$, $1/(1-\lambda)$, $\lambda/(\lambda-1)$. The following should help to understand the group property of these six functions of λ (cf. Chap. I, §2, ex. 12).

(a) The cross ratio is invariant (unchanged) if x_1, x_2, x_3, x_4 are subjected to the permutations of the four-group $\mathfrak{B}$.

(b) All permutations of a coset of $\mathfrak{B}$ in $\mathfrak{S}_4$ change λ into one and the same of the six values.

(c) Since $\mathfrak{B} \lhd \mathfrak{S}_4$ the cosets of $\mathfrak{B}$ form a group, namely the factor group $\mathfrak{S}_4/\mathfrak{B}$; therefore the six functions of λ form a group which is isomorphic to $\mathfrak{S}_4/\mathfrak{B} \simeq \mathfrak{S}_3$ [cf. ex. 2(b)].

15. Find a subset $\mathbf{A} \subset \mathfrak{S}_3$ so that $\mathbf{A} \cap \mathfrak{N}(\mathbf{A}) = \varnothing$.

16. Determine the group of all linear homogeneous transformations which leave the conic section $c_1x_1{}^2 + c_2x_2{}^2 = 1$ invariant:

(a) If $c_1 > 0$, $c_2 > 0$ show that the group is isomorphic to the group $\bar{\mathfrak{D}}_2$ (cf. Chap. I, §4, *a.*). More concisely: The two groups are conjugate subgroups of the linear group $\mathfrak{L}_2$ of all real non-singular 2×2-matrices. Determine the conjugating element.

(b) Let $c_1c_2 < 0$, in particular $c_1 = 1$, $c_2 = -1$. Are these groups isomorphic to those defined in part (a)?

17. (a) Find the conjugacy classes of the icosahedral rotation group $\mathfrak{J}$ (cf. §2, ex. 12).

This is an opportunity to discuss the geometrical interpretation of the conjugation process; the answer will therefore be given in detail.

Let P and P^* be two vertices of the icosahedron $\mathbf{J}$ and ρ an element of $\mathfrak{J}$ which turns $P \to P^*$; we write $P^* = \rho P$. Further let α be the rotation about the axis $\overrightarrow{OP}$ with the angle $2\pi/5$; thus $\alpha \in \mathfrak{J}$ and $\alpha P = P$. Hence $\rho\alpha P = \rho P = P^*$. On the other hand, since $P = \rho^{-1}P^*$, we have $\rho\alpha P = \rho\alpha\rho^{-1}P^*$. Consequently the conjugate $\rho\alpha\rho^{-1}$ of α represents a rotation about the axis $\overrightarrow{OP^*}$.

In a suitable coordinate system (namely the system Σ_0 with $\overline{OP}$ as x_1-axis) the rotation α is represented by the matrix A [cf. §2, ex. 12, (9)] and the rotation ρ by a certain orthogonal matrix R. The rotation $\rho\alpha\rho^{-1}$ about the axis $\overline{OP}{}^*$ will then be represented by the matrix RAR^{-1}; therefore the angle of the rotation $\rho\alpha\rho^{-1}$ is the same as the angle of α (cf. Chap. I, §4, *c.*).

If we choose P^* to be the vertex opposite to P we find that $\rho\alpha\rho^{-1}$, a rotation in the positive sense about $\overrightarrow{OP^*}$, appears as a rotation in the negative sense about $\overrightarrow{OP}$; hence $\rho\alpha\rho^{-1} = \alpha^{-1}$.

From this it is evident that the 6 subgroups $\mathfrak{L}_0$ of $\mathfrak{J}$ are conjugate and that the 12 rotations with the angles $\pm 2\pi/5$ form the class $\mathscr{C}(\alpha)$, those with the angles $\pm 4\pi/5$ from the class $\mathscr{C}(\alpha^2)$. Thus the 24 rotations of the order 5 are divided into two classes, each consisting of 12 elements.

Similarly it is seen that the 15 rotations of the order 2 form a class, and so do the 20 rotations of the order 3. Thus altogether there are 5 classes: $60 = 1+12+12+15+20$.

(*b*) Show that the centre of the icosahedral group $\tilde{\mathfrak{J}}$ ($|\tilde{\mathfrak{J}}| = 120$) is not trivial.

18. *Geometry in the group* $\mathfrak{G}_{\mathbb{F}}$. (Cf. Chap. I, §3, ex. 1; Chap. II, §3, ex. 2, and §4, ex. 5.) The elements of the group $\mathfrak{G}_{\mathbb{F}}$ are represented by the symbols $A = (a, \alpha)$, $a, \alpha \in \mathbb{F}$, $a \neq 0$. Let $B = (b, \beta) \in \mathfrak{G}_{\mathbb{F}}$. One has the multiplication law $AB = (ab, a\beta+\alpha)$, the unit element $I = (1, 0)$ and the inverse $A^{-1} = (a^{-1}, -a^{-1}\alpha)$. If $\mathbb{F}$ is the field of the real numbers it is natural to consider the group elements A as points in a cartesian a, α-plane. Conversely to all points except those on the α-axis ($a = 0$) correspond elements of the group $\mathfrak{G}_{\mathbb{F}}$. Thus we call the cartesian plane from which the points of the α-axis have been removed, "the group plane $\mathfrak{G}_{\mathbb{F}}$". In a more abstract way, the idea of a group plane may be adopted also in the case of an arbitrary field $\mathbb{F}$.

As in elementary analytic geometry a straight line in the group plane $\mathfrak{G}_{\mathbb{F}}$ is represented by an equation, linear in the coordinates of a point on the line, with coefficients in $\mathbb{F}$. The basic facts for a linear geometry in $\mathfrak{G}_{\mathbb{F}}$ are the following (cf. Fig. 7).

(a) Each straight line through I represents a subgroup of the group $\mathfrak{G}_{\mathbb{F}}$. If $A(\neq I)$ is a point on such a line, then the normalizer $\mathfrak{N}(A)$ is the subgroup represented by the line.

(b) Let $\mathfrak{H} = \{(1, \tau) \mid \tau \in \mathbb{F}\}$; then $\mathfrak{H} \lhd \mathfrak{G}_{\mathbb{F}}$ and the set of the normalizers $\mathfrak{N}(A)$, $A \notin \mathfrak{H}$, is a complete set of conjugate subgroups of $\mathfrak{G}_{\mathbb{F}}$.

(c) Every straight line in the group plane $\mathfrak{G}_{\mathbb{F}}$ represents a left coset $B\mathfrak{N}(A)$ of a normalizer $\mathfrak{N}(A)$; it is parallel to the line representing $\mathfrak{N}(A)$ and passes through the point B.

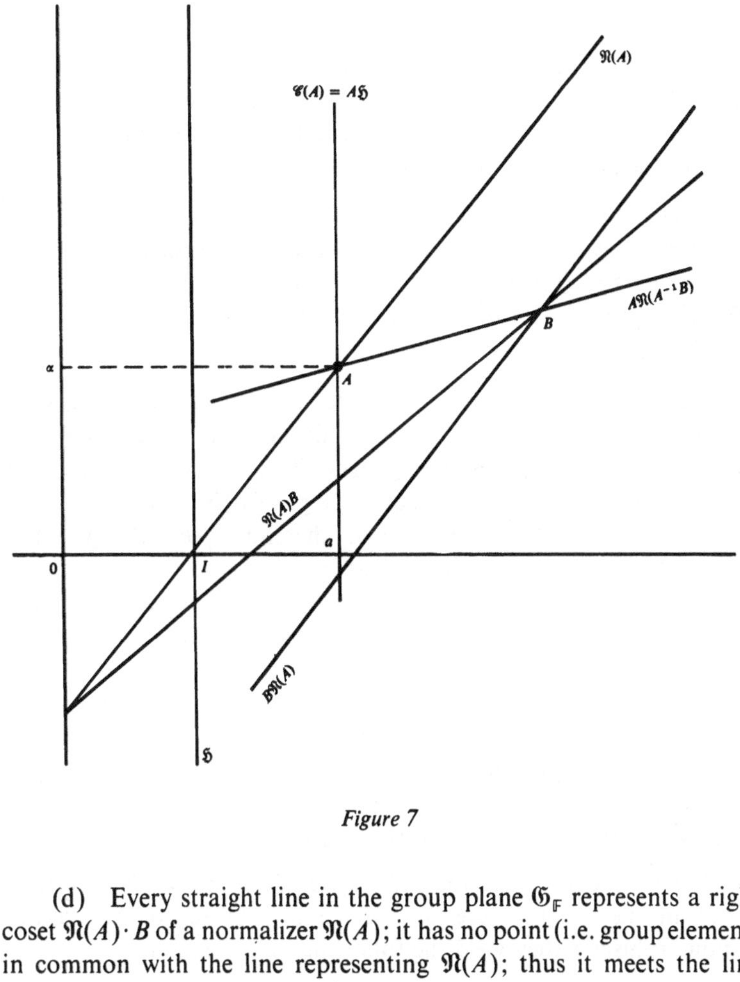

Figure 7

(d) Every straight line in the group plane $\mathfrak{G}_{\mathbb{F}}$ represents a right coset $\mathfrak{N}(A)\cdot B$ of a normalizer $\mathfrak{N}(A)$; it has no point (i.e. group element) in common with the line representing $\mathfrak{N}(A)$; thus it meets the line representing $\mathfrak{N}(A)$ on the α-axis.

So we can state that in the geometry of the group plane $\mathfrak{G}_{\mathbb{F}}$, for every line not parallel to the α-axis there are exactly two different parallel lines through every given point.

(e) If A, B are two group elements, not on one and the same line parallel to the line representing the normal subgroup $\mathfrak{H}$ (that is: not in one and the same coset of $\mathfrak{H}$) then the line through A and B is represented by $A\mathfrak{N}(A^{-1}B)$. Find other representations of this line.

§5 Correspondence theorems. Direct products

a. In §3, Theorem 6 it has been shown that the homomorphic image of a subgroup of a group $\mathfrak{G}$ is a subgroup of the homomorphic image $\mathfrak{G}'$ of $\mathfrak{G}$. Now we shall establish a correspondence between the cosets of the subgroups of $\mathfrak{G}$ and $\mathfrak{G}'$ respectively.

THEOREM 1. *Let $\mathfrak{H} < \mathfrak{G}$ and $f : \mathfrak{G} \to \mathfrak{G}'$ an epimorphism,* $\ker f \leqslant \mathfrak{H}$. *Then $f(\mathfrak{H}) = \mathfrak{H}' < \mathfrak{G}'$ and there is a one–one correspondence between the left cosets of $\mathfrak{H}$ in $\mathfrak{G}$ and the left cosets of $\mathfrak{H}'$ in $\mathfrak{G}'$, namely $a\mathfrak{H} \leftrightarrow a'\mathfrak{H}'$, $a' = f(a)$, for all $a \in \mathfrak{G}$. If the index $(\mathfrak{G} : \mathfrak{H})$ is finite, then so is $(\mathfrak{G}' : \mathfrak{H}')$ and both indices are equal.*

Proof. Let $\mathfrak{H}, a\mathfrak{H}, b\mathfrak{H}, \ldots$ be the distinct left cosets of $\mathfrak{H}$ in $\mathfrak{G}$. Let $a' = f(a), b' = f(b), \ldots$. We shall show that $\mathfrak{H}', a'\mathfrak{H}', b'\mathfrak{H}', \ldots$ then are the distinct left cosets of $\mathfrak{H}'$ in $\mathfrak{G}'$. Proof indirect: Suppose that two of these cosets were equal, e.g. $f(a)\mathfrak{H}' = f(b)\mathfrak{H}'$. Then $f(b)^{-1}f(a)\mathfrak{H}' = f(b^{-1}a)\mathfrak{H}' = \mathfrak{H}'$ and therefore $f(b^{-1}a) \in \mathfrak{H}'$; thus $b^{-1}a \in f^{-1}(\mathfrak{H}')$, the set of all x for which $f(x) \in \mathfrak{H}'$.

By supposition $\ker f = \mathfrak{K} \leqslant \mathfrak{H}$. It will be shown that this implies that $f^{-1}(\mathfrak{H}') = \mathfrak{H}$, thus $b^{-1}a \in \mathfrak{H}$ and $a\mathfrak{H} = b\mathfrak{H}$ which is contrary to the initial assumption.

Instead of f we consider now the natural homomorphism $\bar{f} : \mathfrak{G} \to \mathfrak{G}/\mathfrak{K} \simeq \mathfrak{G}'$. Clearly $\bar{f}$ maps $\mathfrak{H} \to \mathfrak{H}/\mathfrak{K} \simeq \mathfrak{H}'$. By $\bar{f}^{-1}(\mathfrak{H}/\mathfrak{K})$ we denote the set of all elements x of $\mathfrak{G}$ for which $\bar{f}(x) \in \mathfrak{H}/\mathfrak{K}$. These x are exactly the elements of $\mathfrak{G}$ which are contained in $\mathfrak{H}$; thus $\bar{f}^{-1}(\mathfrak{H}/\mathfrak{K}) = \mathfrak{H}$. Hence also $f^{-1}(\mathfrak{H}') = \mathfrak{H}$.

The facts concerning the indices $(\mathfrak{G} : \mathfrak{H})$ and $(\mathfrak{G}' : \mathfrak{H}')$ as mentioned in the theorem are now evident. Applied to the natural homomorphism $\bar{f}$ they yield the following index relation

(1) $$(\mathfrak{G}/\mathfrak{K} : \mathfrak{H}/\mathfrak{K}) = (\mathfrak{G} : \mathfrak{H})$$

or

(2) $$\frac{|\mathfrak{G}/\mathfrak{K}|}{|\mathfrak{H}/\mathfrak{K}|} = \frac{|\mathfrak{G}|}{|\mathfrak{H}|} \quad \text{(cf. §2, Theorem 2')}.$$

The argument used in the preceding proof enables us to decide another important question: As before, let $f : \mathfrak{G} \to \mathfrak{G}'$ denote an epimorphism with the kernel $\mathfrak{K}$. Given a subgroup $\mathfrak{H}' \leqslant \mathfrak{G}'$, is there a subgroup $\mathfrak{H} \leqslant \mathfrak{G}$ such that $f(\mathfrak{H}) = \mathfrak{H}'$? If so, is the subgroup $\mathfrak{H}$ unique in $\mathfrak{G}$?

In the case $\mathfrak{H}' = \mathfrak{e}'$ the question is answered by $f^{-1}(\mathfrak{e}') = \mathfrak{K}$, the kernel of the epimorphism f. In general we consider the inverse image $f^{-1}(\mathfrak{H}') \subseteq \mathfrak{G}$. We know that its elements satisfy the group axioms: Since $e' \in \mathfrak{H}'$ we have $\mathfrak{K} \leqslant f^{-1}(\mathfrak{H}')$, hence $e \in f^{-1}(\mathfrak{H}')$. Further, if $a, b \in f^{-1}(\mathfrak{H}')$, i.e. $f(a), f(b) \in \mathfrak{H}'$ if follows that $f(a)f(b)^{-1} = f(a)f(b^{-1}) = f(ab^{-1}) \in \mathfrak{H}'$ and therefore $ab^{-1} \in f^{-1}(\mathfrak{H}')$. Hence $f^{-1}(\mathfrak{H}') \leqslant \mathfrak{G}$.

As to the uniqueness: A subgroup of which $\mathfrak{e}'$ is the image is in general not unique: Every subgroup of the kernel $\mathfrak{K}$ has $\mathfrak{e}'$ as image. But by $f^{-1}(\mathfrak{H}')$ we denote the set of *all* $x \in \mathfrak{G}$ such that $f(x) \in \mathfrak{H}'$ and this group is obviously unique.

Thus we have the *subgroup correspondence theorem*:

THEOREM 2. *Every epimorphism $f : \mathfrak{G} \to \mathfrak{G}'$ defines a one–one correspondence between the system of all subgroups of $\mathfrak{G}'$ and the system of all subgroups of $\mathfrak{G}$ which contain the kernel of f.*

In the case that f is the natural homomorphism with the kernel $\mathfrak{K}$ we obtain the following result:

COROLLARY. *If $\mathfrak{K} \lhd \mathfrak{G}$ and the factor group $\mathfrak{G}/\mathfrak{K}$ has a subgroup $\mathfrak{H}'$, then $\mathfrak{G}$ has a subgroup $\mathfrak{H}$ such that $\mathfrak{H}/\mathfrak{K} = \mathfrak{H}'$.*

b. Now we shall assume that one of the subgroups $\mathfrak{H}$ or $\mathfrak{H}'$ is normal in $\mathfrak{G}$ or $\mathfrak{G}'$ respectively; it will be seen that then the other one is also normal.

THEOREM 3. *Let $f : \mathfrak{G} \to \mathfrak{G}'$ be an epimorphism and* $\ker f = \mathfrak{K}$. *If $\mathfrak{H}$ is a normal subgroup of $\mathfrak{G}$ and $\mathfrak{K} \leqslant \mathfrak{H}$ then it follows that $\mathfrak{H}' = f(\mathfrak{H}) \trianglelefteq \mathfrak{G}'$ and $\mathfrak{G}'/\mathfrak{H}' \simeq \mathfrak{G}/\mathfrak{H}$.*

Proof. Because $t\mathfrak{H}t^{-1} = \mathfrak{H}$ for all t in $\mathfrak{G}$ we have

$$f(t)f(\mathfrak{H})f(t)^{-1} = f(t\mathfrak{H}t^{-1}) = f(\mathfrak{H})$$

and therefore $t'\mathfrak{H}'t'^{-1} = \mathfrak{H}'$ for all t' in $\mathfrak{G}'$. Moreover the epimorphism f, applied to the cosets of $\mathfrak{H}$ in $\mathfrak{G}$ is a one–one mapping $a\mathfrak{H} \to a'\mathfrak{H}'$ (cf. theorem 1) and $f(a\mathfrak{H}\cdot b\mathfrak{H}) = f(a\mathfrak{H})\cdot f(b\mathfrak{H}) = a'\mathfrak{H}'\cdot b'\mathfrak{H}'$; thus $f : \mathfrak{G}/\mathfrak{H} \to \mathfrak{G}'/\mathfrak{H}'$ is an isomorphism.

If f is the natural homomorphism and $\mathfrak{G}' = \mathfrak{G}/\mathfrak{K}$ we have immediately the isomorphism theorem:

COROLLARY. *In the case that* $\mathfrak{K} \trianglelefteq \mathfrak{H} \trianglelefteq \mathfrak{G}$ *and* $\mathfrak{K} \trianglelefteq \mathfrak{G}$ *one has the isomorphism*

$$\mathfrak{G}/\mathfrak{H} \simeq \frac{\mathfrak{G}/\mathfrak{K}}{\mathfrak{H}/\mathfrak{K}}.$$

Compare this result with the index formula of §2, theorem 2′.

THEOREM 4. *Let* $f : \mathfrak{G} \to \mathfrak{G}'$ *be an epimorphism; let* $\mathfrak{H}'$ *be a normal subgroup of* $\mathfrak{G}'$. *Then the inverse image* $\mathfrak{H} = f^{-1}(\mathfrak{H}')$ *is a normal subgroup of* $\mathfrak{G}$ *which contains* $\mathfrak{K} = \ker f$.

Proof. By supposition $t'\mathfrak{H}'t'^{-1} = \mathfrak{H}'$ for all t' in $\mathfrak{G}'$. Choose an arbitrary t in $\mathfrak{G}$ for which $f(t) = t'$; then $f(t)f(\mathfrak{H})f(t)^{-1} = f(t\mathfrak{H}t^{-1})$ and also $= f(\mathfrak{H})$. As $\mathfrak{K} = \ker f \leqslant \mathfrak{H}$ and also $\mathfrak{K} = t\mathfrak{K}t^{-1} \leqslant t\mathfrak{H}t^{-1}$ it follows from theorem 2 that $t\mathfrak{H}t^{-1} = \mathfrak{H}$.

If f is the natural homomorphism with the kernel $\mathfrak{K}$ we have the following result:

COROLLARY. *If* $\mathfrak{H}'$ *is a normal subgroup of* $\mathfrak{G}/\mathfrak{K}$ *then there exists a normal subgroup* $\mathfrak{H} \trianglelefteq \mathfrak{G}$ *such that* $\mathfrak{H}' = \mathfrak{H}/\mathfrak{K}$.

Finally there is an isomorphism theorem which corresponds to the index formula of §2, *c.*, theorem 3′.

THEOREM 5. *Let* $\mathfrak{H}$ *be an arbitrary subgroup of* $\mathfrak{G}$ *and* $\mathfrak{K} \trianglelefteq \mathfrak{G}$ *so that* $\mathfrak{H}\mathfrak{K} \leqslant \mathfrak{G}$. *Then*

$$\text{(i)}\quad \mathfrak{D} = \mathfrak{H} \cap \mathfrak{K} \trianglelefteq \mathfrak{H}, \qquad \text{(ii)}\quad \mathfrak{H}\mathfrak{K}/\mathfrak{K} \simeq \mathfrak{H}/\mathfrak{H} \cap \mathfrak{K}.$$

Proof. (i) Let $b \in \mathfrak{H} \leqslant \mathfrak{H}\mathfrak{K}$ and $d \in \mathfrak{D}$. Then $bdb^{-1} \in \mathfrak{K}$; but $bdb^{-1} \in \mathfrak{H}$ because $\mathfrak{D} \leqslant \mathfrak{H}$. Thus $bdb^{-1} \in \mathfrak{H} \cap \mathfrak{K} = \mathfrak{D}$. Hence $\mathfrak{D} \trianglelefteq \mathfrak{H}$.

(ii) The one–one correspondence of the cosets $b_\lambda\mathfrak{D}$ of $\mathfrak{D}$ in $\mathfrak{H}$ and the cosets $b_\lambda\mathfrak{K}$ of $\mathfrak{K}$ in $\mathfrak{H}\mathfrak{K}$ as established in §2, *c.*, theorem 3′, is now recognized as an isomorphism between the factor-groups $\mathfrak{H}\mathfrak{K}/\mathfrak{K}$ and $\mathfrak{H}/\mathfrak{D}$, namely $b_\lambda\mathfrak{K} \to b_\lambda\mathfrak{D}$ (where the b_λ represent a transversal of $\mathfrak{D}$ in $\mathfrak{H}$). This completes the proof of the theorem.

The assumption $\mathfrak{K} \trianglelefteq \mathfrak{G}$ in theorem 5 can be replaced by a slightly weaker assumption:

COROLLARY. *If* $\mathfrak{H}, \mathfrak{K} \leqslant \mathfrak{G}$ *and* $\mathfrak{H} \leqslant \mathfrak{N}(\mathfrak{K})$, *it follows that*

$$\text{(i)}\quad \mathfrak{H} \cap \mathfrak{K} \trianglelefteq \mathfrak{H}, \qquad \text{(ii)}\quad \mathfrak{H}\mathfrak{K}/\mathfrak{K} \simeq \mathfrak{H}/\mathfrak{H} \cap \mathfrak{K}.$$

Indeed $\mathfrak{H} \leqslant \mathfrak{N}(\mathfrak{K})$ implies that $x\mathfrak{K}x^{-1} = \mathfrak{K}$ for all $x \in \mathfrak{H}$, that is $\mathfrak{H}\mathfrak{K} = \mathfrak{K}\mathfrak{H}$, whence it follows that $\mathfrak{H}\mathfrak{K}$ is a subgroup of $\mathfrak{G}$.

Remark. In the case of finite subgroups $\mathfrak{H}$, $\mathfrak{K}$ where $|\mathfrak{H}| = h$, $|\mathfrak{K}| = k$, $|\mathfrak{H}\mathfrak{K}| = v$, $|\mathfrak{D}| = d$, under the assumptions of the corollary we have the relation

(3) $\quad hk = vd$

which coincides with the well-known relation between two positive integers h, k, their greatest common divisor $d = (h, k)$ and their least common multiple $v = [h, k]$. (Cf. Chap. I, §3, *a.*, II'.)

c. The internal direct product. A group $\mathfrak{G}$ is said to be the (internal) direct product of two of its subgroups $\mathfrak{A}$, $\mathfrak{B}$ if

(i) $\mathfrak{G} = \mathfrak{A}\mathfrak{B}$; (ii) $\mathfrak{A} \cap \mathfrak{B} = e$; (iii) $\mathfrak{A} \lhd \mathfrak{G}, \mathfrak{B} \lhd \mathfrak{G}$.

If these conditions are satisfied one writes $\mathfrak{G} = \mathfrak{A} \times \mathfrak{B}$.

For some applications it will be of advantage to have the three conditions in another form. Condition (i) can be expressed as follows:

(α) Every element x of $\mathfrak{G}$ can be represented in the form $x = ab$ where $a \in \mathfrak{A}$ and $b \in \mathfrak{B}$.

The following condition is equivalent to (ii):

(β) For every x in $\mathfrak{G}$ the factors a, b are uniquely defined elements of $\mathfrak{A}$, $\mathfrak{B}$ respectively.

Indeed if x can be represented in two ways: $x = ab = a'b'$, then $a'^{-1}a = b'b^{-1} \in \mathfrak{A} \cap \mathfrak{B}$. Hence by (ii) these products are equal to e and therefore $a' = a$ and $b' = b$.

Conversely, (ii) follows immediately from (β). For if (ii) is not valid then there is an element $c \in \mathfrak{A} \cap \mathfrak{B}$, $c \neq e$, and one has for x the second representation $x = ac^{-1} \cdot cb$, $ac^{-1} \in \mathfrak{A}$, $cb \in \mathfrak{B}$. Thus (β) is not valid.

(γ) $ab = ba$ for all $a \in \mathfrak{A}$, $b \in \mathfrak{B}$.

To show that this follows from (iii) and (ii) we consider the *commutator* of a and b, that is

$$\left.\begin{aligned} aba^{-1}b^{-1} &= (aba^{-1})b^{-1} \in \mathfrak{B} \\ &= a\cdot(ba^{-1}b^{-1}) \in \mathfrak{A} \end{aligned}\right\} \text{ by (iii).}$$

Hence by (ii) $aba^{-1}b^{-1} = e$ which implies $ab = ba$.

Conversely (iii) is a consequence of (γ) and (α): Let $a_1 \in \mathfrak{A}$ and $x = ab$ $(a \in \mathfrak{A}, b \in \mathfrak{B})$; then $xa_1x^{-1} = aba_1b^{-1}a^{-1} = aa_1a^{-1} \in \mathfrak{A}$; hence $\mathfrak{A} \lhd \mathfrak{G}$. Similarly $\mathfrak{B} \lhd \mathfrak{G}$.

THEOREM 6. *From* $\mathfrak{G} = \mathfrak{A} \times \mathfrak{B}$ *it follows that*

$$\mathfrak{G}/\mathfrak{A} \simeq \mathfrak{B} \quad \textit{and} \quad \mathfrak{G}/\mathfrak{B} \simeq \mathfrak{A}.$$

Proof. Let $\mathfrak{A}x$ be a coset of $\mathfrak{A}$ in $\mathfrak{G}$. Since $x = ab$, $(a \in \mathfrak{A}, b \in \mathfrak{B})$ we have $\mathfrak{A}x = \mathfrak{A}ab = \mathfrak{A}b$. This suggests that

$$\mathfrak{G} = \mathfrak{A} \cup \mathfrak{A}b \cup \mathfrak{A}b' \cup \ldots = \bigcup_{b \in \mathfrak{B}} \mathfrak{A}b$$

is the decomposition of $\mathfrak{G}$ into cosets (mod $\mathfrak{A}$). Indeed if $b \neq b'$ it follows that $\mathfrak{A}b \neq \mathfrak{A}b'$, for if $\mathfrak{A}b = \mathfrak{A}b'$ we should have $b'b^{-1} \in \mathfrak{A}$ and thus $b'b^{-1} \in \mathfrak{A} \cap \mathfrak{B}$ implying $b'b^{-1} = e$, i.e. $b' = b$.

Thus we have a one–one correspondence between the elements of $\mathfrak{B}$ and those of $\mathfrak{G}/\mathfrak{A}$, namely $b \leftrightarrow \mathfrak{A}b$, $b' \leftrightarrow \mathfrak{A}b'$, ... which obviously is a homomorphism: $bb' \rightarrow \mathfrak{A} \cdot bb' = \mathfrak{A}\mathfrak{A} \cdot bb' = \mathfrak{A}b \cdot \mathfrak{A}b'$ and therefore an isomorphism.

COROLLARY. *If* $\mathfrak{G} = \mathfrak{A} \times \mathfrak{B}$ *is a finite group then* $|\mathfrak{G}| = |\mathfrak{A}|\,|\mathfrak{B}|$.

If $\mathfrak{G} = \mathfrak{A} \times \mathfrak{B}$ we call $\mathfrak{A}$ and $\mathfrak{B}$ "*direct factors*" of $\mathfrak{G}$. A simple example shows that in general direct factors of a group are not uniquely defined by the group (cf. ex. 3, b.). Improper subgroups are not considered as direct factors and not every group has proper direct factors, as e.g. a *simple group*, that is, by definition, a group which has no proper normal subgroup. On the other hand, a normal subgroup of $\mathfrak{G}$ is not always a direct factor of $\mathfrak{G}$ (cf. ex. 2).

THEOREM 7. *The centre of a direct product* $\mathfrak{G} = \mathfrak{A} \times \mathfrak{B}$ *is the direct product of the centres of the direct factors*: $\mathfrak{Z}(\mathfrak{A} \times \mathfrak{B}) = \mathfrak{Z}(\mathfrak{A}) \times \mathfrak{Z}(\mathfrak{B})$.

Proof. (a) To show that $\mathfrak{Z}(\mathfrak{A} \times \mathfrak{B}) \leqslant \mathfrak{Z}(\mathfrak{A}) \times \mathfrak{Z}(\mathfrak{B})$ let $z = z'z''$, $z' \in \mathfrak{A}$, $z'' \in \mathfrak{B}$. For all $x \in \mathfrak{G}$, $x = ab$, $a \in \mathfrak{A}$, $b \in \mathfrak{B}$, we have $xz = zx$, i.e. $abz'z'' = z'z''ab$. Hence by (γ) $az'bz'' = z'az''b$ whence by (β) $az' = z'a$ and $bz'' = z''b$. Thus $z' \in \mathfrak{Z}(\mathfrak{A})$ and $z'' \in \mathfrak{Z}(\mathfrak{B})$. Therefore $z \in \mathfrak{Z}(\mathfrak{A}) \times \mathfrak{Z}(\mathfrak{B})$.

(b) Let $z = z'z''$, $z' \in \mathfrak{Z}(\mathfrak{A})$, $z'' \in \mathfrak{Z}(\mathfrak{B})$. If again $x = ab$, $a \in \mathfrak{A}$, $b \in \mathfrak{B}$ we have $az' = z'a$, $bz'' = z''b$; hence $az'bz'' = z'az''b$ and finally $abz'z'' = z'z''ab$ i.e. $xz = zx$. Thus $z \in \mathfrak{Z}(\mathfrak{A} \times \mathfrak{B})$ and $\mathfrak{Z}(\mathfrak{A}) \times \mathfrak{Z}(\mathfrak{B}) \leqslant \mathfrak{Z}(\mathfrak{A} \times \mathfrak{B})$.

d. It may happen that in the direct product $\mathfrak{G} = \mathfrak{A} \times \mathfrak{B}$ the group $\mathfrak{A}$ is a direct product $\mathfrak{H}_1 \times \mathfrak{H}_2$ of two of it subgroups, and therefore $\mathfrak{G} = (\mathfrak{H}_1 \times \mathfrak{H}_2) \times \mathfrak{B}$. Thus every $x \in \mathfrak{G}$ can be written uniquely in the form $x = (a_1 a_2)b, a_1 \in \mathfrak{H}_1, \; a_2 \in \mathfrak{H}_2, \; b \in \mathfrak{B}$. Since also $x = a_1(a_2 b)$ it follows that $\mathfrak{G} = \mathfrak{H}_1 \times (\mathfrak{H}_2 \times \mathfrak{B})$. Thus the notion of the internal direct product can be extended to three or more direct factors and within a system of direct factors the operation "direct product" is associative. In a general way we define the direct product of r factors as follows:

(i) $\quad \mathfrak{G} = \mathfrak{H}_1 \mathfrak{H}_2 \dots \mathfrak{H}_r$

(ii) $\quad \mathfrak{H}_1 \cap \mathfrak{H}_2 = \mathfrak{e}, \quad \mathfrak{H}_1 \mathfrak{H}_2 \cap \mathfrak{H}_3 = \mathfrak{e}, \dots, \mathfrak{H}_1 \mathfrak{H}_2 \dots \mathfrak{H}_{r-1} \cap \mathfrak{H}_r = \mathfrak{e}$

(iii) $\quad \mathfrak{H}_\rho \lhd \mathfrak{G} \quad (\rho = 1, \dots, r)$.

Clearly (i) states that every $x \in \mathfrak{G}$ can be represented in the form

(α) $\quad x = a_1 a_2 \dots a_r, \quad a_\rho \in \mathfrak{H}_\rho \quad (\rho = 1, \dots, r)$.

Moreover (in analogy with *c.*)

(β) $\quad$ The factors $a_\rho \in \mathfrak{H}_\rho$ are uniquely defined by $x \in \mathfrak{G}$.

(γ) $\quad a_\rho a_\sigma = a_\sigma a_\rho \quad \text{if} \quad \rho \neq \sigma \quad (\rho, \sigma = 1, \dots, r)$.

The relations (γ) follow from (ii) and (iii). By means of the argument used in *c.* we find that the commutators

$$(4) \quad \begin{cases} a_1 a_2 a_1^{-1} a_2^{-1} \in \mathfrak{H}_1 \cap \mathfrak{H}_2 \\ (a_1 a_2) a_3 (a_1 a_2)^{-1} a_3^{-1} \in \mathfrak{H}_1 \mathfrak{H}_2 \cap \mathfrak{H}_3 \\ \dots\dots\dots\dots\dots\dots \\ (a_1 a_2 \dots a_{r-1}) a_r (a_1 a_2 \dots a_{r-1})^{-1} a_r^{-1} \in \mathfrak{H}_1 \mathfrak{H}_2 \dots \mathfrak{H}_{r-1} \cap \mathfrak{H}_r; \end{cases}$$

therefore all these commutators are equal to $\mathfrak{e}$. Hence $a_1 a_2 = a_2 a_1$, $a_1 a_2 a_3 = a_3 a_1 a_2$. For $a_2 = e$ we have $a_1 a_3 = a_3 a_1$ and for $a_1 = e$ likewise $a_2 a_3 = a_3 a_2$. Proceeding in this way we obtain from the other relations (4) all the conditions (γ).

The statement (β) is proved by induction with respect to r. Let us assume that the uniqueness of the factors for all elements of a direct product of $r-1$ direct factors has been established. Then the same follows for a direct product of r direct factors: Let

$$a_1 \ldots a_{r-1} a_r = b_1 \ldots b_{r-1} b_r, \quad a_\rho, b_\rho \in \mathfrak{H}_\rho.$$

Then

$$a_r b_r^{-1} = a_{r-1}^{-1} \ldots a_1^{-1} b_1 \ldots b_{r-1}$$

whence by (γ) and (ii)

$$(5) \qquad a_1^{-1} b_1 \ldots a_{r-1}^{-1} b_{r-1} = e,$$

$$(6) \qquad a_r b_r^{-1} = e.$$

According to the induction assumption we conclude from (5) that $a_1 = b_1, \ldots, a_{r-1} = b_{r-1}$; from (6) follows that $a_r = b_r$. Thus (β) is proved.

As in *c.* one can derive (i), (ii), (iii) from (α), (β), (γ). Obviously (i) and (α) are equivalent. As to (ii), assume e.g. that $\mathfrak{H}_1 \mathfrak{H}_2 \cap \mathfrak{H}_3$ contains an element $c \neq e$. Then we have for $x = a_1 a_2 a_3$ the second representation $x = a_1 a_2 c^{-1} \cdot c a_3$; similarly in the other cases. Finally (iii) follows from (γ).

Theorems 6 and 7 are easy to generalize; in particular the latter, i.e.

$$(7) \qquad \mathfrak{Z}(\mathfrak{G}) = \mathfrak{Z}(\mathfrak{H}_1 \times \mathfrak{H}_2 \times \ldots \times \mathfrak{H}_r) = \mathfrak{Z}(\mathfrak{H}_1) \times \mathfrak{Z}(\mathfrak{H}_2) \times \ldots \times \mathfrak{Z}(\mathfrak{H}_r)$$

follows readily by induction.

e. The external direct product. The idea of the direct product can be applied to solve the following problem: For two arbitrary given groups $\mathfrak{G}_1$ and $\mathfrak{G}_2$ construct a group $\mathfrak{G}$ which has as direct factors two of its subgroups $\mathfrak{A}$ and $\mathfrak{B}$, $\mathfrak{G} = \mathfrak{A} \times \mathfrak{B}$, which are isomorphic to $\mathfrak{G}_1$ and $\mathfrak{G}_2$ respectively.

Elements of $\mathfrak{G}_i$ will be denoted by a_i, b_i $(i = 1, 2)$, the unit element by e_i. The elements of the cartesian product $(\mathfrak{G}_1, \mathfrak{G}_2)$, namely the couples

$$x = (a_1, a_2), y = (b_1, b_2), \ldots$$

form a group $\mathfrak{G}$ with the unit element $e = (e_1, e_2)$, provided we define the multiplication of the couples componentwise, that is

$$xy = (a_1, a_2)(b_1, b_2) = (a_1 b_1, a_2 b_2).$$

We call this group $\mathfrak{G}$ the *external direct* product of $\mathfrak{G}_1$ and $\mathfrak{G}_2$.

The connection with the internal direct product is evident: We notice that the couples

$$a = (a_1, e_2), a_1 \in \mathfrak{G}_1 \quad \text{form a group} \quad \mathfrak{A} \simeq \mathfrak{G}_1,$$

$$b = (e_1, a_2), a_2 \in \mathfrak{G}_2 \quad \text{form a group} \quad \mathfrak{B} \simeq \mathfrak{G}_2,$$

so that

$$\mathfrak{G} = \mathfrak{A} \times \mathfrak{B}$$

in the sense defined in *c*.

Indeed the three conditions (α), (β), (γ) of *c*. are easily verified: (α) For all $x \in \mathfrak{G}$ we have $x = (a_1, a_2) = (a_1, e_2)(e_1, a_2) = ab$; ($\beta$) Since a_1, a_2 are uniquely defined by x, so are a, b; (γ) $ba = (e_1, a_2)(a_1, e_2) = (a_1, a_2) = (a_1, e_2)(e_1, a_2) = ab$.

A natural and descriptive notation for the external direct product of $\mathfrak{G}_1$ and $\mathfrak{G}_2$ is $\mathfrak{G} = (\mathfrak{G}_1, \mathfrak{G}_2)$. In this way we understand without elaborate definition the generalization of the external direct product to more than two direct factors $\mathfrak{G}_1, \mathfrak{G}_2, \ldots, \mathfrak{G}_r$:

$$\mathfrak{G} = (\mathfrak{G}_1, \mathfrak{G}_2, \ldots, \mathfrak{G}_r) = \mathfrak{H}_1 \times \mathfrak{H}_2 \times \ldots \times \mathfrak{H}_r$$

where

$$\mathfrak{H}_1 = (\mathfrak{G}_1, e_2, \ldots, e_r), \quad \mathfrak{H}_2 = (e_1, \mathfrak{G}_2, e_3, \ldots, e_r), \ldots.$$

The groups $\mathfrak{G}_1, \ldots, \mathfrak{G}_r$ may only be defined up to isomorphism; in particular it is admissible that they are all identical, so that $\mathfrak{G} = (\mathfrak{G}_1, \ldots, \mathfrak{G}_1)$ which, of course, does not imply that the subgroups $\mathfrak{H}_1, \ldots, \mathfrak{H}_r$ are equal; they are disjoint, but isomorphic.

Remark 1. We have shown in *d.* that the internal direct product is associative within a system of disjoint direct factors of a group. As an operation within a system of abstract groups $\mathfrak{G}_j$ the external direct product is associative up to isomorphism:

$$(\mathfrak{G}_1, (\mathfrak{G}_2, \mathfrak{G}_3)) \simeq (\mathfrak{G}_1, \mathfrak{G}_2, \mathfrak{G}_3) \simeq ((\mathfrak{G}_1, \mathfrak{G}_2), \mathfrak{G}_3).$$

Remark 2. It is often convenient to use for the external direct product the notation of the internal direct product, that means, instead of $\mathfrak{G} = (\mathfrak{G}_1, \mathfrak{G}_2, \ldots, \mathfrak{G}_r)$ we may write

$$\mathfrak{G} = \mathfrak{G}_1 \times \mathfrak{G}_2 \times \ldots \times \mathfrak{G}_r.$$

THEOREM 8. *Let $\mathfrak{G}$ be the direct product of r ($\geqslant 2$) non-abelian simple groups $\mathfrak{H}_1, \ldots, \mathfrak{H}_r$. Then every normal subgroup $\mathfrak{H}$ of $\mathfrak{G}$ is the direct product of a subcollection of the direct factors $\mathfrak{H}_i$ of $\mathfrak{G}$.*

Proof. Let $\mathfrak{G} = \mathfrak{H}_1 \times \mathfrak{H}_2 \times \ldots \times \mathfrak{H}_r$ and $\mathfrak{H} \lhd \mathfrak{G}$. Assume $b \in \mathfrak{H}$, $b = a_1 a_2 \ldots a_r$, $a_i \in \mathfrak{H}_i$ $(i = 1, 2, \ldots, r)$ and $a_1 \neq e$. Since $\mathfrak{H}_1$ is simple, its centre $\mathfrak{Z}(\mathfrak{H}_1) = \mathfrak{e}$ and there is an element x in $\mathfrak{H}_1$ for which $xa_1 \neq a_1 x$ and

$$bx = a_1 a_2 \ldots a_r x = a_1 \cdot x \cdot a_2 \ldots a_r \neq x \cdot a_1 a_2 \ldots a_r = xb$$

so that the commutator $bxb^{-1}x^{-1} = c \neq e$.

Now $\mathfrak{H}_1 \lhd \mathfrak{G}$; therefore $c = (bxb^{-1})x^{-1} \in \mathfrak{H}_1$. Also $\mathfrak{H} \lhd \mathfrak{G}$; therefore $c = b(xb^{-1}x^{-1}) \in \mathfrak{H}$. Thus $c \in \mathfrak{H} \cap \mathfrak{H}_1 \trianglelefteq \mathfrak{H}_1$ and again because $\mathfrak{H}_1$ is simple: $\mathfrak{e} < \mathfrak{H} \cap \mathfrak{H}_1 = \mathfrak{H}_1 \lhd \mathfrak{H}$. Similarly, if the factor a_i of an element b in $\mathfrak{H}$ is different from e, we conclude that $\mathfrak{H}_i \trianglelefteq \mathfrak{H}$.

Remark 3. The theorem is no longer true if the direct factors $\mathfrak{H}_i$ of $\mathfrak{G}$ include abelian simple groups; cf. ex. 3(b).

f. The external direct product can be defined also in the case of an infinite number of factors $\mathfrak{G}_1, \mathfrak{G}_2, \ldots$. As in the case of a finite number of factors we have in the first place the *cartesian product* $\mathfrak{G} = (\mathfrak{G}_1, \mathfrak{G}_2, \ldots)$, i.e. the system of all infinite sequences $(a_1, a_2, \ldots)$, $a_\rho \in \mathfrak{G}_\rho$, $\rho = 1, 2, \ldots$, a natural generalization of the product

THEOREM 3. *Let $\mathfrak{H}$ be a group, $\mathfrak{L}$ a subgroup, and $\mathfrak{G}$ a "supergroup"* of $\mathfrak{H} : \mathfrak{L} \leqslant \mathfrak{H} \leqslant \mathfrak{G}$.

(a) If $\mathfrak{L} \trianglelefteq\!\!\!\!\lhd \mathfrak{H} \trianglelefteq\!\!\!\!\lhd \mathfrak{G}$ then $\mathfrak{L} \trianglelefteq\!\!\!\!\lhd \mathfrak{G}$.

(b) If $\mathfrak{L} \trianglelefteq\!\!\!\!\lhd \mathfrak{H} \lhd \mathfrak{G}$ then $\mathfrak{L} \lhd \mathfrak{G}$.

Proof. (a) Every automorphism α in $\mathfrak{G}$ leaves $\mathfrak{H}$ invariant and induces an automorphism in $\mathfrak{H}$; therefore it leaves $\mathfrak{L}$ invariant.

(b) Every inner automorphism of $\mathfrak{G}$ leaves $\mathfrak{H}$ invariant and induces an automorphism of $\mathfrak{H}$; therefore it leaves $\mathfrak{L}$ invariant; thus $\mathfrak{L} \lhd \mathfrak{G}$.

This implies that every characteristic subgroup $\mathfrak{L}$ of $\mathfrak{H}$ is normal in every group $\mathfrak{G}$ which contains $\mathfrak{H}$ as a normal subgroup.

Remark. Part (b) of theorem 3 can be inverted in the following way: In §2 it will be shown that for a given group $\mathfrak{H}$ a supergroup $\mathfrak{G}$, the so-called holomorph of $\mathfrak{H}$, can be constructed such that every automorphism of $\mathfrak{H}$ is an inner automorphism of $\mathfrak{G}$ restricted to $\mathfrak{H}$. Thus we can state:

(c) If $\mathfrak{L} \lhd \mathfrak{H} \lhd \mathfrak{G}$ and $\mathfrak{L} \lhd \mathfrak{G}$ then $\mathfrak{L} \lhd\!\!\!\!| \, \mathfrak{H}$.

THEOREM 4. *If $\mathfrak{L} \leqslant \mathfrak{H} \leqslant \mathfrak{G}$, $\mathfrak{L} \lhd\!\!\!\!| \, \mathfrak{G}$ and $\mathfrak{H}/\mathfrak{L} \lhd\!\!\!\!| \, \mathfrak{G}/\mathfrak{L}$ it follows that* $\mathfrak{H} \lhd\!\!\!\!| \, \mathfrak{G}$.

Proof. For every $\alpha \in \Gamma(\mathfrak{G})$ by supposition $\alpha(\mathfrak{L}) = \mathfrak{L}$. Moreover α induces an automorphism $\hat{\alpha}$ of $\mathfrak{G}/\mathfrak{L}$, namely $\hat{\alpha}(x\mathfrak{L}) = \alpha(x)\mathfrak{L}$. Indeed $\hat{\alpha}(x\mathfrak{L}\, y\mathfrak{L}) = \hat{\alpha}(xy\mathfrak{L}) = \alpha(xy)\mathfrak{L} = \alpha(x)\cdot\alpha(y)\mathfrak{L} = \alpha(x)\mathfrak{L}\cdot\alpha(y)\mathfrak{L} = \hat{\alpha}(x\mathfrak{L})\cdot\hat{\alpha}(y\mathfrak{L})$. As an automorphism of $\mathfrak{G}/\mathfrak{L}$ the automorphism $\hat{\alpha}$ preserves the subgroup $\mathfrak{H}/\mathfrak{L}$ (by supposition); thus $\hat{\alpha}$ causes a permutation of the cosets $x\mathfrak{L}$, the elements of $\mathfrak{H}/\mathfrak{L}$, leaving $\mathfrak{L}$ fixed. Hence it follows that $\alpha(\mathfrak{H}) = \mathfrak{H}$.

d. Characteristically simple groups. A group $\mathfrak{G}$ has been called simple (cf. Chap. II, §5, *c.*) when it has no proper normal subgroup. All cyclic groups of prime order are simple. No other finite abelian group $\mathfrak{G}$ is simple. Indeed let $|\mathfrak{G}| = g$, not a prime number. If $\mathfrak{G}$ is cyclic then every proper divisor d of g is the order of a normal subgroup of $\mathfrak{G}$. If $\mathfrak{G}$ is not cyclic then there is an element x in $\mathfrak{G}$ of a certain order $d < g$, $d|g$, and $\langle x \rangle_d \lhd \mathfrak{G}$.

There are simple non-abelian groups, e.g. the icosahedral group $\mathfrak{J}$ (cf. Chap. II, §5, ex. 10) and the alternating groups $\mathfrak{A}_n$ ($n = 5, 6, \ldots$) (cf. ex. 13).

(b) Show that $\mathfrak{G}^{(4)}$ is the direct product of two of its subgroups of order 2 in three different ways.

(c) The group $\mathfrak{G}^{(4)}$ is the union of its three proper subgroups cf. §1, Theorem 3, corollary.

4. Let $\mathfrak{G}$ be a group where all elements except e have order 2. Show that $\mathfrak{G}$ is abelian, and if $\mathfrak{G}$ is finite, that it is the direct product of groups of order 2.

5. Describe up to isomorphism all abstract groups of order 8 which are direct products of proper subgroups.

6. (a) Let $\mathfrak{H}, \mathfrak{K}$ be cyclic subgroups of a larger group $\mathfrak{G}$ and $|\mathfrak{H}| = h$, $|\mathfrak{K}| = k$. If $(h, k) = 1$, then the direct product of $\mathfrak{H}$ and $\mathfrak{K}$ is a cyclic group of order hk.

(b) If $\mathfrak{G}$ is a cyclic group of order $v = [h, k]$, then it contains exactly one cyclic subgroup $\mathfrak{H}$ of order h and one cyclic subgroup $\mathfrak{K}$ of order k. Show that $\mathfrak{G} = \mathfrak{H}\mathfrak{K}$, $|\mathfrak{H} \cap \mathfrak{K}| = (h, k) = d$. In this case the group order formula (3) *is* the arithmetical relation between g.c.d. and l.c.m.

(c) Let p be a prime number. Show that a cyclic group of order p^α is not a direct product of cyclic groups of lower order.

(d) Let $p_1, \ldots, p_r$ be r distinct prime numbers. Then a cyclic group of the order $p_1^{\alpha_1} p_2^{\alpha_2} \ldots p_r^{\alpha_r}$ is the direct product of r uniquely defined cyclic groups of the orders $p_1^{\alpha_1}, \ldots, p_r^{\alpha_r}$.

7. Let $\mathfrak{G} = \mathfrak{A} \times \mathfrak{B}$ and $\mathfrak{L} \trianglelefteq \mathfrak{A}$. Show that $\mathfrak{L} \triangleleft \mathfrak{G}$. [If $\mathfrak{L} \triangleleft \mathfrak{H} \triangleleft \mathfrak{G}$ it does not in general follow that $\mathfrak{L} \triangleleft \mathfrak{G}$, cf. §4, ex. 2(c)].

8. (a) The additive group $\mathbb{C}^+$ of the field of the complex numbers is isomorphic to the direct product of two copies of the additive group of the real number field.

(b) Give two different representations of $\mathbb{C}^+$ as an internal direct product of subgroups.

(c) The additive group of an n-dimensional vector space over a field $\mathbb{F}$ is isomorphic to the external direct product of n copies of the addition group of $\mathbb{F}$.

9. If $\mathfrak{G} = (\mathfrak{G}_1, \mathfrak{G}_2, \ldots)$ is the cartesian product of $\mathfrak{G}_1, \mathfrak{G}_2, \ldots$ and $\mathfrak{H} = \mathfrak{G}_1 \times \mathfrak{G}_2 \times \ldots$ the restricted direct product, show that $\mathfrak{H} \trianglelefteq \mathfrak{G}$ with equality if and only if the number of factors is finite.

automorphism is the inverse of an automorphism we have $\mathfrak{Z} \leqslant \alpha(\mathfrak{Z})$. Thus $\alpha(\mathfrak{Z}) = \mathfrak{Z}$.

Another important characteristic subgroup of a group $\mathfrak{G}$ is the so-called *commutator subgroup* $\mathfrak{G}' \leqslant \mathfrak{G}$, also called the *derived group*. Let us denote by **C** the set of all commutators in $\mathfrak{G}$:

$$\mathbf{C} = \{[x, y] = xyx^{-1}y^{-1} \mid x, y \in \mathfrak{G}\}.$$

By definition

$$\mathfrak{G}' = \langle \mathbf{C} \rangle.$$

For every α in Γ we have

$$\alpha([x, y]) = \alpha(x)\,\alpha(y)\,\alpha(x)^{-1}\alpha(y)^{-1} = [\alpha(x), \alpha(y)];$$

thus $\alpha([x, y])$ is a commutator and therefore $\alpha(\mathbf{C}) \subseteq \mathbf{C}$. Also $\alpha^{-1}(\mathbf{C}) \subseteq \mathbf{C}$; hence $\alpha(\mathbf{C}) = \mathbf{C}$ and therefore $\alpha(\mathfrak{G}') = \mathfrak{G}'$. This proves the first part of the following theorem:

THEOREM 2. *The commutator subgroup $\mathfrak{G}'$ of a group $\mathfrak{G}$ is a characteristic subgroup*: $\mathfrak{G}' \trianglelefteq \mathfrak{G}$ *and the factor group* $\mathfrak{G}/\mathfrak{G}'$ *is abelian. Moreover if* $\mathfrak{H} \trianglelefteq \mathfrak{G}$ *then the factor group* $\mathfrak{G}/\mathfrak{H}$ *is abelian if and only if* $\mathfrak{G}' \trianglelefteq \mathfrak{H} \trianglelefteq \mathfrak{G}$.

Proof. The commutator of two elements $x\mathfrak{G}', y\mathfrak{G}'$ of $\mathfrak{G}/\mathfrak{G}'$, that is

$$[x\mathfrak{G}', y\mathfrak{G}'] = x\mathfrak{G}'y\mathfrak{G}'x^{-1}\mathfrak{G}'y^{-1}\mathfrak{G}' = [x, y]\mathfrak{G}' = \mathfrak{G}',$$

the unit element of $\mathfrak{G}/\mathfrak{G}'$. Thus $x\mathfrak{G}'$ and $y\mathfrak{G}'$ commute.

If $\mathfrak{G}' \trianglelefteq \mathfrak{H} \trianglelefteq \mathfrak{G}$ the argument can be repeated:

$$[x\mathfrak{H}, y\mathfrak{H}] = [x, y]\mathfrak{H} = \mathfrak{H} \quad \text{because} \quad [x, y] \in \mathfrak{G}' \trianglelefteq \mathfrak{H}.$$

Conversely, if $\mathfrak{G}/\mathfrak{H}$ is abelian, it follows that every $[x, y] \in \mathfrak{H}$ therefore $\mathfrak{G}' \trianglelefteq \mathfrak{H}$.

It has been pointed out [Chap. II, §4, ex. 2(c)] that if $\mathfrak{L} \lhd \mathfrak{H} \lhd \mathfrak{G}$ then one cannot conclude that $\mathfrak{L} \lhd \mathfrak{G}$. But we can state the following facts:

Proof. Let $\mathfrak{G}$ be finite and $|\mathfrak{A} \cap \mathfrak{B}| = d$; then by (1) $ld = |\mathfrak{A}| = |\mathfrak{B}| = rd$. Hence $l = r$. If $\mathfrak{G}$ is infinite and $\mathfrak{A}\mathfrak{B} \leqslant \mathfrak{G}$ one has by (2) $l = (\mathfrak{G} : \mathfrak{B})/(\mathfrak{G} : \mathfrak{A}\mathfrak{B}) = (\mathfrak{G} : \mathfrak{A})/(\mathfrak{G} : \mathfrak{A}\mathfrak{B}) = r$.

b. Double cosets. Carrying on with the previously introduced notations we now consider for an arbitrary $x \in \mathfrak{G}$ the so-called double coset mod $(\mathfrak{A}, \mathfrak{B})$ in $\mathfrak{G}$, that is, the subset

$$\mathfrak{A}x\mathfrak{B} = axb, \quad a \in \mathfrak{A}, b \in \mathfrak{B}, \quad x \text{ fixed in } \mathfrak{G}.$$

Double cosets have an important property in common with ordinary left (or right) cosets: *Two double cosets* $\mathfrak{A}x\mathfrak{B}$ *and* $\mathfrak{A}y\mathfrak{B}$ *coincide if they have a common element*. Indeed, let $axb = a'yb'$, $a' \in \mathfrak{A}$, $b' \in \mathfrak{B}$. Then $\mathfrak{A}axb\mathfrak{B} = \mathfrak{A}a'yb'\mathfrak{B}$, thus $\mathfrak{A}x\mathfrak{B} = \mathfrak{A}y\mathfrak{B}$. Hence two different double cosets are disjoint. Moreover every element x of $\mathfrak{G}$ is contained in one and only one double coset, namely in $\mathfrak{A}x\mathfrak{B}$. The group $\mathfrak{G}$ can therefore be represented as a union of the double cosets

$$(3) \qquad \mathfrak{G} = \mathfrak{A}\mathfrak{B} \cup \underset{x \notin \mathfrak{A}\mathfrak{B}}{\mathfrak{A}x\mathfrak{B}} \cup \underset{\substack{y \notin \mathfrak{A}\mathfrak{B} \\ y \notin \mathfrak{A}x\mathfrak{B}}}{\mathfrak{A}y\mathfrak{B}} \cup \ldots$$

The elements $x, y, \ldots$ form a system $\mathbf{R}$ of representatives, that is, a transversal, of the double cosets mod $(\mathfrak{A}, \mathfrak{B})$. In $\mathbf{R}$ the element x may be replaced by $x' = axb$, $a \in \mathfrak{A}$, $b \in \mathfrak{B}$; indeed $\mathfrak{A}x'\mathfrak{B} = \mathfrak{A}x\mathfrak{B}$. Instead of (3) we can write

$$(3') \qquad \mathfrak{G} = \bigcup_{x \in \mathbf{R}} \mathfrak{A}x\mathfrak{B}.$$

In the case of a finite group $\mathfrak{G}$

$$(4) \qquad |\mathfrak{G}| = \sum_{x \in \mathbf{R}} |\mathfrak{A}x\mathfrak{B}|.$$

If $\mathfrak{A}$ and $\mathfrak{B}$ are finite

$$\begin{aligned} |\mathfrak{A}x\mathfrak{B}| &= |\mathfrak{A}x\mathfrak{B} \cdot x^{-1}| = |\mathfrak{A} \cdot \mathfrak{B}^x| \\ &= |x^{-1} \cdot \mathfrak{A}x\mathfrak{B}| = |\mathfrak{A}^{x^{-1}} \cdot \mathfrak{B}| \end{aligned}$$

so that from (1′)

$$(5)\qquad |\mathfrak{A}x\mathfrak{B}| = \frac{|\mathfrak{A}|\cdot|\mathfrak{B}|}{|\mathfrak{A}\cap\mathfrak{B}^x|} = \frac{|\mathfrak{A}|\cdot|\mathfrak{B}|}{|\mathfrak{A}^{x^{-1}}\cap\mathfrak{B}|}.$$

Since $\mathfrak{A}\cap\mathfrak{B}^x \leqslant \mathfrak{B}^x$ and $\mathfrak{A}^{x^{-1}}\cap\mathfrak{B} \leqslant \mathfrak{A}^{x^{-1}}$ we have

$$(6)\qquad \begin{cases} (\mathfrak{G}:\mathfrak{A}) = \sum\limits_{x\in\mathbf{R}} \dfrac{|\mathfrak{B}|}{|\mathfrak{A}^{x^{-1}}\cap\mathfrak{B}|} = \sum\limits_{x\in\mathbf{R}} (\mathfrak{B}:\mathfrak{A}^{x^{-1}}\cap\mathfrak{B}) \\[2ex] (\mathfrak{G}:\mathfrak{B}) = \sum\limits_{x\in\mathbf{R}} \dfrac{|\mathfrak{A}|}{|\mathfrak{A}\cap\mathfrak{B}^x|} = \sum\limits_{x\in\mathbf{R}} (\mathfrak{A}:\mathfrak{A}\cap\mathfrak{B}^x). \end{cases}$$

If one of the subgroups $\mathfrak{A}$ or $\mathfrak{B}$ is normal, e.g. $\mathfrak{A} \trianglelefteq \mathfrak{G}$, we have $\mathfrak{A}\mathfrak{B} = \mathfrak{B}\mathfrak{A} \leqslant \mathfrak{G}$ and $\mathfrak{A}x\mathfrak{B} = x\mathfrak{A}\mathfrak{B}$ so that the double cosets in $\mathfrak{G}$ [mod $(\mathfrak{A}, \mathfrak{B})$] are ordinary left cosets of $\mathfrak{G}$ (mod $\mathfrak{A}\mathfrak{B}$).

The double coset $\mathfrak{A}x\mathfrak{B}$ is the union of certain left cosets $ax\mathfrak{B}$, ($a \in \mathfrak{A}$, x fixed in $\mathfrak{G}$) of $\mathfrak{B}$. Let their number l_x be finite. Then

$$\mathfrak{A}x\mathfrak{B} = a_1x\mathfrak{B} \cup a_2x\mathfrak{B} \cup \ldots \cup a_{l_x}x\mathfrak{B}, \quad a_\lambda \in A,\ \lambda = 1, \ldots, l_x.$$

Multiplication of this equation by x^{-1} from the right side yields

$$\mathfrak{A}\mathfrak{B}^x = a_1\mathfrak{B}^x \cup a_2\mathfrak{B}^x \cup \ldots \cup a_{l_x}\mathfrak{B}^x.$$

This is a coset expansion of $\mathfrak{A}\mathfrak{B}^x$ in cosets of $\mathfrak{B}^x$. These cosets are distinct since otherwise two cosets in the expansion of $\mathfrak{A}x\mathfrak{B}$ would coincide. We consider now the coset expansion of $\mathfrak{A}$ in cosets of $\mathfrak{D}_x = \mathfrak{A}\cap\mathfrak{B}^x$. By (1) $(\mathfrak{A}\mathfrak{B}^x : \mathfrak{B}^x) = (\mathfrak{A} : \mathfrak{D}_x)$. Hence both expansions have the same number of terms. One of these terms is $a_1\mathfrak{D}_x$. Suppose that $a_1\mathfrak{D}_x = a_2\mathfrak{D}_x$; this means that $a_2^{-1}a_1 \in \mathfrak{D}_x$ and since $\mathfrak{D}_x \subseteq \mathfrak{B}^x$ we conclude that $a_1\mathfrak{B}^x = a_2\mathfrak{B}^x$ which contradicts the assumption (cf. §2, *c*.). Hence

$$\mathfrak{A} = a_1\mathfrak{D}_x \cup a_2\mathfrak{D}_x \cup \ldots \cup a_{l_x}\mathfrak{D}_x$$

with disjoint $a_\lambda\mathfrak{D}_x$ for distinct λ. Hence by (2) and (5)

$$(7)\qquad \begin{cases} l_x = (\mathfrak{A}:\mathfrak{A}\cap\mathfrak{B}^x) = \dfrac{|\mathfrak{A}x\mathfrak{B}|}{|\mathfrak{B}|}, \\[2ex] r_x = (\mathfrak{B}:\mathfrak{B}\cap\mathfrak{A}^{x^{-1}}) = \dfrac{|\mathfrak{A}x\mathfrak{B}|}{|\mathfrak{A}|}. \end{cases}$$

Thus by (6)

$$(8)\qquad (\mathfrak{G}:\mathfrak{A}) = \sum_{x\in\mathbf{R}} r_x,\quad (\mathfrak{G}:\mathfrak{B}) = \sum_{x\in\mathbf{R}} l_x.$$

Moreover we note that with regard to (7)

$$(9)\qquad l_x = r_x \quad \text{if} \quad |\mathfrak{A}| = |\mathfrak{B}|.$$

c. Double transversals. In §2 ex. 6(b) a proper subgroup $\mathfrak{H}$ of $\mathfrak{S}_4$ has been mentioned for which, although $\mathfrak{H}$ is not normal in $\mathfrak{S}_4$, a right transversal is also a left transversal. G. A. Miller has first shown in 1910 that such a "double transversal" exists for every subgroup of every finite group $\mathfrak{G}$.

We shall consider a slightly more general situation. Let $\mathfrak{A}$ and $\mathfrak{B}$ be two subgroups of $\mathfrak{G}$ with the same finite index $(\mathfrak{G}:\mathfrak{A}) = (\mathfrak{G}:\mathfrak{B})$ and $\mathfrak{A}\mathfrak{B} \leqslant \mathfrak{G}$ if $\mathfrak{G}$ is infinite. Let $\mathfrak{A}x\mathfrak{B}$ be one of the double cosets in the expansion (3). Split it into its right cosets mod $\mathfrak{A}$:

$$\mathfrak{A}x\mathfrak{B} = \mathfrak{A}xb_1 \cup \mathfrak{A}xb_2 \cup \ldots \cup \mathfrak{A}xb_{r_x} \quad (b_\rho \in \mathfrak{B})$$

as well as into its left cosets mod $\mathfrak{B}$:

$$\mathfrak{A}x\mathfrak{B} = a_1x\mathfrak{B} \cup a_2x\mathfrak{B} \cup \ldots \cup a_{l_x}x\mathfrak{B} \quad (a_\lambda \in \mathfrak{A})$$

With regard to Theorem 1 we have $l_x = r_x$ for all $x \in \mathfrak{G}$ and since $\mathfrak{A}a = \mathfrak{A}$ for all $a \in \mathfrak{A}$, and $b\mathfrak{B} = \mathfrak{B}$ for all $b \in \mathfrak{B}$ we can write

$$\mathfrak{A}x\mathfrak{B} = \mathfrak{A}a_1xb_1 \cup \ldots \cup \mathfrak{A}a_{r_x}xb_{r_x} = a_1xb_1\mathfrak{B} \cup \ldots \cup a_{r_x}xb_{r_x}\mathfrak{B}.$$

Thus $c_1(x) = a_1xb_1, \ldots, c_{r_x}(x) = a_{r_x}xb_{r_x}$ is part of a transversal of the right cosets of $\mathfrak{A}$ as well as part of a transversal of the left cosets of $\mathfrak{B}$ in $\mathfrak{G}$, namely of those cosets contained in $\mathfrak{A}x\mathfrak{B}$.* By applying the argument to all double cosets mod $(\mathfrak{A}, \mathfrak{B})$ in $\mathfrak{G}$ we obtain by (3) the following result:

* The idea of this proof was suggested to the author by H. Zassenhaus.

THEOREM 2. *If* $\mathfrak{A}$ *and* $\mathfrak{B}$ *are two subgroups of equal finite index in* $\mathfrak{G}$, *and* $\mathfrak{A}\mathfrak{B} \leqslant \mathfrak{G}$ *if* $\mathfrak{G}$ *is infinite, then it is possible to choose in* $\mathfrak{G}$ *a* ***double transversal*** *for* $\mathfrak{A}$ *and* $\mathfrak{B}$, *that is, a system of elements* $c_1, c_2, \ldots, c_i \in \mathfrak{G}$ *such that*

$$\mathfrak{G} = \mathfrak{A}c_1 \cup \mathfrak{A}c_2 \cup \ldots \cup \mathfrak{A}c_i = c_1\mathfrak{B} \cup c_2\mathfrak{B} \cup \ldots \cup c_i\mathfrak{B}.$$

In the case that $\mathfrak{A} = \mathfrak{B}$ we have Miller's theorem:

COROLLARY. *If* $\mathfrak{A}$ *is a subgroup of finite index in* $\mathfrak{G}$ *then it is possible to choose a* ***double transversal*** *for* $\mathfrak{A}$ ***in*** $\mathfrak{G}$, *that is, a system of elements* $c_1, \ldots, c_i \in \mathfrak{G}$ *such that*

$$\mathfrak{G} = \mathfrak{A}c_1 \cup \mathfrak{A}c_2 \cup \ldots \cup \mathfrak{A}c_i = c_1\mathfrak{A} \cup c_2\mathfrak{A} \cup \ldots \cup c_i\mathfrak{A}.$$

Remark. Theorem 2, as well as Miller's theorem are special cases of a more general theorem of combinatorial set theory which can be stated in the following form:

THEOREM 3. *Let* $\mathbf{S}$ *be a set of which two partitions are given into n equivalent* disjoint subsets*:

$$\mathbf{S} = \mathbf{A}_1 \cup \mathbf{A}_2 \cup \ldots \cup \mathbf{A}_n = \mathbf{B}_1 \cup \mathbf{B}_2 \cup \ldots \cup \mathbf{B}_n$$

so that no k of the $\mathbf{A}_i$ *are contained in the union of fewer than k of the* $\mathbf{B}_j$ $(k = 1, 2, \ldots, n)$. *Then* ***a common transversal*** *for the* $\mathbf{A}_i$ *and the* $\mathbf{B}_j$ *can be found, that is, for a certain permutation* $\alpha_1, \alpha_2, \ldots, \alpha_n$ *of the indices* $1, 2, \ldots, n$ *the intersections*

$$\mathbf{A}_1 \cap \mathbf{B}_{\alpha_1}, \quad \mathbf{A}_2 \cap \mathbf{B}_{\alpha_2}, \ldots, \mathbf{A}_n \cap \mathbf{B}_{\alpha_n}$$

all are not empty. A set of n elements $a_i \in \mathbf{A}_i \cap \mathbf{B}_{\alpha_i}$ $(i = 1, \ldots, n)$ *represents a common transversal for the two partitions.*

This theorem will not be proved here; it has been rediscovered repeatedly since 1916, sometimes in disguise. In its present form it is due to P. Hall (1935). For a full discussion of its implications we refer

* Two sets $\mathbf{A}_1$, $\mathbf{A}_2$ are *equivalent* if there is a bijective mapping $\mathbf{A}_1 \to \mathbf{A}_2$. Finite sets are equivalent if and only if they have the same number of elements.

to the book by L. Mirsky "*Transversal Theory. An Account of some Aspects of Combinatorial Mathematics*", Academic Press, 1971.

d. A group of double cosets. The cosets of a subgroup $\mathfrak{H}$ in $\mathfrak{G}$ form a group with respect to subset multiplication in $\mathfrak{G}$ if and only if $\mathfrak{H} \trianglelefteq \mathfrak{G}$; this coset group is the factor group $\mathfrak{G}/\mathfrak{H}$. L. Fuchs has dealt with the corresponding problem for double cosets.

To each x in $\mathfrak{G}$ we associate the double coset $\mathfrak{A}x\mathfrak{B}$, $\mathfrak{A} < \mathfrak{G}$, $\mathfrak{B} < \mathfrak{G}$. Clearly $x \in \mathfrak{A}x\mathfrak{B}$ and if also $y \in \mathfrak{G}$ the product xy is an element of the product set $(\mathfrak{A}x\mathfrak{B})(\mathfrak{A}y\mathfrak{B})$; if this is one of the double cosets mod $(\mathfrak{A}, \mathfrak{B})$ then it contains xy and therefore equals $\mathfrak{A}xy\mathfrak{B}$. Hence if the double cosets $\mathfrak{A}x\mathfrak{B}$ form a group $\mathfrak{M}$, the mapping $x \to \mathfrak{A}x\mathfrak{B}$ will be an epimorphism $\mathfrak{G} \to \mathfrak{M}$. All x in $\mathfrak{G}$ for which $x = ab$, $a \in \mathfrak{A}$, $b \in \mathfrak{B}$, are carried into the unit element $\mathfrak{A}\mathfrak{B}$ of $\mathfrak{M}$; hence $\mathfrak{A}\mathfrak{B}$ is the kernel of this homomorphism whence $\mathfrak{A}\mathfrak{B} \trianglelefteq \mathfrak{G}$ and $\mathfrak{A}\mathfrak{B} = \mathfrak{B}\mathfrak{A}$. The homomorphism therefore is the natural homomorphism $\mathfrak{G} \to \mathfrak{G}/\mathfrak{A}\mathfrak{B}$ and the double cosets are the cosets of $\mathfrak{A}\mathfrak{B}$, i.e. $\mathfrak{A}x\mathfrak{B} = x\mathfrak{A}\mathfrak{B}$.

The converse is obviously true. Hence we have the result:

THEOREM 4. *If $\mathfrak{A}$ and $\mathfrak{B}$ are two subgroups of $\mathfrak{G}$ then the double cosets $\mathfrak{A}x\mathfrak{B}$ $(x \in \mathfrak{G})$ form a group $\mathfrak{M}$ if and only if*

$$\mathfrak{A}\mathfrak{B} = \mathfrak{B}\mathfrak{A} \trianglelefteq \mathfrak{G},$$

that is, if and only if $\mathfrak{A}x\mathfrak{B} = x\mathfrak{A}\mathfrak{B} = \mathfrak{A}\mathfrak{B}x$ for every x in $\mathfrak{G}$. The group $\mathfrak{M}$ is a homomorphic image of $\mathfrak{G}$ and coincides with the factor group $\mathfrak{G}/\mathfrak{A}\mathfrak{B}$.

Examples and exercises

1. If $\mathfrak{A}$ and $\mathfrak{B}$ are subgroups of $\mathfrak{G}$ and the indices $(\mathfrak{G} : \mathfrak{A})$ and $(\mathfrak{G} : \mathfrak{B})$ are finite, show that the number of conjugates of $\mathfrak{A}$ obtained by transforming $\mathfrak{A}$ with the elements of $\mathfrak{B}$ equals $(\mathfrak{B} : \mathfrak{N}(\mathfrak{A}) \cap \mathfrak{B})$ where $\mathfrak{N}(\mathfrak{A})$ is the normalizer of $\mathfrak{A}$ in $\mathfrak{G}$.

2. (Miller) Let $\mathfrak{G}$ be finite and $\mathfrak{A} < \mathfrak{G}$. If

$$\mathfrak{G} = \mathfrak{A} \cup x_2\mathfrak{A} \cup \ldots \cup x_i\mathfrak{A} = \mathfrak{A} \cup \mathfrak{A}y_2 \cup \ldots \cup \mathfrak{A}y_i,$$

show that

(a) $s_{\mu\nu} = |x_\mu \mathfrak{A} \cap \mathfrak{A} y_\nu|$ is either zero or a divisor of $|\mathfrak{A}|$;

(b) $s_{\mu\nu} \neq 0$ if and only if the elements of $x_\mu \mathfrak{A}$ transform $\mathfrak{A}$ into a conjugate subgroup $\mathfrak{B}$ such that

$$s_{\mu\nu} = |\mathfrak{A} \cap \mathfrak{B}|.$$

3. Let $\mathfrak{H} < \mathfrak{G}$ and $(\mathfrak{G} : \mathfrak{H})$ finite. Let k be a fixed natural number and $\mathbf{K} \subseteq \mathfrak{G}$, $|\mathbf{K}| = k$. Show that if $k \geqslant (\mathfrak{G} : \mathfrak{H})$ a subset $\mathbf{K}$ of $\mathfrak{G}$ can be found so that $\mathbf{KH} = \mathbf{HK}$.

4. A subgroup $\mathfrak{H} < \mathfrak{G}$ is normal in $\mathfrak{G}$ if and only if every left transversal for $\mathfrak{H}$ is also a right transversal for $\mathfrak{H}$ in $\mathfrak{G}$.

5. Generalize the formula (1′) to a product of more than two finite subgroups.

6. Let $\mathfrak{A}, \mathfrak{B} < \mathfrak{G}$. Define $\mathfrak{A} \circ \mathfrak{B} = \bigcap_{x \in \mathfrak{B}} (x \mathfrak{A} x^{-1} \mathfrak{B})$. Show that $\mathfrak{A} \circ \mathfrak{B}$ is a subsemigroup of $\mathfrak{G}$ with respect to the multiplication law of $\mathfrak{G}$. (L. Fuchs, 1953.)

7. Define $\mathfrak{A} * \mathfrak{B} = (\mathfrak{A} \circ \mathfrak{B}) \cap (\mathfrak{A} \circ \mathfrak{B})^{-1}$ (cf. ex. 6), if $\mathfrak{A}, \mathfrak{B} \leqslant \mathfrak{G}$.

(a) Show that $\mathfrak{A} * \mathfrak{B}$ is a subgroup of $\mathfrak{G}$ contained in $\mathfrak{A}\mathfrak{B}$. If $\mathfrak{A} \trianglelefteq \mathfrak{B}$ or if $\mathfrak{G}$ is finite show that $\mathfrak{A} * \mathfrak{B} = \mathfrak{A} \circ \mathfrak{B}$.

(b) The double cosets $\mathfrak{A} x \mathfrak{B}$ $(x \in \mathfrak{G})$ form a group if and only if $\mathfrak{A} * \mathfrak{B} = \mathfrak{B} * \mathfrak{A} \trianglelefteq \mathfrak{G}$. Show that in general $\mathfrak{A} * \mathfrak{B} \neq \mathfrak{B} * \mathfrak{A}$.

(c) Show that $\mathfrak{A} * (\mathfrak{A} * \mathfrak{B}) = \mathfrak{A} * \mathfrak{B}$.

8. (L. Fuchs). Let $\mathfrak{A}, \mathfrak{B}, \mathfrak{C} \leqslant \mathfrak{G}$, $x, y, \ldots \in \mathfrak{G}$. The expansion $\mathfrak{G} = \mathfrak{A}\mathfrak{C} \cup \mathfrak{A}x\mathfrak{C} \cup \mathfrak{A}y\mathfrak{C} \cup \ldots$ is said to be a refinement of $\mathfrak{G} = \mathfrak{A}\mathfrak{B} \cup \mathfrak{A}x\mathfrak{B} \cup \mathfrak{A}y\mathfrak{B} \cup \ldots$ if for every element x of $\mathfrak{G}$ we have $\mathfrak{A}x\mathfrak{C} \subseteq \mathfrak{A}x\mathfrak{B}$.

Show that this inclusion is satisfied if and only if $\mathfrak{C} \leqslant \mathfrak{A} * \mathfrak{B}$.

Chapter III

AUTOMORPHISMS AND ENDOMORPHISMS

§1 Groups of automorphisms. Characteristic subgroups

a. The identity mapping $\epsilon : \mathfrak{G} \to \mathfrak{G}$, defined by $\epsilon(x) = x$, $x \in \mathfrak{G}$, satisfies the functional equation $\epsilon(xy) = xy = \epsilon(x)\epsilon(y)$. Moreover ϵ is invertible, hence it is an isomorphism. Are there other isomorphisms of $\mathfrak{G}$ onto itself? This is a natural question; it is the fundamental question for an interesting theory.

An isomorphism $\alpha : \mathfrak{G} \to \mathfrak{G}$ is called an *automorphism of* $\mathfrak{G}$. It is represented by an invertible function defined on $\mathfrak{G}$ whose values $\alpha(x)$ are elements of $\mathfrak{G}$ and it satisfies the functional equation

$$(1) \qquad \alpha(xy) = \alpha(x)\,\alpha(y), \quad x, y \in \mathfrak{G}.$$

From what has been shown for isomorphisms it is evident that $\alpha(e) = e$ for every automorphism α of $\mathfrak{G}$; further

$$\alpha(x^{-1}) = (\alpha(x))^{-1}.$$

In the case of a finite group $\mathfrak{G}$, $|\mathfrak{G}| = g$, an automorphism α can be represented by a permutation U of the g group elements

$$(2) \qquad U(\alpha) = \begin{pmatrix} e & x_2 & \dots & x_g \\ e & \alpha(x_2) & \dots & \alpha(x_g) \end{pmatrix}$$

if $x_2, \dots, x_g$ designate the elements of $\mathfrak{G}$ different from e.

It has been noted that the identity mapping is an automorphism ϵ. Also the inverse α^{-1} of an automorphism α is an automorphism: Let $x' = \alpha^{-1}(x)$, $y' = \alpha^{-1}(y)$; by (1) $\alpha(x'y') = \alpha(x')\,\alpha(y') = xy$, hence $x'y' = \alpha^{-1}(x)\,\alpha^{-1}(y) = \alpha^{-1}(xy)$. Further let α, β be two automorphisms of the group $\mathfrak{G}$; then the product $\beta\alpha$ is an automorphism of $\mathfrak{G} : \beta\alpha(xy) = \beta(\alpha(xy)) = \beta(\alpha(x)\alpha(y)) = \beta\alpha(x)\cdot\beta\alpha(y)$. Thus *all the automorphisms of a group* $\mathfrak{G}$ *form a group* $\Gamma = \Gamma(\mathfrak{G})$, the *automorphism group* of $\mathfrak{G}$, often denoted by Aut $\mathfrak{G}$.

If $\mathfrak{G}$ is a finite group of order g let $\mathfrak{S}(\mathfrak{G})$ denote "the symmetric group over $\mathfrak{G}$", i.e. the group of all permutations of the g elements of $\mathfrak{G}$; clearly $\mathfrak{S}(\mathfrak{G}) \simeq \mathfrak{S}_g$. Then Γ is a subgroup of $\mathfrak{S}(\mathfrak{G})$. Since $\alpha(e) = e$ for all $\alpha \in \Gamma$, the group Γ is also isomorphic to a subgroup of the symmetric group $\mathfrak{S}_{g-1}$.

If x is an element of finite order m in $\mathfrak{G}$ then so is the element $\alpha(x)$. Indeed from (1) we conclude that $\alpha(x)^m = \alpha(x^m) = \alpha(e) = e$, and if $\alpha(x)^n = e$ where $0 < n \leqslant m$, then $\alpha(x^n) = e$ whence $x^n = \alpha^{-1}(e) = e$ so that $n = m$.

b. Inner automorphisms. Let x be a variable in the group $\mathfrak{G}$. For a fixed element $t \in \mathfrak{G}$ we consider the mapping $\tau(x) = txt^{-1}$. It satisfies the condition (1): $\tau(xy) = txyt^{-1} = txt^{-1}tyt^{-1} = \tau(x)\cdot\tau(y)$ and it has a unique inverse: $\tau^{-1}(x) = t^{-1}x(t^{-1})^{-1} = t^{-1}xt$. The mapping $\tau(x)$ is called the *inner automorphism* of $\mathfrak{G}$ induced by the element t.

All inner automorphisms of $\mathfrak{G}$ *form a group* $\Delta = \Delta(\mathfrak{G})$. Indeed if $\sigma \in \Delta$, $\sigma(x) = sxs^{-1}$ for a fixed s in $\mathfrak{G}$, then $\sigma\tau(x) = s(txt^{-1})s^{-1} = (st)x(st)^{-1}$ is also an inner automorphism of $\mathfrak{G}$. So are ϵ and τ^{-1}. Clearly $\Delta \leqslant \Gamma$. Indeed

(3) $\quad \Delta \trianglelefteq \Gamma$

since for every automorphism $\alpha \in \Gamma$ we have $\alpha\tau\alpha^{-1}(x) = \alpha(t\alpha^{-1}(x)t^{-1}) = \alpha(t)x\alpha(t)^{-1}$. Thus if τ means transformation of the group elements x of $\mathfrak{G}$ with t (fixed in $\mathfrak{G}$), then $\alpha\tau\alpha^{-1}$ means transformation of x with $\alpha(t)$.

Further we note that there is a homomorphism $\delta : \mathfrak{G} \to \Delta(\mathfrak{G})$ where $\delta(t) = \tau$ is the inner automorphism $\tau(x) = txt^{-1}$, $x \in \mathfrak{G}$. Indeed $\delta(st)(x) = (st)x(st)^{-1} = s(txt^{-1})s^{-1} = \delta(s)(txt^{-1}) = \delta(s)\delta(t)(x)$ and therefore $\delta(st) = \delta(s)\delta(t)$.

The kernel ker $\delta = \delta^{-1}(\epsilon)$ consists of all $z \in \mathfrak{G}$ for which $\delta(z) = \epsilon$, i.e. $zxz^{-1} = x$, $x \in \mathfrak{G}$. Hence

(4) $\quad \ker\delta = \mathfrak{Z}(\mathfrak{G}) = \mathfrak{Z}$ and $\Delta(\mathfrak{G}) \simeq \mathfrak{G}/\mathfrak{Z}$.

Remark. Using the group Δ of the inner automorphisms one can give a fifth definition of a normal subgroup $\mathfrak{H} \triangleleft \mathfrak{G}$:

5. $\mathfrak{H} \triangleleft \mathfrak{G}$ if $\tau(\mathfrak{H}) = \mathfrak{H}$ for all $\tau \in \Delta(\mathfrak{G})$.
(Cf. Chap. II, §4, *b*.)

Similarly for the centre $\mathfrak{Z}(\mathfrak{G})$ we have a definition based on the group $\Delta(\mathfrak{G})$:

$$\mathfrak{Z}(\mathfrak{G}) = \{z \in \mathfrak{G} \mid \tau(z) = z \quad \text{for all} \quad \tau \in \Delta(\mathfrak{G})\}.$$

c. Characteristic subgroups. Two elements a, b of $\mathfrak{G}$ have been called conjugate in $\mathfrak{G}$ if for some $\tau \in \Delta(\mathfrak{G})$ the element $b = \tau(a)$. For the following it will be convenient to extend the concept of conjugacy. Two elements a, b (subsets **A**, **B**) of $\mathfrak{G}$ are said to be *isotype* in $\mathfrak{G}$ if there is an automorphism $\alpha \in \Gamma(\mathfrak{G})$ which carries $a \to b = \alpha(a)$ $[\mathbf{A} \to \mathbf{B} = \alpha(\mathbf{A})]$. Conjugate elements (subsets) are isotype; but the converse is not always true.

Let A be a group of automorphisms of a group $\mathfrak{G}$, i.e. $\mathrm{A} \leqslant \Gamma$. We may call the subgroup $\mathfrak{H}$ of $\mathfrak{G}$ an A-*invariant* subgroup of $\mathfrak{G}$ if $\alpha(\mathfrak{H}) = \mathfrak{H}$ for all $\alpha \in \mathrm{A}$. Thus $\mathfrak{H} \triangleleft \mathfrak{G}$ is equivalent to saying that $\mathfrak{H}$ is a Δ-invariant subgroup of $\mathfrak{G}$. A Γ-invariant subgroup $\mathfrak{H}$ of $\mathfrak{G}$ is called a *characteristic subgroup* of $\mathfrak{G}$; in symbols: $\mathfrak{H} \ntriangleleft \mathfrak{G}$.

We shall call $\mathfrak{Z}_{\mathrm{A}}$ the A-centre of $\mathfrak{G}$ if $z \in \mathfrak{Z}_{\mathrm{A}}$ is equivalent to $\alpha(z) = z$ for all $\alpha \in \mathrm{A}$; that is, $\mathfrak{Z}_{\mathrm{A}}$ consists of all the elements z in $\mathfrak{G}$ which are invariant under all $\alpha \in \mathrm{A}$. In this sense the ordinary centre $\mathfrak{Z} = \mathfrak{Z}(\mathfrak{G})$ is the Δ-centre of $\mathfrak{G}$.

Now suppose that B is another subgroup of the automorphism group $\Gamma(\mathfrak{G})$ and that $\mathrm{A} \leqslant \mathrm{B} \leqslant \Gamma$. Then every B-invariant subgroup is also an A-invariant subgroup, but not conversely. In particular: Every characteristic subgroup is normal. Moreover

$$\mathfrak{Z}_{\mathrm{A}} \geqslant \mathfrak{Z}_{\mathrm{B}} \geqslant \mathfrak{Z}_{\Gamma}.$$

THEOREM 1. *The centre $\mathfrak{Z} = \mathfrak{Z}(\mathfrak{G})$ is a characteristic subgroup of $\mathfrak{G}$.*

Proof. Let $\alpha \in \Gamma$ and let $y = \alpha(x)$, $x \in \mathfrak{G}$. By supposition $xz = zx$ for all $z \in \mathfrak{Z}$ and $x \in \mathfrak{G}$. Hence also $y \cdot \alpha(z) = \alpha(z) \cdot y$ for all $y \in \mathfrak{G}$, that is $\alpha(z) \in \mathfrak{Z}$. Hence $\alpha(\mathfrak{Z}) \leqslant \mathfrak{Z}$. Also $\mathfrak{Z} \leqslant \alpha^{-1}(\mathfrak{Z})$ and since every automorphism is the inverse of an automorphism we have $\mathfrak{Z} \leqslant \alpha(\mathfrak{Z})$. Thus $\alpha(\mathfrak{Z}) = \mathfrak{Z}$.

Another important characteristic subgroup of a group $\mathfrak{G}$ is the so-called *commutator subgroup* $\mathfrak{G}' \leqslant \mathfrak{G}$, also called the *derived group*. Let us denote by **C** the set of all commutators in $\mathfrak{G}$:

$$\mathbf{C} = \{[x, y] = xyx^{-1}y^{-1} \,|\, x, y \in \mathfrak{G}\}.$$

By definition

$$\mathfrak{G}' = \langle \mathbf{C} \rangle.$$

For every α in Γ we have

$$\alpha([x, y]) = \alpha(x)\,\alpha(y)\,\alpha(x)^{-1}\alpha(y)^{-1} = [\alpha(x), \alpha(y)];$$

thus $\alpha([x, y])$ is a commutator and therefore $\alpha(\mathbf{C}) \subseteq \mathbf{C}$. Also $\alpha^{-1}(\mathbf{C}) \subseteq \mathbf{C}$; hence $\alpha(\mathbf{C}) = \mathbf{C}$ and therefore $\alpha(\mathfrak{G}') = \mathfrak{G}'$. This proves the first part of the following theorem:

THEOREM 2. *The commutator subgroup $\mathfrak{G}'$ of a group $\mathfrak{G}$ is a characteristic subgroup: $\mathfrak{G}' \trianglelefteq \mathfrak{G}$ and the factor group $\mathfrak{G}/\mathfrak{G}'$ is abelian. Moreover if $\mathfrak{H} \trianglelefteq \mathfrak{G}$ then the factor group $\mathfrak{G}/\mathfrak{H}$ is abelian if and only if $\mathfrak{G}' \trianglelefteq \mathfrak{H} \trianglelefteq \mathfrak{G}$.*

Proof. The commutator of two elements $x\mathfrak{G}', y\mathfrak{G}'$ of $\mathfrak{G}/\mathfrak{G}'$, that is

$$[x\mathfrak{G}', y\mathfrak{G}'] = x\mathfrak{G}'y\mathfrak{G}'x^{-1}\mathfrak{G}'y^{-1}\mathfrak{G}' = [x, y]\mathfrak{G}' = \mathfrak{G}',$$

the unit element of $\mathfrak{G}/\mathfrak{G}'$. Thus $x\mathfrak{G}'$ and $y\mathfrak{G}'$ commute.

If $\mathfrak{G}' \trianglelefteq \mathfrak{H} \trianglelefteq \mathfrak{G}$ the argument can be repeated:

$$[x\mathfrak{H}, y\mathfrak{H}] = [x, y]\mathfrak{H} = \mathfrak{H} \quad \text{because} \quad [x, y] \in \mathfrak{G}' \trianglelefteq \mathfrak{H}.$$

Conversely, if $\mathfrak{G}/\mathfrak{H}$ is abelian, it follows that every $[x, y] \in \mathfrak{H}$ therefore $\mathfrak{G}' \trianglelefteq \mathfrak{H}$.

It has been pointed out [Chap. II, §4, ex. 2(c)] that if $\mathfrak{L} \triangleleft \mathfrak{H} \triangleleft \mathfrak{G}$ then one cannot conclude that $\mathfrak{L} \triangleleft \mathfrak{G}$. But we can state the following facts:

THEOREM 3. *Let $\mathfrak{H}$ be a group, $\mathfrak{L}$ a subgroup, and $\mathfrak{G}$ a "supergroup" of $\mathfrak{H}$: $\mathfrak{L} \leqslant \mathfrak{H} \leqslant \mathfrak{G}$.*

(a) If $\mathfrak{L} \ntrianglelefteq \mathfrak{H} \ntrianglelefteq \mathfrak{G}$ then $\mathfrak{L} \ntrianglelefteq \mathfrak{G}$.
(b) If $\mathfrak{L} \ntrianglelefteq \mathfrak{H} \lhd \mathfrak{G}$ then $\mathfrak{L} \lhd \mathfrak{G}$.

Proof. (a) Every automorphism α in $\mathfrak{G}$ leaves $\mathfrak{H}$ invariant and induces an automorphism in $\mathfrak{H}$; therefore it leaves $\mathfrak{L}$ invariant.

(b) Every inner automorphism of $\mathfrak{G}$ leaves $\mathfrak{H}$ invariant and induces an automorphism of $\mathfrak{H}$; therefore it leaves $\mathfrak{L}$ invariant; thus $\mathfrak{L} \lhd \mathfrak{G}$.

This implies that every characteristic subgroup $\mathfrak{L}$ of $\mathfrak{H}$ is normal in every group $\mathfrak{G}$ which contains $\mathfrak{H}$ as a normal subgroup.

Remark. Part (b) of theorem 3 can be inverted in the following way: In §2 it will be shown that for a given group $\mathfrak{H}$ a supergroup $\mathfrak{G}$, the so-called holomorph of $\mathfrak{H}$, can be constructed such that every automorphism of $\mathfrak{H}$ is an inner automorphism of $\mathfrak{G}$ restricted to $\mathfrak{H}$. Thus we can state:

(c) If $\mathfrak{L} \lhd \mathfrak{H} \lhd \mathfrak{G}$ and $\mathfrak{L} \lhd \mathfrak{G}$ then $\mathfrak{L} \ntrianglelefteq \mathfrak{H}$.

THEOREM 4. *If $\mathfrak{L} \leqslant \mathfrak{H} \leqslant \mathfrak{G}$, $\mathfrak{L} \ntrianglelefteq \mathfrak{G}$ and $\mathfrak{H}/\mathfrak{L} \ntrianglelefteq \mathfrak{G}/\mathfrak{L}$ it follows that $\mathfrak{H} \ntrianglelefteq \mathfrak{G}$.*

Proof. For every $\alpha \in \Gamma(\mathfrak{G})$ by supposition $\alpha(\mathfrak{L}) = \mathfrak{L}$. Moreover α induces an automorphism $\hat{\alpha}$ of $\mathfrak{G}/\mathfrak{L}$, namely $\hat{\alpha}(x\mathfrak{L}) = \alpha(x)\mathfrak{L}$. Indeed $\hat{\alpha}(x\mathfrak{L}\, y\mathfrak{L}) = \hat{\alpha}(xy\mathfrak{L}) = \alpha(xy)\mathfrak{L} = \alpha(x)\cdot\alpha(y)\mathfrak{L} = \alpha(x)\mathfrak{L}\cdot\alpha(y)\mathfrak{L} = \hat{\alpha}(x\mathfrak{L})\cdot\hat{\alpha}(y\mathfrak{L})$. As an automorphism of $\mathfrak{G}/\mathfrak{L}$ the automorphism $\hat{\alpha}$ preserves the subgroup $\mathfrak{H}/\mathfrak{L}$ (by supposition); thus $\hat{\alpha}$ causes a permutation of the cosets $x\mathfrak{L}$, the elements of $\mathfrak{H}/\mathfrak{L}$, leaving $\mathfrak{L}$ fixed. Hence it follows that $\alpha(\mathfrak{H}) = \mathfrak{H}$.

d. Characteristically simple groups. A group $\mathfrak{G}$ has been called simple (cf. Chap. II, §5, *c.*) when it has no proper normal subgroup. All cyclic groups of prime order are simple. No other finite abelian group $\mathfrak{G}$ is simple. Indeed let $|\mathfrak{G}| = g$, not a prime number. If $\mathfrak{G}$ is cyclic then every proper divisor d of g is the order of a normal subgroup of $\mathfrak{G}$. If $\mathfrak{G}$ is not cyclic then there is an element x in $\mathfrak{G}$ of a certain order $d < g$, $d \mid g$, and $\langle x \rangle_d \lhd \mathfrak{G}$.

There are simple non-abelian groups, e.g. the icosahedral group $\mathfrak{J}$ (cf. Chap. II, §5, ex. 10) and the alternating groups $\mathfrak{A}_n$ ($n = 5, 6, \ldots$) (cf. ex. 13).

A group $\mathfrak{G}$ is said to be *characteristically simple* when it has no proper characteristic subgroup. Every simple group is characteristically simple, but the converse is not true. We shall study only finite characteristically simple groups and show how they can be obtained from simple groups.

THEOREM 5. *Every finite characteristically simple group $\mathfrak{G}$ is either simple or a direct product of isomorphic simple groups.*

*Proof.** In the case of a simple $\mathfrak{G}$ there is nothing to prove. So let us assume that there is a proper non-trivial normal subgroup in $\mathfrak{G}$. Among the normal subgroups of $\mathfrak{G}$ we select one, $\mathfrak{H}$, which contains no non-trivial normal subgroup of $\mathfrak{G}$. We consider all automorphic images $\alpha(\mathfrak{H})$, $\alpha \in \Gamma(\mathfrak{G})$, of $\mathfrak{H}$. By ex. 14(a) we know that all the $\alpha(\mathfrak{H})$ are normal in $\mathfrak{G}$.

We consider now the set **S** of all $\alpha(\mathfrak{H})$, $\alpha \in \Gamma(\mathfrak{G})$. As normal subgroups of $\mathfrak{G}$ the $\alpha(\mathfrak{H})$ commute and therefore $\langle \mathbf{S} \rangle$ equals the product of all these subgroups. Clearly $\langle \mathbf{S} \rangle$ is a characteristic subgroup of $\mathfrak{G}$; hence $\langle \mathbf{S} \rangle = \mathfrak{G}$. Further let $\mathbf{T} = \{\alpha_1(\mathfrak{H}), \ldots, \alpha_k(\mathfrak{H})\}$ represent a subset of **S**, minimal in the sense that still $\langle \mathbf{T} \rangle = \alpha_1(\mathfrak{H}) \cdot \ldots \cdot \alpha_k(\mathfrak{H}) = \mathfrak{G}$, but $\langle \mathbf{T}_i \rangle < \mathfrak{G}$ if $\mathbf{T}_i$ arises from **T** by omitting the subgroup $\alpha_i(\mathfrak{H})$ $(i = 1, \ldots, k)$. For every such i we have $\mathfrak{D}_i = \alpha_i(\mathfrak{H}) \cap \langle \mathbf{T}_i \rangle \lhd \mathfrak{G}$ [cf. Chap. II, §3, ex. 4(a)]. Since $\alpha(\mathfrak{H})$, like $\mathfrak{H}$, contains no proper normal subgroup which is normal in $\mathfrak{G}$, we conclude that either $\mathfrak{D}_i = \mathfrak{e}$ or $\mathfrak{D}_i = \alpha_i(\mathfrak{H})$. Because of the minimality of **T** the latter is excluded. Hence $\mathfrak{D}_i = \mathfrak{e}$ $(i = 1, \ldots, k)$. Therefore we can form the direct product $\alpha_1(\mathfrak{H}) \times \ldots \times \alpha_k(\mathfrak{H})$ and this is equal to $\langle \mathbf{T} \rangle = \mathfrak{G}$.

Finally it is evident that $\mathfrak{H}$ is simple. For, suppose that $\mathfrak{L} \lhd \mathfrak{H}$, $\mathfrak{L} \neq \mathfrak{e}$. As a normal subgroup of a direct factor of $\mathfrak{G}$ also $\mathfrak{L} \lhd \mathfrak{G}$ (cf. Chap. II, §5, ex. 7). Again, with regard to the initial assumption concerning $\mathfrak{H}$ there is no such normal subgroup $\mathfrak{L}$ in $\mathfrak{H}$. Hence there is no non-trivial normal subgroup in $\mathfrak{H}$; thus $\mathfrak{H}$ is simple.

We shall prove now the converse of Theorem 5, using different arguments in the cases of a non-commutative and of a commutative group.

THEOREM 6a. *If $\mathfrak{G}$ is the direct product of r isomorphic non-abelian simple groups then $\mathfrak{G}$ is characteristically simple.*

* A communication by L. Fuchs was helpful for the formulation of this proof.

Proof. By supposition $\mathfrak{G} = (\mathfrak{G}_1, \mathfrak{G}_2, \ldots, \mathfrak{G}_r)$ and the $\mathfrak{G}_i$ are simple, non-abelian, and isomorphic to each other. Every permutation of the $\mathfrak{G}_i$ represents an automorphism of $\mathfrak{G}$. By theorem 8 of Chap. II, §5 every proper normal subgroup $\mathfrak{H} \lhd \mathfrak{G}$ is the direct product of some, but not all, of the $\mathfrak{G}_i$. If $\mathfrak{G}_1$ is a factor of $\mathfrak{H}$ while $\mathfrak{G}_2$ is not a factor of $\mathfrak{H}$, then the automorphism α defined by the transposition (1 2) has the property $\alpha(\mathfrak{H}) \neq \mathfrak{H}$. Thus $\mathfrak{G}$ has no characteristic subgroup.

An abelian group is called an *elementary abelian p-group* if it is the direct product of cyclic groups of prime order p.

THEOREM 6b. *Every elementary abelian p-group $\mathfrak{G}$ is characteristically simple.*

Proof. Let us write $\mathfrak{G}$ as an additive group of the order p^r. Then $\mathfrak{G}$ may be considered to be the additive group of an r-dimensional vector space over the field $\mathbb{R}_p$ (cf. Chap. I, §3, *b.*) and the automorphisms of $\mathfrak{G}$ are the linear homogeneous transformations of this vector space onto itself, that is, all the mappings $\alpha : \mathfrak{G} \to \mathfrak{G}$ which satisfy the conditions

(i) $\quad \alpha(\lambda x + \mu y) = \lambda\alpha(x) + \mu\alpha(y), \quad x, y \in \mathfrak{G}, \quad \lambda, \mu \in \mathbb{R}_p,$

(ii) $\quad \alpha(x) = 0$ implies $x = 0$ (unit element of $\mathfrak{G}$).

Thus $\Gamma(\mathfrak{G})$ is isomorphic to the so-called "general linear group" $GL(r, \mathbb{R}_p)$ (standard notation) of the r-dimensional vector space over the field $\mathbb{R}_p$. According to a fundamental theorem of linear algebra, for every given pair of non-zero points x, y in a vector space an invertible linear mapping α can be found which maps x into $y = \alpha(x)$. Hence if $\mathfrak{H}$ is a proper nontrivial subgroup of $\mathfrak{G}$ the automorphic images of $\mathfrak{H}$ must cover the group $\mathfrak{G}$. Thus $\mathfrak{G}$ has no proper characteristic subgroup.

e. *Γ-invariance.* A subset $\mathbf{S} \subseteq \mathfrak{G}$ is said to be Γ-invariant (cf. *c.*) if $\gamma(\mathbf{S}) = \mathbf{S}$ for all automorphisms $\gamma \in \Gamma(\mathfrak{G})$. The smallest Γ-invariant subset containing the element a of $\mathfrak{G}$ is the set $\mathscr{C}_\Gamma(a)$ of all isotypes $\gamma(a)$ of a in $\mathfrak{G}$. It is called the *characteristic class* (or isotype class) of a in $\mathfrak{G}$. We also define for a subset $\mathbf{A}$ of $\mathfrak{G}$

$$\mathscr{C}_\Gamma(\mathbf{A}) = \{\gamma(\mathbf{A}) \mid \gamma \in \Gamma(\mathfrak{G})\}.$$

In general this is a collection of subsets of $\mathfrak{G}$, but not a subset of $\mathfrak{G}$.

Evidently $\mathscr{C}_\Gamma(e) = e$ and $\mathscr{C}_\Gamma(a) \subset \mathfrak{G}$. Distinct characteristic classes $\mathscr{C}_\Gamma(a), \mathscr{C}_\Gamma(b)$, $b \notin \mathscr{C}_\Gamma(a)$, are disjoint and the classes $\mathscr{C}_\Gamma(x)$, $x \in \mathfrak{G}$, cover the group $\mathfrak{G}$ since $x \in \mathscr{C}_\Gamma(x)$. Because $\Delta \leqslant \Gamma$

$$(5) \qquad \mathscr{C}(a) = \mathscr{C}_\Delta(a) \subseteq \mathscr{C}_\Gamma(a).$$

As in the case of conjugacy (cf. Chap. II, §4, *c.*) it is easy to prove the following theorem:

THEOREM 7. *Every Γ-invariant subset of a group is a union of characteristic classes.*

COROLLARY 1. *A subgroup $\mathfrak{H} \leqslant \mathfrak{G}$ is characteristic* (i.e. Γ-invariant) *if and only if it is a union of characteristic classes.*

COROLLARY 2. *If $\mathfrak{H} \leqslant \mathfrak{G}$ then the intersection of all isotype subgroups $\gamma(\mathfrak{H})$, $\gamma \in \Gamma(\mathfrak{G})$, of $\mathfrak{H}$ in $\mathfrak{G}$ is a characteristic subgroup of $\mathfrak{G}$.*

Now we define the Γ-*normalizer* of a subset $\mathbf{A}$ of $\mathfrak{G}$:

$$\mathrm{N}(\mathbf{A}) = \mathrm{N}_\Gamma(\mathbf{A}) = \{\varphi \in \Gamma \mid \varphi(a) \in \mathbf{A} \quad \text{for all} \quad a \in \mathbf{A}\}.$$

Thus $\mathrm{N}(\mathbf{A})$ is the greatest subgroup of $\Gamma = \Gamma(\mathfrak{G})$ whose elements leave $\mathbf{A}$ invariant.

We begin with two simple theorems which by their analogy with Theorems 4 and 5 in Chap. II, §4, *d.* may justify the name of "Γ-normalizer".

THEOREM 8. *Let $\mathbf{A}$ be a subset of the group $\mathfrak{G}$ such that the index $(\Gamma : \mathrm{N}(\mathbf{A})) = i$ is finite. If $c_\mathbf{A}$ is the number of different isotype sets $\gamma(\mathbf{A})$ ($\gamma \in \Gamma(\mathfrak{G})$) in $\mathscr{C}_\Gamma(\mathbf{A})$, then $c_\mathbf{A} = i$.*

Proof. Let $\beta\mathrm{N}(\mathbf{A})$ be one of the i left cosets of $\mathrm{N}(\mathbf{A})$ in Γ. Then all the automorphisms $\beta\varphi \in \beta\mathrm{N}(\mathbf{A})$, $\varphi \in \mathrm{N}(\mathbf{A})$, map the set $\mathbf{A}$ onto the same isotype set $\beta\varphi(\mathbf{A}) = \beta(\mathbf{A})$; hence there are at most as many distinct $\beta(\mathbf{A})$ as there are cosets of $\mathrm{N}(\mathbf{A})$ in Γ, i.e. $c_\mathbf{A} \leqslant i$.

If, however, for two automorphisms $\alpha, \beta \in \Gamma(\mathfrak{G})$ we have $\alpha(\mathbf{A}) = \beta(\mathbf{A})$, then $\beta^{-1}\alpha(\mathbf{A}) = \mathbf{A}$, i.e. $\beta^{-1}\alpha \in \mathrm{N}(\mathbf{A})$ and therefore $\alpha\mathrm{N}(\mathbf{A}) = \beta\mathrm{N}(\mathbf{A})$. Thus if $\beta\mathrm{N} \neq \alpha\mathrm{N}$ we conclude that $\alpha(\mathbf{A}) \neq \beta(\mathbf{A})$, i.e. $i \leqslant c_\mathbf{A}$. Hence $c_\mathbf{A} = i$.

COROLLARY 1. *For a single element $a \in \mathfrak{G}$ one has*

$$c_a = |\mathscr{C}_\Gamma(a)| = (\Gamma : \mathrm{N}(a)).$$

COROLLARY 2. *If the group $\mathfrak{G}$ has a finite automorphism group $\Gamma(\mathfrak{G})$ the number $c_{\mathbf{A}}$ is a divisor of $|\Gamma(\mathfrak{G})|$.*

THEOREM 9. *Isotype subsets* **A** *and* $\alpha(\mathbf{A})$ *of* $\mathfrak{G}, \alpha \in \Gamma(\mathfrak{G})$, *have* Γ-*normalizers which are conjugate in* $\Gamma(\mathfrak{G})$:

$$\mathrm{N}[\alpha(\mathbf{A})] = \alpha \mathrm{N}(\mathbf{A})\alpha^{-1}, \quad \alpha \in \Gamma(\mathfrak{G}).$$

Proof. According to the definition of N we have

$$\mathrm{N}[\alpha(\mathbf{A})] = \{\psi \in \Gamma(\mathfrak{G}) \mid \psi(\alpha(\mathbf{A})) = \alpha(\mathbf{A})\}.$$

The condition $\psi(\alpha(\mathbf{A})) = \alpha(\mathbf{A})$ implies that $\alpha^{-1}\psi\alpha(\mathbf{A}) = \mathbf{A}$ so that $\alpha^{-1}\mathrm{N}(\alpha(\mathbf{A}))\alpha \leqslant \mathrm{N}(\mathbf{A})$ or $\mathrm{N}(\alpha(\mathbf{A})) \leqslant \alpha\mathrm{N}(\mathbf{A})\alpha^{-1}$. Conversely $\alpha\mathrm{N}(\mathbf{A})\alpha^{-1} = \{\alpha\varphi\alpha^{-1} \mid \varphi(\mathbf{A}) = \mathbf{A}\} = \{\psi \mid \alpha^{-1}\psi\alpha(\mathbf{A}) = \mathbf{A}\}$ and writing the condition in the form $\psi\alpha(\mathbf{A}) = \alpha(\mathbf{A})$ we obtain $\alpha\mathrm{N}(\mathbf{A})\alpha^{-1} \leqslant \mathrm{N}(\alpha(\mathbf{A}))$ which proves the theorem.

Finally we consider together with $\mathrm{N}(\mathbf{A}) = \mathrm{N}_\Gamma(\mathbf{A})$ the subgroup

$$\mathrm{N}_\Delta(\mathbf{A}) = \{\tau \in \Delta \mid \tau(\mathbf{A}) = \mathbf{A}\} \leqslant \mathrm{N}_\Gamma(\mathbf{A})$$

which we call the Δ-normalizer of **A** in $\mathfrak{G}$. It is readily seen that $\mathrm{N}_\Delta(\mathbf{A}) = \Delta \cap \mathrm{N}_\Gamma(\mathbf{A})$ and therefore $\mathrm{N}_\Delta(\mathbf{A}) \trianglelefteq \mathrm{N}_\Gamma(\mathbf{A})$. The homomorphism $\delta : \mathfrak{G} \to \Delta(\mathfrak{G})$ (cf. *b.*) converts the normalizer $\mathfrak{N}(\mathbf{A})$ into $\mathrm{N}_\Delta(\mathbf{A})$ and since the kernel $\ker \delta = \mathfrak{Z} = \mathfrak{Z}(\mathfrak{G}) \trianglelefteq \mathfrak{N}(\mathbf{A})$ we have

$$\mathfrak{N}(\mathbf{A})/\mathfrak{Z} \simeq \mathrm{N}_\Delta(\mathbf{A}).$$

On the other hand let a be a single element of **A**. Then $\mathfrak{N}(a)$ is the set of all elements of $\mathfrak{G}$ which commute with a. The subgroup

$$\mathfrak{C}(\mathbf{A}) = \bigcap_{a \in \mathbf{A}} \mathfrak{N}(a) = \{s \in \mathfrak{G} \mid sas^{-1} = a \quad \text{for all} \quad a \in \mathbf{A}\}$$

is called the *centralizer* of **A** in $\mathfrak{G}$. It is a normal subgroup of $\mathfrak{N}(\mathbf{A})$ and we have

$$\mathfrak{N}(\mathbf{A})/\mathfrak{C}(\mathbf{A}) \simeq \frac{\mathfrak{N}(\mathbf{A})/\mathfrak{Z}}{\mathfrak{C}(\mathbf{A})/\mathfrak{Z}} \simeq \mathrm{N}_\Delta(\mathbf{A}) \Big/ \bigcap_{a \in \mathbf{A}} \mathrm{N}_\Delta(a) = \mathrm{M}(\mathbf{A}).$$

Thus if ω is the homomorphism mapping $\mathfrak{N}(\mathbf{A}) \to \mathrm{M}(\mathbf{A})$ we have $\mathfrak{C}(\mathbf{A}) = \ker \omega$.

Examples and exercises

1. Show that isomorphic groups have isomorphic automorphism groups.

2. Show that a certain automorphism $\alpha : \mathfrak{S}_4 \to \mathfrak{S}_4$ is defined by

$$\alpha(1\ 2) = (2\ 3), \quad \alpha(1\ 3) = (2\ 4), \quad \alpha(1\ 4) = (1\ 2).$$

3. If α is an automorphism of $\mathfrak{G}$ and a an element of $\mathfrak{G}$ of order k (∞) show that $\alpha(a)$ is an element of order k (∞).

4. The automorphism group of an infinite cyclic group is of order 2.

5. (a) Let $\mathfrak{G} = \langle a_1, a_2, \ldots, a_m \rangle$ and $\alpha \in \Gamma(\mathfrak{G})$. Show that also

$$\langle \alpha(a_1), \alpha(a_2), \ldots, \alpha(a_m) \rangle = \mathfrak{G}.$$

(b) Verify that $\Gamma(\mathfrak{S}_3) = \Delta(\mathfrak{S}_3) \simeq \mathfrak{S}_3$. (Hint: Define an automorphism of $\mathfrak{S}_3$ by its action on two transpositions, generators of $\mathfrak{S}_3$.)

6. *The automorphism group of a cyclic p-group.*

Definition. A finite group of order p^k (p a prime, $k \geqslant 0$) is called a p-group.

(a) $\Gamma = \Gamma(\langle a \rangle_p) \simeq \mathfrak{R}_p$.

Proof. Every α in Γ will map the element a into a certain a^i ($1 \leqslant i < p$) and α is defined by the exponent i:

$$\alpha(a) = a^i, \quad \alpha^2(a) = \alpha(a^i) = \alpha(a)^i = a^{i^2}, \ldots, \alpha^r(a) = a^{i^r}, \ldots.$$

In particular if i is a primitive congruence root (mod p) the elements $a, \alpha(a), \alpha^2(a), \ldots, \alpha^{p-2}(a)$ will be all the elements of $\mathfrak{G}$ except e and therefore α is a generating element of $\Gamma = \langle \alpha \rangle_{p-1} \simeq \mathfrak{R}_p$.

(b) The cyclic group $\langle a \rangle_p$ has two isotype classes

$$\mathscr{C}_\Gamma(e) = e, \quad \mathscr{C}_\Gamma(a) = \{a, a^2, \ldots, a^{p-1}\}$$

and the Γ-normalizer of a in $\langle a \rangle_p$ is $\mathrm{N}(a) = \epsilon$.

(c) The group $\Gamma(\langle a \rangle_{p^n})$ is a cyclic group of the order $\varphi(p^n) = p^{n-1}(p-1)$ and isomorphic to the residue class group $\mathfrak{R}_{p^n}$.

As in ex. 6(a) a generating element of Γ is an automorphism α for which $\alpha(a) = a^i$ where i is a primitive congruence root (mod p^n) (cf. the remark in Chap. II, §2, ex. 10).

(d) Let a^j be an element of order p^m $(1 \leqslant m \leqslant n)$ in $\langle a \rangle_{p^n}$; the isotope class $\mathscr{C}_\Gamma(a^j)$ consists of all the elements of order p^m in $\langle a \rangle_{p^n}$ and, therefore, $|\mathscr{C}_\Gamma(a^j)| = \varphi(p^m)$ (cf. Chap. II, §2, *d.*, Theorem 4).

(e) Consider in detail the special case $\langle a \rangle_9$.

It is found that $i = 2$ is a primitive root (mod 9) and if α is defined by $\alpha(a) = a^2$ we have $\alpha^2(a) = a^4$, $\alpha^3(a) = a^8$, $\alpha^4(a) = a^7$ $\alpha^5(a) = a^5$, $\alpha^6(a) = a$. Hence $\Gamma(\langle a \rangle_9) = \langle \alpha \rangle_6$. Further

$$\mathscr{C}_\Gamma(a) = \{a, a^2, a^4, a^5, a^7, a^8\}, \qquad \mathrm{N}(a) = \epsilon$$

$$\mathscr{C}_\Gamma(a^3) = \{a^3, a^6\}, \qquad \mathrm{N}(a^3) = \{\epsilon, \alpha^2, \alpha^4\}.$$

7. The dihedral group $\mathfrak{D}_5$ can be presented in the form

$$\mathfrak{D}_5 = \langle a, b \,|\, a^5 = b^2 = e, aba = b \rangle, \quad |\mathfrak{D}_5| = 10,$$

(cf. Chap. I, §4, *b.*). Let α be an automorphism of $\mathfrak{D}_5$; if its effect on the two generators a, b of $\mathfrak{D}_5$ is known, the effect on the other elements can be calculated by means of the functional equation (1). Thus if $\alpha(a) = a^2$, $\alpha(b) = b$ we find $\alpha(a^2) = a^4$, $\alpha(a^3) = a$, $\alpha(a^4) = a^3$, $\alpha(ba) = \alpha(b)\alpha(a) = ba^2$, $\alpha(ba^2) = ba^4$, $\alpha(ba^3) = ba$, $\alpha(ba^4) = ba^3$. A second automorphism may be defined by $\beta(a) = a$, $\beta(b) = ba$ and these two, α and β, are generators of the automorphism group $\Gamma(\mathfrak{D}_5)$ and

$$\Gamma(\mathfrak{D}_5) = \langle \alpha, \beta \,|\, \alpha^4 = \beta^5 = \epsilon, \alpha\beta = \beta^2\alpha \rangle, \quad |\Gamma(\mathfrak{D}_5)| = 20.$$

In addition verify that

$$\Delta(\mathfrak{D}_5) = \langle \alpha^2, \beta \rangle = \langle \alpha^2 \rangle_2 \langle \beta \rangle_5, \quad |\Delta(\mathfrak{D}_5)| = 10$$

and note that $\langle \beta \rangle_5 \lhd \Gamma(\mathfrak{D}_5)$.

8. If $\mathfrak{G}$ is an elementary abelian p-group of order p^r (cf. *d.*) then the automorphism group $\Gamma(\mathfrak{G})$ has order

$$\begin{aligned}|\Gamma| &= (p^r-1)(p^r-p)(p^r-p^2)\dots(p^r-p^{r-1})\\ &= p^{\frac{1}{2}r(r-1)}(p^r-1)(p^{r-1}-1)\dots(p-1).\end{aligned}$$

9. *The automorphism group of a cyclic group of composite order.*

Let $p_1, \dots, p_r$ be r distinct prime numbers, $r \geqslant 2$, and $\mathfrak{P}_i$ a p_i-group of the order $|\mathfrak{P}_i| = p_i^{k_i}$, $i = 1, 2, \dots, r$. We consider the direct product $\mathfrak{G} = (\mathfrak{P}_1, \mathfrak{P}_2, \dots, \mathfrak{P}_r)$ (cf. Chap. II, §5, *d.* and *e.*). Putting $\mathfrak{H}_1 = (\mathfrak{P}_1, e_2, \dots, e_r)$, $\mathfrak{H}_2 = (e_1, \mathfrak{P}_2, e_3, \dots, e_r)$ etc., denoting by e_i the unit element of $\mathfrak{P}_i$, we may write

$$\mathfrak{G} = \mathfrak{H}_1 \times \mathfrak{H}_2 \times \dots \times \mathfrak{H}_r$$

so that every element x of $\mathfrak{G}$ appears in the form $x = x_1 x_2 \dots x_r$ with uniquely defined x_i in $\mathfrak{H}_i$, $(i = 1, \dots, r)$.

Now let α be an automorphism of $\mathfrak{G}$; then

$$\alpha(x) = \alpha(x_1)\alpha(x_2)\dots\alpha(x_r) \quad \text{with} \quad \alpha(x_i) \in \mathfrak{H}_i;$$

indeed no x_i can be carried by an automorphism of $\mathfrak{G}$ out of the group $\mathfrak{H}_i$; hence $\mathfrak{H}_i \lhd \mathfrak{G}$ and α induces an automorphism in $\mathfrak{H}_i$. Conversely, every automorphism in $\mathfrak{H}_i$ defines an automorphism in $\mathfrak{G}$:

$$\alpha_1(x) = \alpha_1(x_1)\, x_2 \dots x_r, \; \alpha_2(x) = x_1 \alpha_2(x_2)\, x_3 \dots x_r, \dots$$

so that

$$\alpha(x) = \alpha_1(x_1)\alpha_2(x_2)\dots\alpha_r(x_r)$$

and $\alpha = \alpha_1 \alpha_2 \dots \alpha_r$. Clearly

$$\begin{aligned}\alpha_1\alpha_2(x) &= \alpha_1(x_1)\, \alpha_2(x_2)\, x_3 \dots x_r\\ &= \alpha_2(x_2)\, \alpha_1(x_1)\, x_3 \dots x_r = \alpha_2\alpha_1(x).\end{aligned}$$

Likewise $\alpha_i\alpha_j = \alpha_j\alpha_i$ $(i, j = 1, \dots, r)$ and therefore

$$(6) \qquad \Gamma(\mathfrak{G}) \simeq (\Gamma(\mathfrak{H}_1), \Gamma(\mathfrak{H}_2), \dots, \Gamma(\mathfrak{H}_r)).$$

According to Chap. II, §5, ex. 6(d) the cyclic group of order $n = p_1^{k_1} p_2^{k_2} \ldots p_r^{k_r}$ is the direct product of *cyclic* p_i-groups $\mathfrak{P}_i$. Its automorphism group is given by (6) where the $\mathfrak{H}_i$ are the uniquely defined cyclic p_i-subgroups of $\mathfrak{G}$, $|\mathfrak{H}_i| = p_i^{k_i}$.

In the special case

$$\mathfrak{G} = \langle a \rangle_9 \times \langle b \rangle_5 = \langle ab \rangle_{45}, \quad ab = ba$$

we have by (6)

$$\Gamma(\mathfrak{G}) = \langle \alpha \rangle_6 \times \langle \beta \rangle_4$$

with α as in ex. 6(e) and β defined by $\beta(b) = b^3$. This is not a cyclic group, but an abelian group because $\alpha\beta = \beta\alpha$. The group $\Gamma(\mathfrak{G})$ is isomorphic to the residue class group $\mathfrak{R}_{45}$.

As in the cases of ex. 6 it is found that a characteristic class in $\mathfrak{G}$ contains *all* the elements of equal order. This is not so in general.

10. The non-cyclic group of order 4 is characteristically simple. Its automorphism group is isomorphic to the symmetric group $\mathfrak{S}_3$.

11. Let $\mathbf{A}$ be a subset of the group $\mathfrak{G}$. The Δ-normalizer $\mathrm{N}_\Delta(\mathbf{A}) = \{\tau \in \Delta \,|\, \tau(\mathbf{A}) = t\mathbf{A}t^{-1} = \mathbf{A}\}$. Show that

$$|\mathscr{C}(\mathbf{A})| = (\Delta : \mathrm{N}_\Delta(\mathbf{A}))$$

(Cf. Chap. II, §4, Theorem 5, as well as Theorem 8 of the present section.)

12. Every finite abelian group $\mathfrak{G}$ of order $g > 1$ contains a subgroup of prime order.

(Proof by continuing the argument used in the first paragraph of subsection *d*.)

13. THEOREM. *The alternating group* $\mathfrak{A}_n$ $(n > 4)$ *is non-abelian and simple.*

LEMMA 1. *Any permutation in* $\mathfrak{A}_n$ $(n > 2)$ *is either a 3-cycle or a product of several 3-cycles.*

Proof. Every element of $\mathfrak{A}_n$ is the product of an even number of transpositions. Hence it suffices to prove the lemma for a product of two transpositions. These may or may not have a common symbol. In the first case: (12)(23) = (123); in the second case: (12)(34) = (12)(23)(23)(34) = (123)(234).

LEMMA 2. *If* $\mathfrak{H} \trianglelefteq \mathfrak{A}_n$ *(n > 4) and* $\mathfrak{H}$ *contains a 3-cycle then* $\mathfrak{H} = \mathfrak{A}_n$.

Proof. Suppose that $(123) \in \mathfrak{H}$. Let i, j, k be any three of the symbols $1, 2, \ldots, n$ and

$$T = \begin{pmatrix} 1 & 2 & 3 & \ldots \\ i & j & k & \ldots \end{pmatrix}.$$

If this permutation happens to be odd replace it by $(r\ s)T$ where r, s are two of the symbols $1, 2, \ldots, n$, but different from i, j, k (which is always possible if $n \geqslant 5$). The resulting element $T \in \mathfrak{A}_n$ and $T(123)T^{-1} = (ijk) \in \mathfrak{H}$. Therefore $\mathfrak{H}$ contains all 3-cycles. From Lemma 1 it follows that $\mathfrak{H} = \mathfrak{A}_n$.

Proof of the theorem. Again assume that $\mathfrak{H} \trianglelefteq \mathfrak{A}_n$ $(n > 4)$. We shall try to find in $\mathfrak{H}$ a permutation $Q \neq I$ such that Q moves the minimum number m of the symbols $1, 2, \ldots, n$. Clearly $m > 2$. If $m = 3$ it follows from Lemma 2 that $\mathfrak{H} = \mathfrak{A}_n$.

So let us assume that $m = 4$; then Q is a product of two disjoint transpositions, say $Q = (12)(34)$. [We note that (1234) is odd.] We shall show that $\mathfrak{H}$ then contains a 3-cycle which means that m cannot be equal to 4. Indeed let $P = (345)$. The commutator $[P, Q] = PQP^{-1}Q^{-1} = (PQP^{-1})Q^{-1}$, as a product of two elements of $\mathfrak{H}$ is an element of $\mathfrak{H}$; it is found to be

$$[P, Q] = (345)(12)(34)(354)(12)(34) = (345)(34)(354)(34)$$
$$= (345)(345) = (354).$$

If $m > 4$ then there are the following possibilities:

(a) $Q = (12)(34)(5\ldots)\ldots,$ (b) $Q = (12)(345\ldots)\ldots$

(c) $Q = (1234\ldots)\ldots,$ (d) $Q = (123)(45\ldots)\ldots$

In each case one finds that the commutator $[P, Q]$, i.e. another element of $\mathfrak{H}$, leaves fixed at least one symbol more than Q. Thus no number > 3 can be the minimum m; thus $\mathfrak{H} = \mathfrak{A}_n$.

14. (a) If $\mathfrak{H} \lhd \mathfrak{G}$ then $\alpha(\mathfrak{H}) \lhd \mathfrak{G}$ for every $\alpha \in \Gamma(\mathfrak{G})$.

(b) If $\mathfrak{H} < \mathfrak{G}$ the intersection $\bigcap_{\alpha \in \Gamma} \alpha(\mathfrak{H}) \ntriangleleft \mathfrak{G}$.

(c) If $\mathfrak{H}$ is the only subgroup of the order $|\mathfrak{H}|$ in $\mathfrak{G}$ then $\mathfrak{H}$ is a characteristic subgroup of $\mathfrak{G}$.

(d) If $\mathfrak{H} \lhd \mathfrak{G}$ and there is no other non-trivial normal subgroup in $\mathfrak{G}$ then $\mathfrak{H} \ntriangleleft \mathfrak{G}$.

15. The commutator group of the group $\mathfrak{G}_{\mathbb{F}}$ (cf. Chap. II, §3, ex. 2 and §4, ex. 5) coincides with the subgroup $\mathfrak{H} \lhd \mathfrak{G}_{\mathbb{F}}$.

16. Every characteristic class is a union of complete conjugacy classes. Every automorphism effects a permutation of the conjugacy classes within a characteristic class.

17. Definition (R. Baer, 1934). The set of all elements y in a group $\mathfrak{G}$ which satisfy the condition

$$y\mathfrak{H} = \mathfrak{H}y \quad \text{for all} \quad \mathfrak{H} < \mathfrak{G}$$

is called the *nucleus* (German: Kern) of $\mathfrak{G}$.

(a) The nucleus is a characteristic subgroup $\mathfrak{K} = \mathfrak{K}(\mathfrak{G})$ of $\mathfrak{G}$.

(b) The condition of the definition can be weakened:

$$\mathfrak{K} = \{y \in \mathfrak{G} \mid y\langle x\rangle y^{-1} = \langle x\rangle \quad \text{for all} \quad x \in \mathfrak{G}\}.$$

(c) $\mathfrak{K} = \bigcap_{\mathfrak{H} \leqslant \mathfrak{G}} \mathfrak{N}(\mathfrak{H}) = \bigcap_{x \in \mathfrak{G}} \mathfrak{N}(\langle x\rangle)$.

(d) $\mathfrak{Z}(\mathfrak{G}) \leqslant \mathfrak{K}(\mathfrak{G})$.

(e) For all subgroups $\mathfrak{L}$ in $\mathfrak{G}$ the intersection

$$\mathfrak{L} \cap \mathfrak{K}(\mathfrak{G}) \leqslant \mathfrak{K}(\mathfrak{L}).$$

(f) $\mathfrak{K}[\mathfrak{K}(\mathfrak{G})] = \mathfrak{K}(\mathfrak{G})$.

(g) Every subgroup of $\mathfrak{K}(\mathfrak{G})$ is a normal subgroup of $\mathfrak{K}(\mathfrak{G})$.

A group $\mathfrak{G}$ is said to be *characteristically simple* when it has no proper characteristic subgroup. Every simple group is characteristically simple, but the converse is not true. We shall study only finite characteristically simple groups and show how they can be obtained from simple groups.

THEOREM 5. *Every finite characteristically simple group $\mathfrak{G}$ is either simple or a direct product of isomorphic simple groups.*

*Proof.** In the case of a simple $\mathfrak{G}$ there is nothing to prove. So let us assume that there is a proper non-trivial normal subgroup in $\mathfrak{G}$. Among the normal subgroups of $\mathfrak{G}$ we select one, $\mathfrak{H}$, which contains no non-trivial normal subgroup of $\mathfrak{G}$. We consider all automorphic images $\alpha(\mathfrak{H})$, $\alpha \in \Gamma(\mathfrak{G})$, of $\mathfrak{H}$. By ex. 14(a) we know that all the $\alpha(\mathfrak{H})$ are normal in $\mathfrak{G}$.

We consider now the set $\mathbf{S}$ of all $\alpha(\mathfrak{H})$, $\alpha \in \Gamma(\mathfrak{G})$. As normal subgroups of $\mathfrak{G}$ the $\alpha(\mathfrak{H})$ commute and therefore $\langle \mathbf{S} \rangle$ equals the product of all these subgroups. Clearly $\langle \mathbf{S} \rangle$ is a characteristic subgroup of $\mathfrak{G}$; hence $\langle \mathbf{S} \rangle = \mathfrak{G}$. Further let $\mathbf{T} = \{\alpha_1(\mathfrak{H}), \ldots, \alpha_k(\mathfrak{H})\}$ represent a subset of $\mathbf{S}$, minimal in the sense that still $\langle \mathbf{T} \rangle = \alpha_1(\mathfrak{H}) \cdot \ldots \cdot \alpha_k(\mathfrak{H}) = \mathfrak{G}$, but $\langle \mathbf{T}_i \rangle < \mathfrak{G}$ if $\mathbf{T}_i$ arises from $\mathbf{T}$ by omitting the subgroup $\alpha_i(\mathfrak{H})$ $(i = 1, \ldots, k)$. For every such i we have $\mathfrak{D}_i = \alpha_i(\mathfrak{H}) \cap \langle \mathbf{T}_i \rangle \lhd \mathfrak{G}$ [cf. Chap. II, §3, ex. 4(a)]. Since $\alpha(\mathfrak{H})$, like $\mathfrak{H}$, contains no proper normal subgroup which is normal in $\mathfrak{G}$, we conclude that either $\mathfrak{D}_i = \mathfrak{e}$ or $\mathfrak{D}_i = \alpha_i(\mathfrak{H})$. Because of the minimality of $\mathbf{T}$ the latter is excluded. Hence $\mathfrak{D}_i = \mathfrak{e}$ $(i = 1, \ldots, k)$. Therefore we can form the direct product $\alpha_1(\mathfrak{H}) \times \ldots \times \alpha_k(\mathfrak{H})$ and this is equal to $\langle \mathbf{T} \rangle = \mathfrak{G}$.

Finally it is evident that $\mathfrak{H}$ is simple. For, suppose that $\mathfrak{L} \lhd \mathfrak{H}$, $\mathfrak{L} \neq \mathfrak{e}$. As a normal subgroup of a direct factor of $\mathfrak{G}$ also $\mathfrak{L} \lhd \mathfrak{G}$ (cf. Chap. II, §5, ex. 7). Again, with regard to the initial assumption concerning $\mathfrak{H}$ there is no such normal subgroup $\mathfrak{L}$ in $\mathfrak{H}$. Hence there is no non-trivial normal subgroup in $\mathfrak{H}$; thus $\mathfrak{H}$ is simple.

We shall prove now the converse of Theorem 5, using different arguments in the cases of a non-commutative and of a commutative group.

THEOREM 6a. *If $\mathfrak{G}$ is the direct product of r isomorphic non-abelian simple groups then $\mathfrak{G}$ is characteristically simple.*

* A communication by L. Fuchs was helpful for the formulation of this proof.

intersection of all maximal subgroups of $\mathfrak{G}$; hence $x \in \mathfrak{L}_A$ and we have $\langle A, x\rangle \leqslant \mathfrak{L}_A < \mathfrak{G}$ which contradicts the assumption according to which $\langle A, x\rangle = \mathfrak{G}$. Thus $\langle A\rangle < \mathfrak{G}$ is impossible; rather $\langle A\rangle = \mathfrak{G}$ and therefore x is a nongenerator.

Show that $\Phi(\mathfrak{G})$ is a characteristic subgroup of $\mathfrak{G}$.

19. Show that $|\mathfrak{G}'| \geqslant \max\limits_{a \in \mathfrak{G}} |\mathscr{C}(a)|$

§2 The holomorph of a finite group. Complete groups

a. Given a group $\mathfrak{G}$, it is relatively easy and in principle always possible to establish the group $\Delta = \Delta(\mathfrak{G})$ of all inner automorphisms τ of $\mathfrak{G}$ since each of these mappings can be realized by an operation $\tau(x) = txt^{-1}$ involving only elements of $\mathfrak{G}$ and multiplication and inversion in $\mathfrak{G}$. This is no more so in the case of outer (i.e. non-inner) automorphisms of $\mathfrak{G}$. The determination of the group $\Gamma = \Gamma(\mathfrak{G})$ is therefore a much more difficult problem.

We begin with the observation that if $\mathfrak{G}$ is a *normal* subgroup of a certain group $\mathfrak{H}$ then every inner automorphism of the group $\mathfrak{H}$, restricted to $\mathfrak{G}$, will give us an (inner or outer) automorphism of $\mathfrak{G}$. Thus we arrive at the following evidently natural question: Is it possible to find a group $\mathfrak{H}$ containing the given group $\mathfrak{G}$ as a normal subgroup so that every automorphism α of $\mathfrak{G}$ is realized by an inner automorphism of $\mathfrak{H}$ restricted to $\mathfrak{G}$? That is, for every α in $\Gamma(\mathfrak{G})$ there should be at least one u in $\mathfrak{H}$ such that

$$\alpha(x) = uxu^{-1} \quad \text{for all} \quad x \text{ in } \mathfrak{G}.$$

Up to isomorphism we shall determine a group $\mathfrak{H}$ which satisfies this condition and which can be constructed for every given group $\mathfrak{G}$. But the condition is not sufficient to determine uniquely such a group.

A group containing a given group $\mathfrak{G}$ as a subgroup is called an *extension* (or a supergroup) of $\mathfrak{G}$. It may be determined up to isomorphism only, that is, as a group containing a subgroup which is isomorphic to the given group $\mathfrak{G}$. In this sense is the symmetric group $\mathfrak{S}(\mathfrak{G})$, according to Cayley's theorem, an extension of all groups $\mathfrak{G}$ with the same set $\mathbf{G}$ of elements. We may also consider $\mathfrak{S}_g$ as an extension of every group of order g.

A group is called a *normal extension* of $\mathfrak{G}$ if it contains $\mathfrak{G}$ as a normal subgroup. The direct product of $\mathfrak{G}$ with another group is a solution of the normal extension problem for $\mathfrak{G}$. A more general solution will be discussed in §3.

Being at present unprepared for an "abstract" approach we shall begin with a definite representation of the given group $\mathfrak{G}$ by a group $\mathfrak{G}^*$ of permutations, $\mathfrak{G}^* \simeq \mathfrak{G}$, assuming that the given group $\mathfrak{G}$ has finite order $|\mathfrak{G}| = g$. As $\mathfrak{G}^*$ we choose the regular representation of $\mathfrak{G}$ (cf. Chap. I, §2, ex. 10) which is a subgroup of the symmetric group $\mathfrak{S}(\mathfrak{G})$. With a fixed element a in $\mathfrak{G}$ we associate the permutation

$$A = \begin{pmatrix} x \\ ax \end{pmatrix} \quad (x \in \mathfrak{G}).$$

Further we define the group $\mathfrak{H}^* \leqslant \mathfrak{S}(\mathfrak{G})$ to be the normalizer of $\mathfrak{G}^*$ in $\mathfrak{S}(\mathfrak{G})$:

$$\mathfrak{H}^* = \mathfrak{N}_{\mathfrak{S}}(\mathfrak{G}^*) = \{T \in \mathfrak{S}(\mathfrak{G}) \,|\, T\mathfrak{G}^*T^{-1} = \mathfrak{G}^*\}$$

so that $\mathfrak{G}^* \trianglelefteq \mathfrak{H}^*$ and therefore every inner automorphism of $\mathfrak{H}^*$, by restricting it to $\mathfrak{G}^*$, induces an automorphism of $\mathfrak{G}^*$. This well-defined group $\mathfrak{H}^*$ is called the *holomorph* of $\mathfrak{G}^*$ and it defines, up to isomorphism, the holomorph $\mathfrak{H}$ of $\mathfrak{G}$.

From §1, ex. 1 we know that $\Gamma(\mathfrak{G}) \simeq \Gamma(\mathfrak{G}^*)$. This isomorphism is realized by the mapping which sends $\alpha \in \Gamma(\mathfrak{G})$ onto $\alpha^* \in \Gamma(\mathfrak{G}^*)$ defined by

$$\alpha^*(A) = \alpha^* \begin{pmatrix} x \\ ax \end{pmatrix} \overset{\text{def}}{=} \begin{pmatrix} x \\ \alpha(a)x \end{pmatrix}.$$

This mapping $\alpha \to \alpha^*$ is indeed an isomorphism. It is a homomorphism:

$$\alpha^*(BA) = \alpha^* \begin{pmatrix} x \\ bax \end{pmatrix} = \begin{pmatrix} x \\ \alpha(ba)x \end{pmatrix} = \begin{pmatrix} x \\ \alpha(b)\alpha(a)x \end{pmatrix}$$

$$= \begin{pmatrix} x \\ \alpha(b)x \end{pmatrix}\begin{pmatrix} x \\ \alpha(a)x \end{pmatrix} = \alpha^*(B)\,\alpha^*(A).$$

It is an isomorphism because $\alpha^* = \beta^*$ implies $\alpha = \beta$.

b. In §1, *a.* we have learned that the group $\Gamma = \Gamma(\mathfrak{G})$ of all automorphisms α of $\mathfrak{G}$ can be represented isomorphically by a subgroup $\Gamma^* < \mathfrak{S}(\mathfrak{G})$, the elements of Γ^* being the permutations

$$U = U(\alpha) = \begin{pmatrix} x \\ \alpha(x) \end{pmatrix}, \quad x \in \mathfrak{G}, \quad \alpha \in \Gamma \quad \text{[cf. §1, (2)]}.$$

Transformation of an element A in $\mathfrak{G}^*$ by U in Γ^*, i.e. an inner automorphism of $\mathfrak{S}(\mathfrak{G})$, yields

$$(1) \qquad UAU^{-1} = \begin{pmatrix} x \\ \alpha(x) \end{pmatrix}\begin{pmatrix} x \\ ax \end{pmatrix}\begin{pmatrix} \alpha(x) \\ x \end{pmatrix} = \begin{pmatrix} ax \\ \alpha(ax) \end{pmatrix}\begin{pmatrix} \alpha(x) \\ ax \end{pmatrix}$$

$$= \begin{pmatrix} \alpha(x) \\ \alpha(a)\,\alpha(x) \end{pmatrix} = \begin{pmatrix} x \\ \alpha(a)\,x \end{pmatrix},$$

which coincides with the element $\alpha^*(A)$ of $\mathfrak{G}^*$. We conclude that $U \in \mathfrak{N}_{\mathfrak{S}(\mathfrak{G})}(\mathfrak{G}^*) = \mathfrak{H}^*$ for all $\alpha \in \Gamma$. Since $\mathfrak{G}^* \lhd \mathfrak{H}^*$ we have

$$\mathfrak{G}^*\Gamma^* = \Gamma^*\mathfrak{G}^* \leqslant \mathfrak{H}^*.$$

Conversely let us assume that

$$P = \begin{pmatrix} x \\ x' \end{pmatrix}$$

is an element of $\mathfrak{H}^*$. With regard to the definition of this group $PAP^{-1} \in \mathfrak{G}^*$ for every A in $\mathfrak{G}^*$. Since transformation by P (an element of $\mathfrak{H}^*$) induces an automorphism of $\mathfrak{G}^*$, we have for a certain $\alpha \in \Gamma$:

$$PAP^{-1} = \alpha^*(A) = \begin{pmatrix} x \\ \alpha(a)\cdot x \end{pmatrix} = \begin{pmatrix} x' \\ \alpha(a)\cdot x' \end{pmatrix}.$$

On the other hand

$$PAP^{-1} = \begin{pmatrix} x \\ x' \end{pmatrix}\begin{pmatrix} x \\ ax \end{pmatrix}\begin{pmatrix} x' \\ x \end{pmatrix} = \begin{pmatrix} ax \\ (ax)' \end{pmatrix}\begin{pmatrix} x' \\ ax \end{pmatrix} = \begin{pmatrix} x' \\ (ax)' \end{pmatrix},$$

so that

$$(ax)' = \alpha(a)\,x'.$$

In particular for $x = e$ one has $a' = \alpha(a)\, e'$ where e' is the image of e under the permutation $P \in \mathfrak{H}^*$.

If $e' = e$ it follows that $a' = \alpha(a)$, hence $x' = \alpha(x)$ for all $x \in \mathfrak{G}$; thus $P \in \Gamma^*$, i.e. P is representative of the automorphism α of $\mathfrak{G}$.

If $e' \neq e$ we multiply P by the permutation

$$E = \begin{pmatrix} e'x \\ x \end{pmatrix} \in \mathfrak{G}^*$$

which maps $e' \to e$. Thus

$$EP = \begin{pmatrix} e'x \\ x \end{pmatrix}\begin{pmatrix} x \\ x' \end{pmatrix} = \begin{pmatrix} x' \\ e'^{-1}x' \end{pmatrix}\begin{pmatrix} x \\ x' \end{pmatrix} = \begin{pmatrix} x \\ x'' \end{pmatrix}$$

is an element of $\mathfrak{H}^*$ which maps $e \to e'' = e$; consequently

$$(EP)A(EP)^{-1} = \begin{pmatrix} x'' \\ \beta(a)x'' \end{pmatrix} \quad (\beta \in \Gamma)$$

with $EP = Q \in \Gamma^*$. Therefore $P = E^{-1}Q \in \mathfrak{G}^*\Gamma^*$ and $\mathfrak{H}^* \leqslant \mathfrak{G}^*\Gamma^*$.

We summarize the results of this subsection:

THEOREM 1. *The holomorph $\mathfrak{H}$ of a finite group $\mathfrak{G}$ is a normal extension of $\mathfrak{G}$ which is isomorphic to the group $\mathfrak{H}^* = \mathfrak{N}_{\mathfrak{S}}(\mathfrak{G}^*) = \mathfrak{G}^*\Gamma^* \leqslant \mathfrak{S}(\mathfrak{G})$ where $\mathfrak{G}^*$ is the regular Cayley representation of $\mathfrak{G}$, $\mathfrak{G}^* \lhd \mathfrak{H}^*$, and Γ^* is the group of all permutations*

$$U = \begin{pmatrix} x \\ \alpha(x) \end{pmatrix}, \quad \alpha \in \Gamma(\mathfrak{G}).$$

COROLLARY. *Every automorphism of $\mathfrak{G}^*$ is obtained by restriction of an inner automorphism of $\mathfrak{H}^*$ to $\mathfrak{G}^*$* (cf. (1)).

c. The two factors $\mathfrak{G}^*$ and Γ^* of $\mathfrak{H}^*$ have a few properties in common with the direct factors of a direct product. Firstly: $\mathfrak{G}^* \cap \Gamma^* = I$. In fact, suppose that an element

$$A = \begin{pmatrix} x \\ ax \end{pmatrix} \in \mathfrak{G}^*$$

coincides with an element

$$U = \begin{pmatrix} x \\ \alpha(x) \end{pmatrix} \in \Gamma^*;$$

this implies that $\alpha(x) = ax$ for all $x \in \mathfrak{G}$. In particular $\alpha(e) = e = a$. Thus the identity permutation I is the only common element of $\mathfrak{G}^*$ and Γ^*.

Moreover, every $P \in \mathfrak{H}^*$ has a unique representation $P = AU$, $A \in \mathfrak{G}^*$, $U \in \Gamma^*$. Indeed

$$AU = \begin{pmatrix} x \\ ax \end{pmatrix}\begin{pmatrix} x \\ \alpha(x) \end{pmatrix} = \begin{pmatrix} x \\ a\,\alpha(x) \end{pmatrix},$$

and if

$$B = \begin{pmatrix} x \\ bx \end{pmatrix} \quad \text{and} \quad V = \begin{pmatrix} x \\ \beta(x) \end{pmatrix},$$

$\beta \in \Gamma$, and $AU = BV$, then $a\,\alpha(x) = b\,\beta(x)$. By setting $x = e$ we have $a = b$ and thus also $\alpha = \beta$. Hence $A = B$ and $U = V$.

But, in general Γ^* is not a normal subgroup of $\mathfrak{H}^*$. For, $AUA^{-1} \in \Gamma^*$ implies that

$$\begin{pmatrix} x \\ ax \end{pmatrix}\begin{pmatrix} x \\ \alpha(x) \end{pmatrix}\begin{pmatrix} ax \\ x \end{pmatrix} = \begin{pmatrix} ax \\ a\,\alpha(x) \end{pmatrix} = \begin{pmatrix} x \\ a\,\alpha(a^{-1}x) \end{pmatrix}$$

and therefore the mapping $x \to a \cdot \alpha(a)^{-1}\alpha(x)$ is an automorphism of $\mathfrak{G}$. This condition is easily reduced to $a\alpha(a)^{-1} = e$ or $\alpha(a) = a$ which means that a must be invariant under the automorphism α. Thus the holomorph is the direct product of $\mathfrak{G}^*$ and Γ^* if and only if all the elements a of $\mathfrak{G}$ are invariant under all the automorphisms of $\mathfrak{G}$, that is, the identity ϵ is the only automorphism of $\mathfrak{G}$. This requires that $|\mathfrak{G}| = 2$, cf. ex. 2.

*d. Is the holomorph minimal?** The question arises whether for a given group $\mathfrak{G}$ a smaller extension $\mathfrak{K}^*$ of $\mathfrak{G}^*$, $\mathfrak{K}^* \leqslant \mathfrak{H}^*$, can be found which could be substituted for $\mathfrak{H}^*$ itself in the sense that every automorphism of $\mathfrak{G}$ is induced by transforming $\mathfrak{G}^*$ by an element of $\mathfrak{K}^*$. We begin the discussion with the question: Which elements P in

* The following discussion is due to J. Fischer.

$\mathfrak{H}^*$ induce in $\mathfrak{G}^*$ the identity automorphism? The answer to this question is given by the following theorem:

THEOREM 2. *The permutation* $P = AU \in \mathfrak{H}^*$ *transforms every element of* $\mathfrak{G}^*$ *into itself if and only if*

$$P = \bar{A} = \begin{pmatrix} x \\ xa \end{pmatrix} \quad (a \in \mathfrak{G}).$$

Remark. All these permutations $\bar{A}$ form a subgroup $^*\mathfrak{G} < \mathfrak{H}^*$: If

$$a \to \bar{A}, \quad b \to \bar{B} = \begin{pmatrix} x \\ xb \end{pmatrix}$$

then

$$ab \to \begin{pmatrix} x \\ x(ab) \end{pmatrix} = \begin{pmatrix} x \\ xb \end{pmatrix}\begin{pmatrix} x \\ xa \end{pmatrix} = \bar{B}\bar{A}.$$

Although *this* mapping $\mathfrak{G} \to {}^*\mathfrak{G}$ is not an isomorphism, the mapping $a \to \bar{A}^{-1}$ maps $\mathfrak{G}$ isomorphically onto $^*\mathfrak{G}$; hence also $\mathfrak{G}^* \simeq {}^*\mathfrak{G}$.

Proof. Using the notations of subsection *c*.

$$P = AU = \begin{pmatrix} x \\ a \cdot \alpha(x) \end{pmatrix}$$

and the condition $PBP^{-1} = B$ for all $B \in \mathfrak{G}^*$ means

$$(2) \qquad \begin{pmatrix} x \\ a\alpha(x) \end{pmatrix}\begin{pmatrix} x \\ bx \end{pmatrix}\begin{pmatrix} a\alpha(x) \\ x \end{pmatrix} = \begin{pmatrix} a\alpha(x) \\ a\alpha(b)\cdot\alpha(x) \end{pmatrix}$$

$$= \begin{pmatrix} x \\ a\alpha(b)a^{-1}x \end{pmatrix} = \begin{pmatrix} x \\ bx \end{pmatrix}$$

whence $a \cdot \alpha(b) \cdot a^{-1} = b$. Thus α must be the inner automorphism $x \to a^{-1}xa$ and this we can write in the form

$$a\,\alpha(x) = xa$$

which proves the theorem.

As a consequence of theorem 2 we can state that *two elements P and Q of $\mathfrak{H}^*$ induce the same automorphism of $\mathfrak{G}^*$ if and only if $Q = \bar{A}P$, $\bar{A} \in {}^*\mathfrak{G}$, that is, if and only if $Q \in {}^*\mathfrak{G}P$.* Indeed, referring to the first two equations (2) we have

$$PBP^{-1} = \binom{x}{cx}, \quad c = a\,\alpha(b)a^{-1}$$

and therefore

$$QBQ^{-1} = \bar{A}\binom{x}{cx}\bar{A}^{-1} = \binom{x}{cx}.$$

Conversely, let $Q = TP$, $T \in \mathfrak{H}^*$. Then $QBQ^{-1} = PBP^{-1}$ implies that

$$T\binom{x}{cx}T^{-1} = \binom{x}{cx}$$

and, therefore, by theorem 2: $T \in {}^*\mathfrak{G}$. Thus we have the following result:

Corollary 1. *Let $\mathfrak{K}^*$ denote an extension of $\mathfrak{G}^*$, $\mathfrak{K}^* \leqslant \mathfrak{H}^*$. Every automorphism of $\mathfrak{G}^*$ is induced by transformation with a suitable element of $\mathfrak{K}^*$ if and only if $\mathfrak{K}^*$ contains a transversal of the cosets of ${}^*\mathfrak{G}$ in $\mathfrak{H}^*$.*

In ex. 5(a), (b) we exhibit groups $\mathfrak{G}$ where $\mathfrak{K}^*$ *must* equal $\mathfrak{H}^*$ and where we can have $\mathfrak{G}^* < \mathfrak{K}^* < \mathfrak{H}^*$ respectively; the case where $\mathfrak{K}^*$ *may* equal $\mathfrak{G}^*$ will be considered in the next subsection. In connection with the preceding discussion these examples make it evident that the holomorph $\mathfrak{H}$, as an extension of $\mathfrak{G}$ in which all automorphisms of $\mathfrak{G}$ appear as inner automorphisms restricted to $\mathfrak{G}$, can be replaced by a subgroup $\mathfrak{K} \leqslant \mathfrak{H}$.

e. Complete groups. The two subgroups $\mathfrak{G}^*$ and ${}^*\mathfrak{G}$ will be important also in the following discussion. We list some of their properties which are all easy to verify:

Theorem 3. (a) *$\bar{A}B = B\bar{A}$ for all $\bar{A} \in {}^*\mathfrak{G}$, $B \in \mathfrak{G}^*$;*

(b) ${}^*\mathfrak{G} \cap \mathfrak{G}^* = \mathfrak{Z}(\mathfrak{G}^*) = \mathfrak{Z}({}^*\mathfrak{G})$;

(c) $\mathfrak{G}^* \lhd \mathfrak{H}^*$, ${}^*\mathfrak{G} \lhd \mathfrak{H}^*$;

(d) *The elements of the product* $*\mathfrak{G}\cdot\mathfrak{G}^*$ *are exactly those elements of* $\mathfrak{H}^*$ *which induce the inner automorphisms of* $\mathfrak{G}^*$.

(e) $*\mathfrak{G}\cdot\mathfrak{G}^* = *\mathfrak{G}\times\mathfrak{G}^*$ *if and only if* $\mathfrak{Z}(\mathfrak{G}) = \mathfrak{e}$.

(f) *If all automorphisms of* $\mathfrak{G}$ *are inner then* $\mathfrak{H}^* = *\mathfrak{G}\cdot\mathfrak{G}^*$.

DEFINITION. *A group* $\mathfrak{G}$ *is said to be complete if* $\mathfrak{Z}(\mathfrak{G}) = \mathfrak{e}$ *and* $\Gamma(\mathfrak{G}) = \Delta(\mathfrak{G})$; *that is, if* $\mathfrak{G}$ *has "no centre" and if all its automorphisms are inner automorphisms.*

COROLLARY 2. *A group* $\mathfrak{G}$ *is complete if and only if its holomorph* $\mathfrak{H}^*$ *is isomorphic to the direct product* $*\mathfrak{G}\times\mathfrak{G}^*$. *In this case one can choose* $\mathfrak{K}^* = \mathfrak{G}^*$.

We conclude with a simple theorem on complete groups:

THEOREM 4. *Let* $\mathfrak{G}$ *be a complete normal subgroup of a group* $\mathfrak{L}$, *and let* $\mathfrak{C}(\mathfrak{G}) = \mathfrak{C}_{\mathfrak{L}}(\mathfrak{G})$ *denote the centralizer of* $\mathfrak{G}$ *in* $\mathfrak{L}$ (cf. §1, *e.*). *Then*

$$\mathfrak{L} = \mathfrak{G}\times\mathfrak{C}_{\mathfrak{L}}(\mathfrak{G}).$$

Remark. This contains Corollary 2 of Theorem 3: Take $\mathfrak{H}^*$ instead of $\mathfrak{L}$, $\mathfrak{G}^*$ instead of $\mathfrak{G}$. Moreover $\mathfrak{C}_{\mathfrak{H}^*}(\mathfrak{G}^*) = *\mathfrak{G}$ which is another way of stating theorem 2.

Proof. By definition

$$\mathfrak{C}_{\mathfrak{L}}(\mathfrak{G}) = \{s\in\mathfrak{L}\,|\,sxs^{-1} = x,\ x\in\mathfrak{G}\} \tag{3}$$

and therefore $\mathfrak{G}\cap\mathfrak{C}(\mathfrak{G}) = \mathfrak{Z}(\mathfrak{G})$. Since $\mathfrak{G}$ is complete we have $\mathfrak{Z}(\mathfrak{G}) = \mathfrak{e}$ so that

$$\mathfrak{G}\cap\mathfrak{C}(\mathfrak{G}) = \mathfrak{e}.$$

Because all automorphisms of $\mathfrak{G}$ are inner, for every t in $\mathfrak{L}$ an element a can be found in $\mathfrak{G}$ so that $txt^{-1} = axa^{-1}$, i.e. $(a^{-1}t)x(a^{-1}t)^{-1} = x$ for all x in $\mathfrak{G}$. Thus $a^{-1}t\in\mathfrak{C}(\mathfrak{G})$ or $t\in a\,\mathfrak{C}(\mathfrak{G})$ and therefore every t in $\mathfrak{L}$ appears in the form $t = ab$, $a\in\mathfrak{G}$, $b\in\mathfrak{C}(\mathfrak{G})$.

Finally by (3) for every s in $\mathfrak{C}(\mathfrak{G})$ we have

$$tst^{-1} = absb^{-1}a^{-1} = as'a^{-1} = s'\in\mathfrak{C}(\mathfrak{G}).$$

Hence $\mathfrak{C}(\mathfrak{G})\lhd\mathfrak{L}$ and the theorem is proved.

Examples and exercises

1. Isomorphic groups have isomorphic holomorphs.

2. A mapping $\varphi \neq \epsilon$ of a set onto itself is said to be involutory if $\varphi^2 = \epsilon$. Every involutory mapping is invertible.

(a) Show that every abelian group $\mathfrak{G}$ of order >2 admits an involutory automorphism. Hint: Distinguish the two cases

(i) $\mathfrak{G}$ contains at least one element of order >2;

(ii) All elements different from e have order 2.

(b) Every non-abelian group admits an automorphism $\neq \epsilon$.

3. The holomorph of the non-cyclic group of order 4 is isomorphic to the symmetric group $\mathfrak{S}_4$.

4. Determine the holomorphs of the cyclic groups.

(a) The holomorph of $\langle a \rangle_p$ (p a prime) is isomorphic to the group $\mathfrak{G}_{\mathbb{F}_p}$ where $\mathbb{F}_p$ ($\simeq \mathbb{R}_p$) denotes a field with p elements (cf. Chap. II, §3, ex. 2).

(b) Determine the holomorph of $\langle a \rangle_n$ where n is a composite number (cf. §1, ex. 9).

(c) Determine the holomorph of $\langle a \rangle_\infty$ (cf. §1, ex. 4).

5. (a) If $\mathfrak{G}$ is abelian the smallest extension $\mathfrak{K}^*$ of $\mathfrak{G}^*$ which has the holomorph property coincides with the holomorph $\mathfrak{H}^*$.

Proof. Because $\mathfrak{G}$ is abelian we have $\mathfrak{G}^* = {}^*\mathfrak{G}$; thus, according to corollary 1 we consider the system of the cosets of $\mathfrak{G}^*$ in $\mathfrak{H}^*$. Every element of $\mathfrak{H}^*$ can be written in the form $P = AU$, $A \in \mathfrak{G}^*$,

$$U = \begin{pmatrix} x \\ \alpha(x) \end{pmatrix} \in \Gamma^*.$$

Thus $\mathfrak{G}^*P = \mathfrak{G}^*U$. Equality of two cosets: $\mathfrak{G}^*U = \mathfrak{G}^*V$ implies $AU = BV$ and thus $U = V$ (cf. *c.*). Hence $\mathfrak{K}^*$ must contain an element of the form AU for each $U \in \Gamma^*(\mathfrak{G})$. Thus $\Gamma^* < \mathfrak{K}^*$ and since $\mathfrak{G}^* < \mathfrak{K}^*$ we conclude that $\mathfrak{K}^* = \mathfrak{G}^*\Gamma^* = \mathfrak{H}^*$.

(b) If $\mathfrak{G} = \mathfrak{D}_5$ (cf. §1, ex. 7) we may select $\mathfrak{K}^*$ so that $\mathfrak{G}^* < \mathfrak{K}^* < \mathfrak{H}^*$.

We denote the elements of $\mathfrak{D}_5$ as follows:

$$a_0 = e, \quad a_1 = a, \quad a_2 = a^2, \quad a_3 = a^3, \quad a_4 = a^4,$$
$$a_5 = b, \quad a_6 = ba, \quad a_7 = ba^2, \quad a_8 = ba^3, \quad a_9 = ba^4.$$

To the generators a, b of $\mathfrak{D}_5$ then correspond the two permutations

$$A = \begin{pmatrix} x \\ ax \end{pmatrix} = (0\ 1\ 2\ 3\ 4)(5\ 9\ 8\ 7\ 6)$$

$$B = \begin{pmatrix} x \\ bx \end{pmatrix} = (0\ 5)(1\ 6)(2\ 7)(3\ 8)(4\ 9)$$

of the Cayley representation $\mathfrak{D}_5{}^*$ of $\mathfrak{D}_5$. In §1, ex. 7 we have found a presentation of the automorphism group $\Gamma(\mathfrak{D}_5) = \langle \alpha, \beta \rangle$. The isomorphic subgroup $\Gamma_5{}^* < \mathfrak{S}_{10}$ is generated by the permutations

$$\alpha^* = \begin{pmatrix} x \\ \alpha(x) \end{pmatrix} = (1\ 2\ 4\ 3)(6\ 7\ 9\ 8),$$

$$\beta^* = \begin{pmatrix} x \\ \beta(x) \end{pmatrix} = (5\ 6\ 7\ 8\ 9).$$

Then

$$\mathfrak{H}_5{}^* = \langle A, B, \alpha^*, \beta^* \rangle = \mathfrak{D}_5{}^*\Gamma_5{}^*, \quad |\mathfrak{H}_5{}^*| = 200.$$

Since (cf. §1, ex. 7) β is an inner automorphism of $\mathfrak{D}_5$ it is evident that $\mathfrak{K}_5{}^* = \langle A, B, \alpha^* \rangle = \mathfrak{D}_5{}^* \cdot \langle \alpha \rangle_4$, $|\mathfrak{K}_5{}^*| = 40$, is the smallest extension of $\mathfrak{G}^*$ in $\mathfrak{H}^*$ which has the holomorph property.

6. For the symmetric group $\mathfrak{S}_3$ determine the holomorph as a subgroup of $\mathfrak{S}_6$.

7. Let $P = AU$ ($A \in \mathfrak{G}^*$, $U \in \Gamma^*$) be an element of $\mathfrak{H}^*$. Determine the elements $B \in \mathfrak{G}^*$, $V \in \Gamma^*$ for which $P = VB$.

8. If $\mathfrak{G}$ is a group without centre [i.e. $\mathfrak{Z}(\mathfrak{G}) = \mathfrak{e}$] then the automorphism group $\Gamma(\mathfrak{G})$ also has no centre.

9. Theorem. (O. Hölder, 1895). *The symmetric group $\mathfrak{S}_n$ is complete if $n \geqslant 3$, $n \neq 6$.*

Apart from the easily established fact that $\mathfrak{Z}(\mathfrak{S}_n) = \mathfrak{I}$ for all $n \geqslant 3$ the proof of the theorem is based upon the following

Lemma. *Let $n \geqslant 3$, $n \neq 6$, and $\alpha \in \Gamma(\mathfrak{S}_n)$. If $(i\ j)$ is a transposition in $\mathfrak{S}_n$ then so is $\alpha(i\ j) = \alpha((i\ j))$.*

Proof. Referring to Chap. II, §4, ex. 1, (c) the conjugacy class $\mathscr{C}(i\ j)$ is characterized by $k_1 = n-2$, $k_2 = 1$; hence by p. 89, (7) $|\mathscr{C}(i\ j)| = n!/(n-2)!\cdot 2$. Now $\alpha(i\ j)$ is a permutation of order 2, hence a transposition or a product of r disjoint transpositions. Thus its class $\mathscr{C}(\alpha(i\ j))$ is characterized by $k_1 = n-2r$, $k_2 = r$ and $|\mathscr{C}(\alpha(i\ j))| = n!/(n-2r)!\cdot 2^r\cdot r!$

However $\alpha(i\ j)$ has the same number of distinct conjugates as any one transposition in $\mathfrak{S}_n$. Hence $|\mathscr{C}(\alpha(i\ j))| = |\mathscr{C}(i\ j)|$ which implies

$$\text{(4)} \qquad \frac{(n-2)!}{(n-2r)!} = (n-2)(n-3)\ldots(n-2r+1) = r!2^{r-1}.$$

If $r = 1$ this is satisfied for all $n > 2$. If $r = 2$ one has $(n-2)(n-3) = 4$ which is not satisfied by any natural n. For $r = 3$ the equation (4) becomes

$$(n-2)(n-3)(n-4)(n-5) = 24$$

and this has the solution $n = 6$ and none other.

If $r \geqslant 4$ (and obviously $n \geqslant 2r$) we have

$$(n-2)(n-3)\ldots(n-2r+1) \geqslant (2r-2)! > r!2^{r-1}$$

because $(2r-2)(2r-3)\ldots(r+1) > 2^{r-1}$. Hence (4) cannot be satisfied by any other couples r, n except $1, n$ and $3, 6$. Thus only in the case $n = 6$ could $\alpha(ij)$ be either a transposition or a product of three disjoint transpositions. In all other cases $\alpha(ij)$ is a transposition.

Proof of the theorem. The transpositions (12), (13), ..., $(1n)$ represent a system of generators of $\mathfrak{S}_n$. Indeed for a pair i, j out of the set $\{1, 2, \ldots, n\}$, $i \neq j$, we have $(1i)(1j)(1i) = (ij)$ so that by transformation within the system of these elements $(1j)$ all the transpositions of $\mathfrak{S}_n$ can be produced, and since every element of $\mathfrak{S}_n$ is a product of transpositions it is also a product of transpositons out of the system (12), ..., $(1n)$, i.e. $\mathfrak{S}_n = \langle(12), (13), \ldots, (1n)\rangle$. By §1, ex. 5(a) it follows that $\langle\alpha(12), \alpha(13), \ldots, \alpha(1n)\rangle = \mathfrak{S}_n$ for every α in $\Gamma(\mathfrak{S}_n)$.

Further observe that the product of two transpositions is of order 3 if and only if they contain a common symbol; in particular if $\mu \neq \nu$ we have $(1\mu)(1\nu) = (1\ \nu\ \mu)$ so that $\alpha(1\mu)\,\alpha(1\nu) = \alpha(1\ \nu\ \mu)$ has order 3 for every α in $\Gamma(\mathfrak{S}_n)$ and the transpositions $\alpha(1\mu)$ and $\alpha(1\nu)$ have a common symbol, say i. Now if $\rho \neq \mu$, $\rho \neq \nu$ then the transposition $\alpha(1\rho)$ must have a symbol in common with $\alpha(1\mu)$ and with $\alpha(1\nu)$, thus i must be a symbol common to all three.

Hence if $\alpha(1\mu) = (i\ \mu')$ then $\mu \to \mu'$ is an invertible function of μ, defined by the automorphism α for $\mu = 2, \ldots, n$ with values $\mu' \neq i$ in the set $\{1, \ldots, n\}$. So we may complete the function μ' by setting $1' = i$.

Therefore

$$T = \begin{pmatrix} 1 & 2 & 3 & \ldots & n \\ i & 2' & 3' & \ldots & n' \end{pmatrix}$$

is a permutation in $\mathfrak{S}_n$ which is uniquely defined by the automorphism α and

$$T(1\mu)T^{-1} = (i\mu') = \alpha(1\mu).$$

Thus with regard to its effect on the generators (1μ) the automorphism α can be replaced by the inner automorphism $\tau(X) = TXT^{-1}$, $X \in \mathfrak{S}_n$. Hence every automorphism of $\mathfrak{S}_n$ is an inner automorphism.

COROLLARY. $\Gamma(\mathfrak{S}_n) = \Delta(\mathfrak{S}_n) \simeq \mathfrak{S}_n$ *if* $n \geqslant 3$ *and* $n \neq 6$.

As to the case $n = 6$ cf. D. W. Miller, On a theorem of Hölder, *Amer. Math. Monthly*, 65 (1958), 252-254. Cf. also Chap. V, §1, ex. 3.

10. Let $a \in \mathfrak{G}$, $\alpha \in \Gamma(\mathfrak{G})$; show that the centralizer $\mathfrak{C}(\alpha(a)) = \alpha(\mathfrak{C}(a))$.

11. For every $a \in \mathfrak{G}$ let $\mathscr{F}(a)$ denote the set of all $x \in \mathfrak{G}$ which have in $\mathfrak{G}$ the same centralizer: $\mathscr{F}(a) = \{x \mid \mathfrak{C}(x) = \mathfrak{C}(a)\}$. We shall call $\mathscr{F}(a)$ the *equicentralizer system** of a in $\mathfrak{G}$, in short: an EC system. The centre $\mathfrak{Z}(\mathfrak{C}(a))$ will be called the *norm centre* of a in $\mathfrak{G}$. The basic properties

* The study of the EC systems has been originated by M. Cipolla in 1910. He called them "fundamental systems" and the corresponding centralizers "fundamental groups". They are still called so in the Italian literature, the only one where attention has been given to these systems so far. For references cf. the book by G. Zappa, Fondamenti di Teoria dei Gruppi, Vol. I, Chap. II, §6 and Chap. V, §4. Roma 1965.

of the EC systems are stated in the following propositions whose proofs are left to the reader.

(a) The EC systems $\mathscr{F}(a)$, $a \in \mathfrak{G}$, constitute a partition of the group $\mathfrak{G}$ into disjoint classes. The corresponding centralizers are defined by these systems: $\mathfrak{C}(a) = \mathfrak{C}(\mathscr{F}(a))$.

(b) For all $a, b \in \mathfrak{G}$ the three statements $b \in \mathscr{F}(a)$, $\mathfrak{C}(a) = \mathfrak{C}(b)$, $\mathscr{F}(a) = \mathscr{F}(b)$ are equivalent.

(c) For all $a \in \mathfrak{G}$ one has $\mathscr{F}(a) \subseteq \mathfrak{C}(a)$ and $\mathscr{F}(a) \subseteq \mathfrak{Z}(\mathfrak{C}(a))$. If for some a one has $\mathscr{F}(a) = \mathfrak{C}(a)$ then the group $\mathfrak{G}$ is abelian. A group is abelian if and only if it is covered by a single EC system.

(d) The centre $\mathfrak{Z}(\mathfrak{G})$ is an EC system of $\mathfrak{G}$ and every EC system of $\mathfrak{G}$ is a union of cosets of $\mathfrak{Z}(\mathfrak{G})$.

(e) Different EC systems have distinct centralizers and distinct norm centres. Hence:

(i) Two different centralizers of a group cannot have equal centres.

(ii) If $\mathscr{F}(a) = \mathfrak{Z}(\mathfrak{C}(a))$ then $a \in \mathfrak{Z}(\mathfrak{G})$, hence $\mathfrak{C}(a) = \mathfrak{G}$.

(*f*) If $x \in \mathscr{F}(a)$ then $x^{-1} \in \mathscr{F}(a)$. If x has order n and $(n, k) = 1$ then $x^k \in \mathscr{F}(a)$.

(g) If $\mathbf{T}$ is a transversal (i.e. a set of representatives) of the EC systems of $\mathfrak{G}$ then $\mathfrak{Z}(\mathfrak{G}) = \bigcap_{x \in \mathbf{T}} \mathfrak{C}(x) = \bigcap_{x \in \mathbf{T}} \mathfrak{Z}(\mathfrak{C}(x))$.

(h) If $b \in \mathfrak{C}(a)$ then $\mathscr{F}(b) \subseteq \mathfrak{C}(a)$ and if $b \in \mathfrak{Z}(\mathfrak{C}(a))$ then $\mathscr{F}(b) \subseteq \mathfrak{Z}(\mathfrak{C}(a))$. Hence every centralizer and every norm centre is a union of EC systems.

12. Let $a \in \mathfrak{G}$, $\alpha \in \Gamma(\mathfrak{G})$. Show that $\mathscr{F}(\alpha(a)) = \alpha(\mathscr{F}(a))$ (cf. ex. 10). Hence an automorphism of $\mathfrak{G}$ either maps $\mathscr{F}(a)$ onto itself or onto some other EC system.

13. (a) Let $a, x \in \mathfrak{G}$. Show that $xax^{-1} \in \mathscr{F}(a)$ if and only if $x \in \mathfrak{N}(\mathfrak{C}(a))$.

(b) $\mathfrak{N}(\mathfrak{C}(a)) = \mathfrak{N}((\mathscr{F}a))$.

(c) If $\mathfrak{N}(\mathfrak{C}(a)) = \mathfrak{C}(a)$ then $\mathscr{F}(a)$ contains no conjugates of a except a.

14. (a) No group has exactly two EC systems.

(b) No group has exactly three EC systems.

15. Show that the quaternion group of order 8 has exactly four EC systems.

§3 Group extensions

The problem of this section has been formulated in §2, *a*. It will be dealt with in stepwise increasing generality. The special position of the holomorph among the normal extensions of a group will be established. The method of O. Schreier will be described.

a. The semi-direct product. Let $\mathfrak{G}$ be a group with identity e and typical elements x, y, z; let $\mathfrak{A}$ be another group, its identity i, and typical elements a, b, c. Further we assume that there is a homomorphism $\varphi : \mathfrak{A} \to \Gamma(\mathfrak{G})$, i.e. an epimorphism φ of $\mathfrak{A}$ onto a subgroup A of $\Gamma(\mathfrak{G})$. If α, β are typical elements of A, let $\alpha = \varphi(a)$, $\beta = \varphi(b)$; then $\alpha\beta = \varphi(a)\cdot\varphi(b) = \varphi(ab)$. Moreover $\varphi(i)$ is the identity automorphism ϵ.

THEOREM 1. *The set of all couples* (a, x), $a \in \mathfrak{A}$, $x \in \mathfrak{G}$, *forms a group, to be denoted by* $(\mathfrak{A}, \mathfrak{G})_\varphi = (\mathfrak{A}, \mathfrak{G})$ *in which the multiplication of the two couples* (a, x) *and* (b, y) *is defined by*

$$(1) \qquad (a, x)(b, y) = (ab, x\cdot\alpha(y)), \quad \alpha = \varphi(a).$$

The group $(\mathfrak{A}, \mathfrak{G})_\varphi$ is called a *semi-direct product* of $\mathfrak{G}$ by $\mathfrak{A}$ with respect to φ. It will be seen to be a solution of the normal extension problem for $\mathfrak{G}$.

Proof of the theorem. The associative law is readily verified for the multiplication of the couples as defined by (1):

$$((a, x)(b, y))(c, z) = (ab, x\,\alpha(y))(c, z) = (abc, x\cdot\alpha(y)\cdot\alpha\beta(z))$$

$$(a, x)((b, y)(c, z)) = (a, x)(bc, y\cdot\beta(z)) = (abc, x\cdot\alpha(y\cdot\beta(z))).$$

The couple (i, e) represents the right unit element of $(\mathfrak{A}, \mathfrak{G})$ as $(a, x)(i, e) = (ai, x\,\alpha(e)) = (a, x)$, and $(a^{-1}, \alpha^{-1}(x^{-1}))$ is the right inverse of (a, x); indeed

$$(a, x)(a^{-1}, \alpha^{-1}(x^{-1})) = (i, x\,\alpha(\alpha^{-1}(x^{-1}))) = (i, e).$$

Thus Theorem 1 is proved.

The direct product appears as a special case of a semi-direct product $(\mathfrak{A}, \mathfrak{G})$, namely the case in which $A = \{\epsilon\}$. The multiplication rule (1) then is $(a, x)(b, y) = (ab, xy)$, i.e. the rule of the direct product.

At the other end of the scale we have the holomorph $\mathfrak{H}$ of $\mathfrak{G}$ which is also a semi-direct product. Take $\mathfrak{A}$ to be isomorphic—or equal—to the automorphism group $\Gamma = \Gamma(\mathfrak{G})$. In this case we have a one-one correspondence between the elements (α, x) of $(\Gamma, \mathfrak{G})$ and the permutations

$$\begin{pmatrix} t \\ xt \end{pmatrix}\begin{pmatrix} t \\ \alpha(t) \end{pmatrix} = \begin{pmatrix} t \\ x\alpha(t) \end{pmatrix},$$

the elements of $\mathfrak{H}^*$, and this correspondence is indeed an isomorphism. We need only compare the multiplication rule (1) in $(\Gamma, \mathfrak{G})$:

$$(\alpha, x)(\beta, y) = (\alpha\beta, x\alpha(y))$$

with the composition rule in $\mathfrak{H}^*$:

$$\begin{pmatrix} t \\ x\alpha(t) \end{pmatrix}\begin{pmatrix} t \\ y\beta(t) \end{pmatrix} = \begin{pmatrix} t \\ x\alpha(y)\alpha\beta(t) \end{pmatrix}.$$

Thus we may represent the holomorph $\mathfrak{H}$ as the semi-direct product $(\Gamma, \mathfrak{G})$.

b. A semi-direct product as defined above is an "external product" in the sense of Chap. II, §5, *e*. To it corresponds an internal semi-direct product whose factors are subgroups of $(\mathfrak{A}, \mathfrak{G})$ isomorphic to the given groups $\mathfrak{A}$ and $\mathfrak{G}$ respectively. This will become evident now as we establish those properties which a semi-direct product has in common with the direct product.

(i) A semi-direct product $\mathfrak{L} = (\mathfrak{A}, \mathfrak{G})$ contains the elements (i, x), $x \in \mathfrak{G}$; they form a normal subgroup $(i, \mathfrak{G}) \lhd \mathfrak{L}$ which through the mapping $x \to (i, x)$ is isomorphic to $\mathfrak{G}$.

Indeed by (1) $(i, x)(i, y) = (i, xy)$. Moreover transformation of (i, x) with any element $(b, y) \in \mathfrak{L}$ yields another element of $(i, \mathfrak{G})$:

$$\begin{aligned}(b, y)(i, x)(b^{-1}, \beta^{-1}(y^{-1})) &= (b, y\,\beta(x))(b^{-1}, \beta^{-1}(y^{-1})) \\ &= (i, y\,\beta(x)y^{-1}).\end{aligned}$$

(ii) The elements (a, e) form a subgroup (in general not normal) in $\mathfrak{L}$, denoted by $(\mathfrak{A}, e)$ which is isomorphic to $\mathfrak{A}$.

Proof. By (1) $(a, e)(b, e) = (ab, e)$ and $(b, y)(a, e)(b, y)^{-1} = (ba, y)(b^{-1}, \beta^{-1}(y^{-1})) = (bab^{-1}, y\beta\alpha\beta^{-1}(y^{-1}))$, which need not be an element of $(\mathfrak{A}, e)$.

(iii) The external semi-direct product $(\mathfrak{A}, \mathfrak{G})_\varphi$ equals the product $(i, \mathfrak{G})(\mathfrak{A}, e) = (\mathfrak{A}, e)(i, \mathfrak{G})$ and can in this way be considered as the internal semi-direct product of $(i, \mathfrak{G})$ by $(\mathfrak{A}, e)$ with respect to φ.

Indeed $(a, x) = (i, x)(a, e)$, $a \in \mathfrak{A}$, $x \in \mathfrak{G}$.

(iv) The intersection $(\mathfrak{A}, e) \cap (i, \mathfrak{G}) = (i, e)$, the unit element of $\mathfrak{L} = (\mathfrak{A}, \mathfrak{G})$ (cf. ex. 6).

(v) The factor group $(\mathfrak{A}, \mathfrak{G})/(i, \mathfrak{G})$—briefly $\mathfrak{L}/\mathfrak{G}$—is isomorphic to the given group $\mathfrak{A}$.

Its elements are the cosets $(i, \mathfrak{G})(a, x) = (a, \mathfrak{G}x) = (a, \mathfrak{G}) = (i, \mathfrak{G})(a, e)$. They multiply according to the rule

$$(i, \mathfrak{G})(a, e)\cdot(i, \mathfrak{G})(b, e) = (i, \mathfrak{G})(ab, e),$$

which is the rule in $(\mathfrak{A}, e)$, by (ii) isomorphic to $\mathfrak{A}$.

An external direct product $(\mathfrak{A}, \mathfrak{G})$ can be formed for two arbitrary groups $\mathfrak{A}$ and $\mathfrak{G}$. To be distinct from the direct product for a semi-direct product it is required that there is an epimorphism $\varphi : \mathfrak{A} \to \mathrm{A} \leqslant \Gamma(\mathfrak{G})$ where $\mathrm{A} > \epsilon$. There may be more than one epimorphism of $\mathfrak{A}$ onto one and the same subgroup A, as well as epimorphisms of $\mathfrak{A}$ onto other subgroups of $\Gamma(\mathfrak{G})$. Thus there may be several different semi-direct product solutions to the extension problem.

c. Are there other solutions? We shall not be able to arrive at a general decision on this question; but we shall discuss a construction which in the case of two given groups $\mathfrak{G}$ and $\mathfrak{A}$ might lead to another solution of the normal extension problem, namely a group $\mathfrak{P}$ such that $\mathfrak{G} \lhd \mathfrak{P}$ and $\mathfrak{P}/\mathfrak{G} \simeq \mathfrak{A}$.

First let us assume that $\mathfrak{P}$ is a normal extension of the given group $\mathfrak{G}$ by the given group $\mathfrak{A}$. We will deduce some conditions that $\mathfrak{P}$ must satisfy. Denote by f^* one of the isomorphisms $\mathfrak{A} \to \mathfrak{P}/\mathfrak{G}$ and by **T** a transversal of the cosets of $\mathfrak{G}$ in $\mathfrak{P}$ (cf. Chap. II, §2, *a.*). If u, v are typical elements* of **T** let

$$f^*(i) = \mathfrak{G}, \quad f^*(a) = \mathfrak{G}u, \quad f^*(b) = \mathfrak{G}v, \quad a, b \in \mathfrak{A},$$

so that

$$f^*(ab) = f^*(a)f^*(b) = \mathfrak{G}u\mathfrak{G}v = \mathfrak{G}uv.$$

* We point out that in general the elements of **T** do not form a group. Cf. ex. 3.

The isomorphism f^* can be represented uniquely by an invertible function f which is defined in the group $\mathfrak{A}$ and whose values are elements of the transversal **T**, namely $f(i) = t \in \mathfrak{G}$ (taking $f(i) = e$ would simplify some of the subsequent relations, but might introduce a restriction), $f(a) = u$, $f(b) = v$. Briefly we can write $f(\mathfrak{A}) = \mathbf{T}$. Evidently $f(ab)$ represents the element of **T** which defines the coset $\mathfrak{G}uv$; thus for every pair of elements a, b in $\mathfrak{A}$ there is an element $g(a, b)$ in $\mathfrak{G}$ for which

(2) $$uv = f(a)f(b) = g(a, b)f(ab)$$

whence

(3) $$g(a, b) = f(a)f(b)f(ab)^{-1}.$$

This function of the two variables a and b in $\mathfrak{A}$, whose values are elements of $\mathfrak{G}$, is called the *factor system* of the extension $\mathfrak{P}$ of $\mathfrak{G}$ by $\mathfrak{A}$. The function $g(a, b)$ is determined by the isomorphism $f^* : \mathfrak{A} \to \mathfrak{P}/\mathfrak{G}$ and by the transversal **T** of $\mathfrak{G}$ in $\mathfrak{P}$, since f^* and **T** define the mapping $f : \mathfrak{A} \to \mathbf{T}$.

For $g(a, b)$ it is now easy to formally verify the relation

$$g(a, b)g(ab, c) = g(b, c)^{f(a)}g(a, bc)$$

by expressing each of the g-factors in terms of f-factors according to (3).

The right side of the last equation involves an inner automorphism of $\mathfrak{P}$, realized by transformation by $f(a)$. This inner automorphism of $\mathfrak{P}$ induces an automorphism of $\mathfrak{G}$ which we shall denote by α:

$$\alpha(x) = f(a)xf(a)^{-1} = x^{f(a)};$$

similarly

$$\beta(x) = f(b)xf(b)^{-1} = x^{f(b)}, \text{ etc.}$$

The functional equation for $g(a, b)$ thus assumes the form

(4) $$g(a, b)g(ab, c) = \alpha(g(b, c))\, g(a, bc).$$

Since

$$\mathfrak{P} = \bigcup_{u \in \mathbf{T}} \mathfrak{G}u = \mathfrak{G} \cdot \mathbf{T}$$

any two elements of $\mathfrak{P}$ appear in the form of products xu, yv for some x, y in $\mathfrak{G}$ and u, v in $\mathbf{T}$. The multiplication formula for two such products follows from the trivial identity $xu \cdot yv = xuyu^{-1} \cdot uv$; putting $u = f(a)$, $v = f(b)$ we have by (3)

$$xf(a) \cdot yf(b) = xf(a)yf(a)^{-1} \cdot f(a)f(b)$$

$$= x\,\alpha(y)g(a, b) \cdot f(ab). \tag{5}$$

Now let $a = i$ in (2). Thus $f(i) = g(i, b)$ for all b in $\mathfrak{A}$ and in particular $f(i) = g(i, i)$ so that the unit element of $\mathfrak{P}$ appears in the form

$$e = g(i, i)^{-1}f(i).$$

Taking this into account it is readily verified that

$$(xf(a))^{-1} = \alpha^{-1}(x^{-1}g(i, i)^{-1}g(a, a^{-1})^{-1}) \cdot f(a^{-1}) \tag{6}$$

is the product representation of the inverse of $xf(a)$. [Simply multiply (6) from the left by $xf(a)$ making use of (5).] For $x = e$ it follows that

$$f(a)^{-1} = \alpha^{-1}(g(i, i)^{-1}g(a, a^{-1})^{-1})f(a^{-1}). \tag{6'}$$

The product of two of the automorphisms α, β can be expressed in terms of the functions $f(a)$ and $g(a, b)$. Let π denote the automorphism associated with the product ab, that is

$$\pi(x) = f(ab)xf(ab)^{-1} = x^{f(ab)}.$$

Thus

$$\alpha\beta(x) = \alpha(\beta(x)) = x^{f(a)f(b)} = x^{g(a,b)f(ab)}$$

whence

$$\alpha\beta(x) = \pi(x)^{g(a,b)},$$

a relation involving only the automorphisms and the factor system.

d. Construction of a normal extension. Based on the preceding analysis of an extension $\mathfrak{P}$ of $\mathfrak{G}$ by $\mathfrak{A}$ we shall now demonstrate how for two given groups $\mathfrak{G}$ and $\mathfrak{A}$ a solution of the normal extension problem can be constructed if we assume the existence of a function $g(a, b)$ of the two variables a, b in $\mathfrak{A}$ with values $g(a, b) \in \mathfrak{G}$, and the existence of a mapping $\sigma : \mathfrak{A} \to \Gamma(\mathfrak{G})$ (we shall write $\sigma(a) = \alpha$, $\sigma(b) = \beta$, $\sigma(ab) = \pi$, $a, b \in \mathfrak{A}$) for which the following conditions are satisfied:

$$(7) \qquad g(a, b)g(ab, c) = \alpha(g(b, c))g(a, bc), \quad a, b, c \in \mathfrak{A}$$

$$(7') \qquad \alpha(\beta(x)) = \pi(x)^{g(a,b)}, \quad x \in \mathfrak{G}.$$

Further let $\mathbf{T}$ denote a set of symbols which has with $\mathfrak{G}$ exactly one element in common: $\mathfrak{G} \cap \mathbf{T} = g(i, i)$. Moreover we assume the existence of a one-one correspondence

$$f : \mathfrak{A} \to \mathbf{T}, \quad f(i) = g(i, i).$$

We consider the set $\mathbf{P}$ of all formal or symbolic products

$$\mathbf{P} = \{xf(a) \mid x \in \mathfrak{G}, a \in \mathfrak{A}\},$$

"formal" in the sense that the multiplication of an element x of $\mathfrak{G}$ by an element $f(a)$ of $\mathbf{T}$ is not yet defined if $a \neq i$. Then we can define a composition law in $\mathbf{P}$ as follows

$$(7'') \qquad xf(a) \cdot yf(b) = x\,\alpha(y)g(a, b)f(ab)$$

for all a, b in $\mathfrak{A}$ and x, y in $\mathfrak{G}$, a formula clearly suggested by the relation (5).

We shall show that the set $\mathbf{P}$ of the symbolic products is a group with respect to the law (7″). For this purpose we have to verify the associativity, the existence of a right unit element and of a right inverse in $\mathbf{P}$.

Evidently by (7″) the product of two symbolic products $xf(a)$ and $yf(b)$ is a symbolic product. By repeated application of (7″) and (7′) we find

$$\begin{aligned}(xf(a)\cdot yf(b))\cdot zf(c) &= (x\,\alpha(y)\,g(a,b)\,f(ab))\cdot z\,f(c)\\ &= x\,\alpha(y)\,g(a,b)\,\pi(z)\,g(ab,c)\,f(abc)\\ &= x\,\alpha(y)\,\pi(z)^{g(a,b)}\underline{g(a,b)g(ab,c)}\,f(abc)\end{aligned}$$

$$\begin{aligned}xf(a)\cdot(yf(b)\cdot zf(c)) &= x\,f(a)\cdot(y\,\beta(z)\,g(b,c)\,f(bc))\\ &= x\,\alpha(y)\,\alpha\beta(z)\,\alpha(g(b,c))\,g(a,bc)\,f(abc)\\ &= x\,\alpha(y)\,\pi(z)^{g(a,b)}\underline{\alpha(g(b,c))g(a,bc)}\,f(abc).\end{aligned}$$

With regard to (7) these two expressions are equal. Thus the symbolic products multiply associatively.

As to the unit element in **P** we recall that $f(i) = g(i,i)$. This suggests to try the unit element of $\mathfrak{G}$, namely

$$e = g(i,i)^{-1}f(i)$$

as unit element of **P**. Indeed, putting $b = c = i$ in (7) we have $g(a,i)\,g(a,i) = \alpha(g(i,i))\,g(a,i)$ whence $\alpha(g(i,i)) = g(a,i)$; therefore

$$\begin{aligned}x\,f(a)\cdot g(i,i)^{-1}f(i) &= x\,\alpha(g(i,i)^{-1})g(a,i)f(a)\\ &= x\,g(a,i)^{-1}g(a,i)f(a) = x\,f(a).\end{aligned}$$

Finally it is easy to verify that

$$(x\,f(a))^{-1} = \alpha^{-1}(x^{-1}g(i,i)^{-1}g(a,a^{-1})^{-1})f(a^{-1}).$$

Thus **P** is in fact a group $\mathfrak{P}$.

e. It remains to verify that the group $\mathfrak{P}$ actually solves the normal extension problem. The elements $x = x\,g(i,i)^{-1}f(i)$ form a subgroup $\mathfrak{G}$ of $\mathfrak{P}$:

$$xy = x\,g(i,i)^{-1}f(i)\cdot y\,g(i,i)^{-1}f(i) = xy\,g(i,i)^{-1}f(i)$$

and also $\mathfrak{G} \lhd \mathfrak{P}$:

$$\begin{aligned} &xf(a)\cdot y\cdot (xf(a))^{-1} \\ &= xf(a)\cdot yg(i,i)^{-1}f(i)\cdot \alpha^{-1}(x^{-1}g(i,i)^{-1}g(a,a^{-1})^{-1})f(a^{-1}) \\ &= x\,\alpha(y)\,\alpha(g(i,i)^{-1})g(a,i)f(a)\cdot \alpha^{-1}(\ldots)f(a^{-1}) \\ &= x\,\alpha(y)x^{-1}\cdot g(i,i)^{-1}g(a,a^{-1})^{-1}\,g(a,a^{-1})\,f(i) = x\,\alpha(y)x^{-1}. \end{aligned}$$

Now we consider the factor group $\mathfrak{P}/\mathfrak{G}$. Its elements are the cosets

$$\begin{aligned} \mathfrak{G}xf(a) &= \{y\,g(i,i)^{-1}f(i)\cdot xf(a) \mid y \in \mathfrak{G},\ a \text{ fixed in } \mathfrak{A},\ x \text{ fixed in } \mathfrak{G}\} \\ &= \{y\,g(i,i)^{-1}x^{g(i,i)}g(i,a)f(a) \mid \ldots\} \\ &= \{yx\,g(i,i)^{-1}g(i,a)f(a) \mid \ldots\} = \mathfrak{G}f(a) \end{aligned}$$

which multiply according to the rule

$$\mathfrak{G}f(a)\cdot \mathfrak{G}f(b) = \mathfrak{G}g(a,b)f(ab) = \mathfrak{G}f(ab).$$

Thus the mapping $\omega : a \to \mathfrak{G}f(a)$, i.e. $\mathfrak{A} \to \mathfrak{P}/\mathfrak{G}$ is a homomorphism. But $\mathfrak{G}f(a) = \mathfrak{G}f(b)$ implies by (6′) that

$$f(b)f(a^{-1}) = \alpha^{-1}(g(i,i)^{-1}g(b,a^{-1}))f(ba^{-1}) \in \mathfrak{G}$$

i.e. $f(ba^{-1}) \in \mathfrak{G}$ whence $b = a$. Thus the two cosets $\mathfrak{G}f(a)$ and $\mathfrak{G}f(b)$ are distinct so that ω is an isomorphism: $\mathfrak{P}/\mathfrak{G} \simeq \mathfrak{A}$.

The theory of sections *c.–e.* is due to O. Schreier (1926).

Examples and exercises

1. (a) Rewrite the definition of the semi-direct product $(\mathfrak{A}, \mathfrak{G})_\varphi$ assuming that the group $\mathfrak{G}$ is additive, the group $\mathfrak{A}$ is multiplicative.

(b) Show that the group $\mathfrak{G}_{\mathbb{F}}$ (cf. Chap. II, §3, ex. 2) is isomorphic to a semi-direct product $(\mathfrak{A}, \mathfrak{G})$ where $\mathfrak{G} = \mathbb{F}^+$ is the addition group of the field $\mathbb{F}$ and $\mathfrak{A} = \mathfrak{F}$ the multiplicative group of $\mathbb{F}$.

2. Is the alternating group $\mathfrak{A}_4$ isomorphic to a semi-direct product of the four group $\mathfrak{V}$ and the alternating group $\mathfrak{A}_3$?

3. An extension $\mathfrak{P}$ of a group $\mathfrak{G}$ is not always a semi-direct product of $\mathfrak{G}$ by a group $\mathfrak{A}$. Choose for $\mathfrak{P}$ the cyclic group of order 4:

$$\mathfrak{P} = \langle x \rangle_4, \quad \mathfrak{G} = \langle x^2 \rangle_2, \quad \mathfrak{A} = \langle a \rangle_2, \quad (x^4 = e, a^2 = i).$$

There are four possible transversals for $\mathfrak{G}$ in $\mathfrak{P}$:

T:	$\{e, x\}$	$\{e, x^3\}$	$\{x^2, x\}$	$\{x^2, x^3\}$
$f(i)$:	e	e	x^2	x^2
$f(a)$:	x	x^3	x	x^3
$g(i, i)$:	e	e	x^2	x^2
$g(a, i)$:	e	e	x^2	x^2
$g(i, a)$:	e	e	x^2	x^2
$g(a, a)$:	x^2	x^2	x^2	x^2

4. Which are the functions $f(a)$, $g(a, b)$, and the automorphisms $\alpha, \beta, \ldots$ if in ex. 3 the extension of $\mathfrak{G}$ by $\mathfrak{A}$ is to be a semi-direct product $(\mathfrak{A}, \mathfrak{G})$?

5. The dihedral group $\mathfrak{D}_n$ is isomorphic to a semi-direct product of a cyclic group $\langle x \rangle_n$ by a cyclic group $\langle a \rangle_2$.

6. Let $\mathfrak{L}$ be a group and $\mathfrak{G} \lhd \mathfrak{L}$. Then $\mathfrak{L}$ is said *to split over* $\mathfrak{G}$ if there exists a subgroup $\mathfrak{A} < \mathfrak{L}$ such that

$$\mathfrak{A} \cap \mathfrak{G} = \mathfrak{e} \quad \text{and} \quad \mathfrak{L} = \mathfrak{A} \cdot \mathfrak{G}.$$

The subgroup $\mathfrak{A}$ is called a *complement* to $\mathfrak{G}$ in $\mathfrak{L}$.

(a) If $\mathfrak{L}$ is a semi-direct product of two of its subgroups, $\mathfrak{G} \lhd \mathfrak{L}$ and $\mathfrak{H} < \mathfrak{L}$, then $\mathfrak{L}$ splits over $\mathfrak{G}$. Hence the semi-direct product is also called a *splitting extension* of $\mathfrak{G}$.

(b) If $\mathfrak{L}$ splits over one of its normal subgroups $\mathfrak{G}$, is it then a splitting extension of $\mathfrak{G}$?

(c) If $\mathfrak{L}$ splits over one of its normal subgroups $\mathfrak{G}$ and $\mathfrak{A}$ is the complement of $\mathfrak{G}$ in $\mathfrak{L}$, then the elements of $\mathfrak{A}$ represent a transversal of $\mathfrak{G}$ in $\mathfrak{L}$. (Do the elements of $\mathfrak{G}$ form a transversal of $\mathfrak{A}$ in $\mathfrak{L}$?)

§4 A problem of Burnside

After the preceding example of a general theory we proceed to deal with a special question: *Are there finite groups where outer automorphisms leave the conjugacy classes $\mathscr{C}(x)$ invariant*? This question was set and decided by W. Burnside in 1913; he established a family of groups of the order p^6 (p is an odd prime number) which have outer automorphisms leaving the classes $\mathscr{C}(x)$ invariant. The smallest group in this family has the order $3^6 = 729$. We shall show that another family of groups, of the orders $8n \cdot \varphi(8n)$ ($n = 1, 2, \ldots$), has the property in question.

The problem can be formulated as follows: Given a group $\mathfrak{G}$; let $\mathrm{A} \leqslant \Gamma(\mathfrak{G})$, $\mathrm{A} = \{\alpha \in \Gamma(\mathfrak{G}) \mid \alpha(\mathscr{C}(x)) = \mathscr{C}(x) \text{ for all } x \in \mathfrak{G}\}$. Clearly $\Delta(\mathfrak{G}) \leqslant \mathrm{A}$. Are there groups $\mathfrak{G}$ for which $\Delta < \mathrm{A}$? We shall call such a group a "($\Delta < \mathrm{A}$)-group".

THEOREM 1. (G. E. Wall, 1947). *Let $\mathfrak{L}_m$ be the group of all mappings $\xi \to a\xi + x$, $a \in \mathfrak{R}_m$, $x \in \mathbb{R}_m{}^+$* (cf. Chap. I, §3, ex. 4). *If $m = 8n$ ($n = 1, 2, \ldots$) the group $\mathfrak{L}_m$ is a ($\Delta < \mathrm{A}$)-group.*

a. Preliminaries. The group $\mathfrak{L}_m$ ($m > 2$) is isomorphic to the holomorph of the additive group $\mathbb{R}_m{}^+$ of the residues modulo m (cf. §2, ex. 4) which we represent by the numbers $0, 1, \ldots, m-1$ (instead of $\bar{0}, \bar{1}, \ldots, \overline{m-1}$). Thus $\mathfrak{L}_m$ also equals the semi-direct product $(\mathfrak{R}_m, \mathbb{R}_m{}^+)$ [cf. §3, ex. 1(b)] where $\mathfrak{R}_m$ represents the multiplicative group of the residues $a \pmod{m}$, $(a, m) = 1$.

As in the simplest case $\mathfrak{L}_p = \mathfrak{G}_{\mathbb{R}_p}$ (cf. §1, ex. 6(a); §2, ex. 4(a)) the elements of $\mathfrak{L}_m$ may be written as couples $A = (a, x)$, $B = (b, y)$, $\ldots$ where $a, b, \ldots \in \mathfrak{R}_m$; $x, y, \ldots \in \mathbb{R}_m{}^+$. The couples $(1, x)$ form a normal subgroup $\mathfrak{H} = (1, \mathbb{R}_m{}^+)$ in $\mathfrak{L}_m$; this can be shown as in Chap. II, §3, ex. 2.

Further it is easily seen that $\mathfrak{H}$ contains the commutator subgroup of $\mathfrak{L}_m$ (cf. §1, ex. 15 and §4, ex. 1).

We also note that if $A \notin \mathfrak{H}$ the conjugacy class $\mathscr{C}(A)$ in $\mathfrak{L}_m$ is a subset of the coset $\mathfrak{H}A$ (cf. Chap. II, §4, ex. 5). Indeed

$$\mathfrak{H}A = \{TA \mid T = (1, t) \in \mathfrak{H}\} = \{(a, x+t) \mid t \in \mathbb{R}_m{}^+\};$$
$$\mathscr{C}(A) = \{BAB^{-1} \mid B \in \mathfrak{L}_m\} = \{(ba, bx+y)(b^{-1}, -b^{-1}y)\}$$
$$= \{(a, (1-a)y + bx)\}.$$

For $\mathfrak{H}A$ and $\mathscr{C}(A)$ to be equal it must be possible for every $a \in \mathfrak{R}_m$, $x, s \in \mathbb{R}_m{}^+$ to solve the congruence

$$(1) \qquad (1-a)y+bx \equiv s+x, \text{ i.e. } (1-a)y-(1-b)x \equiv s \pmod{m}$$

in the unknowns $b, y, (b, m) = 1$.

Suppose that $m = 2n$; then a and b are odd, $1-a$ and $1-b$ even; therefore (1) cannot have a solution if s is odd. Thus $\mathscr{C}(A) \subset \mathfrak{H}A$.

b. LEMMA 1. *Let* $m = 4n$. *The elements of* $\mathfrak{R}_m$ *are then the residues represented by*

$$a = 2\lambda+1, \quad 0 \leqslant \lambda < 2n, \quad (a, n) = 1.$$

Those with even $\lambda = 2\mu$ *form a subgroup* $\mathfrak{R}_m{}'$ *of index* 2 *in* $\mathfrak{R}_m$.

Proof. Represent the $\varphi(m)$ residues by

$$a = 2\lambda+1, \quad -n \leqslant \lambda < n, \quad (a, n) = 1,$$

so that if $a \in \mathfrak{R}_m$ then also $-a \in \mathfrak{R}_m$. Let $\lambda \geqslant 0$ and λ even (odd). Since $-(2\lambda+1) = 2\lambda'+1$ where $\lambda' = -(\lambda+1) < 0$ and λ' odd (even), there are equal numbers of residues with even and with odd λ. Those with even λ are $\equiv 1 \pmod 4$; they form a subgroup $\mathfrak{R}_m{}'$ of index 2 in $\mathfrak{R}_m$.

LEMMA 2. *Let* $m = 8n$; *the elements of* $\mathfrak{R}_m$ *appear in the form* $a = 4\lambda \pm 1, 0 \leqslant \lambda \leqslant 2n$ (*-1 and $4 \cdot 2n+1$ excluded*), $(a, n) = 1$. *Those with even* λ (*i.e. the residues* $\equiv \pm 1 \pmod 8$)) *form a subgroup* $\mathfrak{R}_m{}''$ *of index* 2 *in* $\mathfrak{R}_m$.

Proof. Clearly all the residues $\equiv \pm 1 \pmod 8$ form a proper subgroup $\mathfrak{R}_m{}''$ in $\mathfrak{R}_m$. The product of two residues $\equiv \pm 3 \pmod 8$ (which are not contained in $\mathfrak{R}_m{}''$) equals a residue $\equiv \pm 1 \pmod 8$, i.e. an element of $\mathfrak{R}_m{}''$. Thus by Chap. II, §2, ex. 8 we have $(\mathfrak{R}_m : \mathfrak{R}_m{}'') = 2$.

Remark. Because not every $4\mu+1 \equiv 8\lambda \pm 1$ the two subgroups $\mathfrak{R}_m{}'$ and $\mathfrak{R}_m{}''$ are distinct. Cf. ex. 2.

Now let $\mathfrak{G}$ be an arbitrary subgroup of index 2 in $\mathfrak{R}_m$, $m = 8n$. In $\mathfrak{R}_m$ we define a function $\delta = \delta_{\mathfrak{G}} : \mathfrak{R}_m \to \mathbb{R}_2{}^+$, namely

$$\delta(a) = \delta_{\mathfrak{G}}(a) \equiv \begin{cases} 0 \pmod 2 & \text{if } \quad a \in \mathfrak{G} \\ 1 \pmod 2 & \text{if } \quad a \notin \mathfrak{G}, \end{cases}$$

which may be called *the characteristic function of the subgroup* $\mathfrak{G}$. It represents a homomorphism of $\mathfrak{G}$ onto the additive residue class group (mod 2):

$$\delta(ab) \equiv \delta(a)+\delta(b) \pmod 2.$$

Since all residues in $\mathfrak{R}_m$ are odd, this congruence is equivalent to

$$(2) \qquad \delta(ab) \equiv \delta(a)+a\delta(b) \pmod 2.$$

For $\mathfrak{G} = \mathfrak{R}_m'$ and $a = 2\lambda+1$, one has

$$(3) \qquad \delta(a) = \delta_{\mathfrak{R}_m'}(a) \equiv \lambda = \frac{a-1}{2} \pmod 2.$$

c. For a subgroup $\mathfrak{G} < \mathfrak{R}_m$, $(\mathfrak{R}_m : \mathfrak{G}) = 2$, $m = 8n$, we introduce the mapping

$$\omega = \omega_{\mathfrak{G}} : \mathfrak{L}_m \to \mathfrak{L}_m,\ (a, x) \to \omega(a, x) = \left(a, x+\frac{m}{2}\,\delta_{\mathfrak{G}}(a)\right).$$

This mapping is involutory: $\omega^2 = \epsilon$; indeed (dropping the subscript $\mathfrak{G}$):

$$\omega(\omega(a, x)) = \omega\left(a, x+\frac{m}{2}\cdot\delta(a)\right) = (a, x+m\cdot\delta(a)) = (a, x).$$

Hence ω is invertible and $\omega^{-1} = \omega$.

Further ω is an *automorphism* of $\mathfrak{L}_m$:

$$\omega((a, x)(b, y)) = \omega(ab, ay+x) = \left(ab, ay+x+\frac{m}{2}\,\delta(ab)\right)$$

and

$$\omega(a, x)\,\omega(b, y) = \left(a, x+\frac{m}{2}\,\delta(a)\right)\left(b, y+\frac{m}{2}\,\delta(b)\right)$$
$$= \left(ab, x+ay+\frac{m}{2}\,(\delta(a)+a\,\delta(b))\right)$$

so that by (2)

$$\omega((a, x)(b, y)) = \omega(a, x)\,\omega(b, y) \quad \text{q.e.d.}$$

For the following put $m = 2^\mu r$ assuming r to be odd and $\mu \geqslant 3$ so that at any rate 8 is a divisor of m.

LEMMA 3. *The automorphism* $\omega = \omega_{\mathfrak{G}}$ *is an inner automorphism if and only if* $\mathfrak{G} = \mathfrak{R}_m'$.

Proof. The mapping ω is an inner automorphism if and only if there exists a fixed $C = (c, z) \in \mathfrak{L}_m$ so that $\omega(A) = CAC^{-1}$ for all $A = (a, x) \in \mathfrak{L}_m$, that is

$$\left(a, x+\frac{m}{2}\delta(a)\right) = (a, (1-a)z+cx)$$

or equivalently, if and only if for all $a \in \mathfrak{R}_m$ and $x \in \mathbb{R}_m^+$ the congruence

$$(4) \qquad (1-a)z+cx \equiv x+2^{\mu-1}r\delta(a) \pmod{2^\mu r}$$

has a solution c, z.

(i) Suppose that $\omega = \omega_{\mathfrak{G}}$ is an inner automorphism of $\mathfrak{L}_m$. Put $x = 0$ in the congruence (4) and use that r is odd:

$$(5) \qquad (1-a)z \equiv 2^{\mu-1}r\ \delta(a) \pmod{2^\mu r}, \quad \text{thus also} \quad \pmod{2^\mu}.$$

Put $x = 1$ in (4) and apply (5):

$$(5') \qquad c \equiv 1 \pmod{2^\mu r}.$$

Again since r is odd: $2^{\mu-1}r \equiv 2^{\mu-1} \pmod{2^\mu}$; hence by (5)

$$(6) \qquad (1-a)z \equiv 2^{\mu-1}\ \delta(a) \pmod{2^\mu}.$$

In order to determine z let $a = 4r-1$ which is an element of $\mathfrak{R}_m$ [because $(4r-1, 2^\mu r) = 1$], but not an element of $\mathfrak{R}_m'$ [because $4r-1 = 2(2r-1)+1$]. Thus by (6)

$$2(1-2r)z \equiv 2^{\mu-1}\ \delta(4r-1) \pmod{2^\mu}.$$

This implies that $2^{\mu-2} \mid z$; hence there is an integer z_1, for which

$$z = 2^{\mu-2}z_1.$$

We substitute this result into (6) and obtain

$$(1-a)2^{\mu-2}\, z_1 \equiv \frac{1-a}{2} 2^{\mu-1} z_1 \equiv 2^{\mu-1}\, \delta(a) \quad (\text{mod } 2^{\mu})$$

whence

$$(7) \qquad \delta(a) \equiv \frac{1-a}{2} z_1 \quad (\text{mod } 2),$$

so that for an arbitrary odd z_1 the function $\delta_{\mathfrak{G}}(a)$ becomes the characteristic function of the subgroup $\mathfrak{G} = \mathfrak{R}_m'$ as given by (3). (For an even z_1 we have $\delta_{\mathfrak{G}}(a) \equiv 0$ for all a which contradicts our assumption.) Thus if $\omega = \omega_{\mathfrak{G}}$ is an inner automorphism of $\mathfrak{L}_m$ then $\mathfrak{G} = \mathfrak{R}_m'$.

(ii) Conversely, if $\mathfrak{G} = \mathfrak{R}_m'$, i.e. $\delta_{\mathfrak{G}}(a) \equiv (a-1)/2$ (mod 2), then (7) is valid for every odd z_1 so that for $z = 2^{\mu-2} z_1\, r$ the conditions (6) and (5) are satisfied; combining this with (5′) we obtain (4) for all x in $\mathbb{R}_m^+$. Thus $\omega(A) = CAC^{-1}$, $C = (1, 2^{\mu-2} z_1 r)$.

d. Now we come to the final step in the proof of the theorem.

LEMMA 4. *The automorphism* $\omega = \omega_{\mathfrak{G}}$ *leaves the classes* $\mathscr{C}(A)$, $A \in \mathfrak{L}_m$, *invariant if* $\mathfrak{G} = \mathfrak{R}_m''$.

Proof. Let $\mathfrak{G}$ denote again an arbitrary subgroup of index 2 in $\mathfrak{L}_m$. As before we shall write δ and ω instead of $\delta_{\mathfrak{G}}$ and $\omega_{\mathfrak{G}}$.

For a fixed $A = (a, x)$ in $\mathfrak{L}_m$ we form the subclass

$$\begin{aligned} \mathscr{C}'(A) &= \{TAT^{-1} \mid T = (1, t) \in \mathfrak{H}\} \quad \text{(cf. subsection } a.) \\ &= \{[a, (1-a)t + x] \mid t \in \mathbb{R}_m^+\} \end{aligned}$$

which includes at most m distinct elements. Let $A_2 \in \mathscr{C}(A)$, but $A_2 \notin \mathscr{C}'(A)$. Then the subclass $\mathscr{C}'(A_2) = \{TA_2T^{-1} \mid T \in \mathfrak{H}\} \subset \mathscr{C}(A)$ and $\mathscr{C}'(A) \cap \mathscr{C}'(A_2) = 0$. Maybe $\mathscr{C}(A) = \mathscr{C}'(A) \cup \mathscr{C}'(A_2)$; if not, there is an A_3 in $\mathscr{C}(A)$, $A_3 \notin \mathscr{C}'(A) \cup \mathscr{C}'(A_2)$ so that the subclass $\mathscr{C}'(A_3)$ is disjoint from the other two subclasses. Eventually we have

$$\mathscr{C}(A) = \mathscr{C}'(A) \cup \mathscr{C}'(A_2) \cup \ldots \cup \mathscr{C}'(A_k),$$

a union of k disjoint subclasses.

From the definition of ω it is clear that if $a \in \mathfrak{G}$ then $\omega(A) = A$. Thus we need only consider the case that $a \notin \mathfrak{G}$. In subsection *a.* it has been established that $\mathscr{C}(A) \subseteq \mathfrak{H}A$; hence all the elements of $\mathscr{C}(A)$ [including of course those of $\mathscr{C}'(A)$] have a unique representation of the form (a, x) with a fixed a in $\mathfrak{R}_m$ and $x \in \mathbb{R}_m{}^+$. Evidently

$$\omega(A) = \left(a, x+\frac{m}{2}\delta(a)\right) \in \mathfrak{H}A.$$

Now let d be the g.c.d. of $1-a$ and m. Then $2 \mid d$ and for appropriate integers ξ and η we have

$$d = (1-a)\xi + m\eta.$$

Consequently

$$d \equiv (1-a)\xi \pmod{m}.$$

We consider the two possibilities

(i) $d \mid m/2$; it will be shown that then there is an element $T = (1, t) \in \mathfrak{H}$ for which $\omega(A) = TAT^{-1}$ and therefore $\omega(A) \in \mathscr{C}'(A)$.

(ii) $d \nmid m/2$; in this case $a \in \mathfrak{R}_m{}''$.

Assuming that these two facts have been established, we see that the automorphism $\omega = \omega_{\mathfrak{G}}$, $\mathfrak{G} = \mathfrak{R}_m{}''$, leaves either $\mathscr{C}'(A)$ invariant or A itself fixed. Hence for $\mathfrak{G} = \mathfrak{R}_m{}''$ the automorphism $\omega_{\mathfrak{G}}$ maps every conjugacy class $\mathscr{C}(A)$, $A \in \mathfrak{L}_m$, onto itself and, by Lemma 3 and the Remark following Lemma 2, $\omega_{\mathfrak{G}}$ is not an inner automorphism of $\mathfrak{L}_m$.

We now verify the two statements (i) and (ii).

(i) By supposition $m/2d$ is an integer. If $t = m\xi/2d$ then $T = (1, t)$ has the required property:

$$TAT^{-1} = (a, (1-a)t + x) = \left(a, \frac{dm}{2d} + x\right) = \left(a, \frac{m}{2} + x\right)$$

and since for $a \notin \mathfrak{G}$ we have $(m/2)\,\delta_{\mathfrak{G}}(a) \equiv m/2 \pmod{m}$, we conclude that $TAT^{-1} = \omega(A)$.

(ii) Since $d \nmid m/2$, but $d \mid m$, it follows that $d = 2^{\mu}r'$ and $r' \mid r$. Because $d \mid (a-1)$ and $a-1 < m$ we have $r' < r$. Let $a-1 = qd$. Then $a = 1+qd = 1+u\cdot 2^{\mu} = 1 \pm 8\lambda$ for some integer u and $\mu \geqslant 3$. Thus in fact $a \in \mathfrak{R}_m{}''$ which completes the proof.

Examples and exercises

1. Show that the subgroup $\mathfrak{H} = (1, \mathbb{R}_m{}^+) \lhd \mathfrak{L}_m$ contains the commutator subgroup of $\mathfrak{L}_m$. When does the commutator subgroup coincide with $\mathfrak{H}$?

2. Determine all subgroups of index 2 in $\mathfrak{R}_8$, $\mathfrak{R}_{16}$, $\mathfrak{R}_{40}$ and tabulate the corresponding characteristic functions δ.

3. For $\mathfrak{L}_8$ find explicitly the outer automorphisms which preserve the conjugate classes.

4. Construct a representation of $\mathfrak{L}_8$ by permutations of degree 8.

5. Show that all the automorphisms of $\mathfrak{L}_p = \mathfrak{G}_{\mathbb{R}_p}$ are inner automorphisms.

§5 Endomorphisms and operators

a. An endomorphism is, by definition, a homomorphism of a group $\mathfrak{G}$ into itself, i.e. a function α defined in $\mathfrak{G}$ with values in $\mathfrak{G}$ satisfying the functional equation

$$\alpha(xy) = \alpha(x)\alpha(y), \quad x, y \in \mathfrak{G}.$$

By Chap. II, §3, *a.* the image of $\mathfrak{G}$ under an endomorphism is a subgroup $\mathfrak{G}^* \leqslant \mathfrak{G}$. If $\mathfrak{G}^* = \mathfrak{G}$ and α is one-one, then α is an automorphism. There are, however, endomorphisms $\alpha : \mathfrak{G} \to \mathfrak{G}$ where α is not injective, cf. ex. 1.

The example of the infinite cyclic group shows that in the case of an infinite group $\mathfrak{G}$ a monomorphism $\mathfrak{G} \to \mathfrak{G}$ is not necessarily an automorphism, cf. ex. 2. An endomorphism $\mathfrak{G} \to \mathfrak{G}$ which is a monomorphism is sometimes called a *meromorphism*.

For any two endomorphisms α, β in $\mathfrak{G}$ one can form the mapping $\beta\alpha$ which is again an endomorphism in $\mathfrak{G}$. Since the composition of mappings is associative we can state that the system of all endomorphisms of a group $\mathfrak{G}$ is a *semigroup* $\mathrm{E}(\mathfrak{G})$ with respect to functional composition. This semigroup contains every automorphism of the group $\mathfrak{G}$, in particular the identity ϵ. It also contains the *trivial endomorphism* o which maps every $x \in \mathfrak{G}$ into the unit element e

(into the zero element 0 in the case of an additive group). The mapping o is also called the *null-* (or *zero-*) *endomorphism* because from $o(x) = e$ for all $x \in \mathfrak{G}$ it follows that $o(\alpha(x)) = \alpha(o(x)) = \alpha(e) = e = o(x)$. We write

$$o\alpha = \alpha o = o$$

for all endomorphisms α and note that o has the typical property of a zero element with respect to multiplication.

The elements $x \in \mathfrak{G}$ which by a fixed endomorphism α are mapped onto the unit element e form a normal subgroup of $\mathfrak{G}$, the *kernel* of the endomorphism: $\ker \alpha = \alpha^{-1}(e)$ (cf. Chap. II, §3, *b.*). Is every normal subgroup $\mathfrak{H} \lhd \mathfrak{G}$ the kernel of an endomorphism? Here is the answer to this question:

THEOREM 1. *A normal subgroup $\mathfrak{H} \lhd \mathfrak{G}$ is the kernel of an endomorphism $\mathfrak{G} \to \mathfrak{G}$ if and only if $\mathfrak{G}$ contains a subgroup $\tilde{\mathfrak{G}}$ which is isomorphic to the factor group $\mathfrak{G}/\mathfrak{H}$.*

Proof. 1. Let α be an endomorphism of $\mathfrak{G}$ and $\mathfrak{H}$ its kernel. By $f_{\mathfrak{H}}$ we denote the natural homomorphism $\mathfrak{G} \to \mathfrak{G}/\mathfrak{H}$ (cf. Chap. II, §3, *c.*). The endomorphism maps the distinct cosets $a\mathfrak{H}$ ($a \in \mathfrak{G}$) onto the distinct elements $\alpha(a\mathfrak{H}) = \alpha(a) \in \mathfrak{G}$ and these $\alpha(a)$ form a subgroup $\tilde{\mathfrak{G}} < \mathfrak{G}$. The group $\tilde{\mathfrak{G}}$ is an isomorphic image of $\mathfrak{G}/\mathfrak{H}$. Denoting this isomorphism by δ we have $\delta(f_{\mathfrak{H}}(x)) = \alpha(x)$, $x \in \mathfrak{G}$.

2. Suppose that there is an isomorphism δ of $\mathfrak{G}/\mathfrak{H}$ onto a subgroup $\tilde{\mathfrak{G}} < \mathfrak{G}$. Then $\alpha = \delta f_{\mathfrak{H}}$ is an endomorphism of $\mathfrak{G}$ which has $\mathfrak{H}$ as its kernel.

COROLLARY. *The condition $\mathfrak{G}/\mathfrak{H} \simeq \tilde{\mathfrak{G}} < \mathfrak{G}$ is satisfied if the cosets of $\mathfrak{H}$ have a transversal whose elements form a subgroup of $\mathfrak{G}$.*

For example, if $\mathfrak{G} = \mathfrak{H}\mathfrak{A}$ is the internal direct or semidirect product of $\mathfrak{H}$ by a subgroup $\mathfrak{A} < \mathfrak{G}$, then $\mathfrak{A} = \tilde{\mathfrak{G}}$. The elements of $\mathfrak{A}$ represent indeed a transversal of the cosets of $\mathfrak{H}$ in $\mathfrak{G}$ (cf. §3, *b.*).

In the quaternion group $\mathfrak{Q}$ it is not possible to select a group transversal for the subgroup $\mathfrak{H}$ of order 2 (cf. ex. 3).

b. Let $\mathfrak{G}$ be an additive abelian group. The composition of two of its endomorphisms α, β yields an endomorphism $\beta\alpha$. They can also be added: The endomorphism $\alpha+\beta$ is defined by

$$(\alpha+\beta)(x) = \alpha(x)+\beta(x), \quad x \in \mathfrak{G}.$$

Addition in $\mathfrak{G}$ is commutative and associative; therefore addition of endomorphisms is commutative and associative. Moreover multiplication and addition of endomorphisms are distributive:

$$\gamma(\alpha+\beta)(x) = \gamma(\alpha(x)+\beta(x)) = \gamma\alpha(x)+\gamma\beta(x) = (\gamma\alpha+\gamma\beta)(x),$$

$$(\alpha+\beta)\gamma(x) = \alpha\gamma(x)+\beta\gamma(x) = (\alpha\gamma+\beta\gamma)(x), \quad x \in \mathfrak{G};$$

briefly we may write: $\gamma(\alpha+\beta) = \gamma\alpha+\gamma\beta$, $(\alpha+\beta)\gamma = \alpha\gamma+\beta\gamma$. Multiplication of endomorphisms is not in general commutative. The neutral element for the addition is the null-endomorphism o which maps every $x \to o(x) = 0$, the neutral element of $\mathfrak{G}$. For every endomorphism α of $\mathfrak{G}$ we have the unique negative $-\alpha$ which satisfies the condition $\alpha+(-\alpha) = \alpha-\alpha = o$; it is defined by $-\alpha(x) = \alpha(-x)$, $x \in \mathfrak{G}$. Thus we have the result:

THEOREM 2. *The endomorphisms of an abelian group $\mathfrak{G}$ form a ring.*

This ring will be denoted by $\mathrm{E} = \mathrm{E}(\mathfrak{G})$.

Theorem 2 is of particular interest in the case of an *elementary abelian group* (cf. §1, *d.*), that is, the direct product of cyclic groups of order p where p is a fixed prime. An elementary abelian group $\mathfrak{A}$ can be considered as a vector space over the residue class field $\mathbb{R}_p$. An endomorphism appears as a *linear homogeneous transformation* of $\mathfrak{A} \to \mathfrak{A}$; in fact if $\alpha \in \mathrm{E}(\mathfrak{A})$ then

$$(1) \qquad \alpha(x+y) = \alpha(x)+\alpha(y), \quad \alpha(\rho x) = \rho\alpha(x), \quad x, y \in \mathfrak{G}, \quad \rho \in \mathbb{R}_p.$$

Here ρ may be interpreted as an integer so that $\rho x = x+\ldots+x$ (ρ terms). In particular, the multiplier kp represents the null-endomorphism o for every integer k [cf. ex. 5(d)].

For the sake of simplicity let $\mathfrak{A}$ be the direct product of n cyclic subgroups of order p, say $\mathfrak{A}_1, \mathfrak{A}_2, \ldots, \mathfrak{A}_n$. The direct product $\mathfrak{A} = (\mathfrak{A}_1, \ldots, \mathfrak{A}_n)$ then represents an n-dimensional vector space over the field $\mathbb{R}_p$. In the direct summand $\mathfrak{A}_\nu$ we choose an element $e_\nu \neq 0$. Every element of $\mathfrak{A}_\nu$ then appears in the form ξe_ν ($\xi \in \mathbb{R}_p$) and every element $x \in \mathfrak{A}$ can be written in the form

$$(2) \qquad x = \xi_1 e_1 + \xi_2 e_2 + \ldots + \xi_n e_n$$

with uniquely defined $\xi_1, \ldots, \xi_n \in \mathbb{R}_p$. Hence the system $e_1, \ldots, e_n$ is called a *basis* of $\mathfrak{A}$.

Now let α be an endomorphism of $\mathfrak{A}$. By (1) and (2)

$$(3)\qquad \alpha(x) = \xi_1\alpha(e_1)+\ldots+\xi_n\alpha(e_n) = \eta_1 e_1+\ldots+\eta_n e_n.$$

As elements of $\mathfrak{A}$ the $\alpha(e_\nu)$ $(\nu = 1, \ldots, n)$ appear in the form

$$\alpha(e_\nu) = \alpha_{1\nu}e_1+\ldots+\alpha_{n\nu}e_n, \quad \alpha_{\mu\nu}\in \mathbb{R}_p.$$

Therefore by (2)

$$\alpha(x) = \sum_\nu \xi_\nu\alpha(e_\nu) = \sum_\nu \xi_\nu \sum_\mu \alpha_{\mu\nu}e_\mu = \sum_\mu \eta_\mu e_\mu$$

so that with regard to the uniqueness of the coefficients of the e_μ we have

$$(4)\qquad \eta_\mu = \sum_{\nu=1}^{n} \alpha_{\mu\nu}\xi_\nu \quad (\mu = 1, \ldots, n).$$

Thus the matrix $(\alpha_{\mu\nu})$ represents the endomorphism α with respect to the basis $e_1, \ldots, e_n$.

Conversely every $n\times n$-matrix $(\alpha_{\mu\nu})$, $\alpha_{\mu\nu} \in \mathbb{R}_p$ induces an endomorphism in $\mathfrak{A}$ since the equations (3) and (4) are equivalent.

c. We call a set Ω of symbols ω an *operator domain* for the group $\mathfrak{G}$ if to every $\omega \in \Omega$ is associated a well defined endomorphism $\alpha_\omega \in \mathrm{E}(\mathfrak{G})$; thus Ω is an *index set for a subset of* $\mathrm{E}(\mathfrak{G})$. To different $\omega, \omega' \in \Omega$ may correspond one and the same endomorphism: $\alpha_\omega = \alpha_{\omega'}$ [cf. ex. 5(d)].

The combination of $\mathfrak{G}$ and Ω is often considered as a new entity, denoted by $(\mathfrak{G}, \Omega)$. A subgroup $\mathfrak{H} \leqslant \mathfrak{G}$ is said to be Ω-*admissible* if for every $\omega \in \Omega$ one has $\alpha_\omega(\mathfrak{H}) \leqslant \mathfrak{H}$. If $\mathfrak{H}$ is Ω-admissible we can say that $(\mathfrak{H}, \Omega)$ is a group with the operator domain Ω (cf. ex. 5).

A subgroup $\mathfrak{H}$ which is admissible with respect to, that is invariant under, all endomorphisms of $\mathfrak{G}$ is said to be *fully invariant in* $\mathfrak{G}$, in symbols: $\mathfrak{H} \lll \mathfrak{G}$. Clearly $\mathfrak{G} \lll \mathfrak{G}$, and $\mathfrak{e} \ll \mathfrak{G}$ if $\mathfrak{G}$ contains more than one element. As in the case of the characteristic subgroups, the property of being fully invariant is transitive: From $\mathfrak{H}_1 \lll \mathfrak{H}_2$ and $\mathfrak{H}_2 \lll \mathfrak{H}_3 \leqslant \mathfrak{G}$ it follows that $\mathfrak{H}_1 \lll \mathfrak{H}_3$ (cf. §1, *c*.).

Every fully invariant subgroup is characteristic and therefore normal. It is Ω-admissible for every set Ω of operators. Every subgroup is Ω-admissible if Ω is empty or if $\alpha_\omega = \epsilon$ for every $\omega \in \Omega$.

All normal subgroups are Ω-admissible if Ω indexes a subset of $\Delta(\mathfrak{G})$. All characteristic subgroups are admissible if Ω indexes a subset of $\Gamma(\mathfrak{G})$.

A "refined" group theory can be constructed making use of the notion of "group with an operator domain" $(\mathfrak{G}, \Omega)$. We shall develop the beginnings of this theory.

Let $\mathfrak{G}$ and $\tilde{\mathfrak{G}}$ be two groups with the same operator domain Ω. Let α_ω and $\tilde{\alpha}_\omega$ be the endomorphisms associated with Ω in $E(\mathfrak{G})$ and $E(\tilde{\mathfrak{G}})$ respectively. The groups $\mathfrak{G}$ and $\tilde{\mathfrak{G}}$ are said to be Ω-homomorphic (Ω-isomorphic) if there is a homomorphism (isomorphism) $f : \mathfrak{G} \to \tilde{\mathfrak{G}}$, $\tilde{x} = f(x) \in \tilde{\mathfrak{G}}$ if $x \in \mathfrak{G}$, which satisfies the condition

$$f(\alpha_\omega(x)) = \tilde{\alpha}_\omega(f(x))$$

for all $x \in \mathfrak{G}$ and $\omega \in \Omega$.

It follows now that if $\mathfrak{H} \leqslant \mathfrak{G}$ is an Ω-admissible subgroup of $\mathfrak{G}$ then its Ω-homomorphic (Ω-isomorphic) image $\tilde{\mathfrak{H}} = f(\mathfrak{H})$ is an Ω-admissible subgroup of $\tilde{\mathfrak{G}}$. Indeed if $\alpha_\omega(\mathfrak{H}) \leqslant \mathfrak{H}$ then

$$\tilde{\alpha}_\omega(\tilde{\mathfrak{H}}) = \tilde{\alpha}_\omega(f(\mathfrak{H})) = f(\alpha_\omega(\mathfrak{H})) \leqslant f(\mathfrak{H}) = \tilde{\mathfrak{H}}.$$

In particular we consider the case of a natural homomorphism $f_{\mathfrak{K}} : \mathfrak{G} \to \mathfrak{G}/\mathfrak{K}$ with the kernel $\mathfrak{K} \lhd \mathfrak{G}$. We shall have to arrange that the operator domain Ω of $\mathfrak{G}$ is also operator domain for the factor group $\mathfrak{G}/\mathfrak{K}$ so that to every ω is associated an endomorphism $\bar{\alpha}_\omega \in E(\mathfrak{G}/\mathfrak{K})$. Clearly we need that $\bar{\alpha}_\omega(\mathfrak{K}) = \mathfrak{K}$ since $\mathfrak{K}$ is the unit element of $\mathfrak{G}/\mathfrak{K}$. Thus, we adopt the natural definition of $\bar{\alpha}_\omega$, namely

$$(5) \qquad \bar{\alpha}_\omega(xK) = \alpha_\omega(x)\mathfrak{K}, \quad x \in \mathfrak{G}.$$

This implies that for $y \in \mathfrak{K}$:

$$\bar{\alpha}_\omega(y\mathfrak{K}) = \bar{\alpha}_\omega(\mathfrak{K}) = \alpha_\omega(y) \cdot \mathfrak{K} = \mathfrak{K}.$$

Hence $\alpha_\omega(y) \in \mathfrak{K}$ or $\alpha_\omega(\mathfrak{K}) \leqslant \mathfrak{K}$ and $\mathfrak{K}$ is Ω-admissible.

It is now readily seen that $\bar{\alpha}_\omega$ is indeed well defined in $\mathfrak{G}/\mathfrak{K}$. If $y \in \mathfrak{K}$ we have $\bar{\alpha}_\omega(xy\mathfrak{K}) = \alpha_\omega(xy)\mathfrak{K} = \alpha_\omega(x)\,\alpha_\omega(y)\mathfrak{K} = \alpha_\omega(x) \cdot \mathfrak{K} = \bar{\alpha}_\omega(x\mathfrak{K})$.

Further $\bar{\alpha}_\omega$ is an endomorphism in $\mathfrak{G}/\mathfrak{K}$ since $\bar{\alpha}_\omega(x\mathfrak{K}\cdot y\mathfrak{K}) = \alpha_\omega(xy)\mathfrak{K} = \alpha_\omega(x)\mathfrak{K}\cdot\alpha_\omega(y)\mathfrak{K} = \bar{\alpha}_\omega(x\mathfrak{K})\,\bar{\alpha}_\omega(y\mathfrak{K})$ for all $x, y \in \mathfrak{G}$.

Finally we show that the natural homomorphism $f_\mathfrak{K}$ is an Ω-homomorphism. We use the fact that $f_\mathfrak{K}(x) = x\mathfrak{K}$ whence

$$f_\mathfrak{K}(\alpha_\omega(x)) = \alpha_\omega(x)\mathfrak{K} = \bar{\alpha}_\omega(x\mathfrak{K}) = \bar{\alpha}_\omega(f_\mathfrak{K}(x)).$$

We state the result of this discussion:

THEOREM 3. *The natural homomorphism* $f_\mathfrak{K} : \mathfrak{G} \to \mathfrak{G}/\mathfrak{K}$ *is an* Ω-*homomorphism if the kernel* $\mathfrak{K}$ *is an* Ω-*admissible normal subgroup of* $\mathfrak{G}$ (cf. ex. 6).

On the basis of these considerations the correspondence theorems of Chap. II, §§3–5 can be "refined" for groups $(\mathfrak{G}, \Omega)$ with an operator domain Ω. The corresponding refined theorems are obtained by substituting for "normal subgroup" and "homomorphism" the words "Ω-admissible subgroup" and "Ω-homomorphism".

Examples and exercises

1. Let $\mathfrak{G}$ be a group and $\mathfrak{H}$ a proper normal subgroup of $\mathfrak{G}$ for which the factorgroup $\mathfrak{G}/\mathfrak{H}$ is isomorphic to $\mathfrak{G}$. Denote this isomorphism by φ. Let $f : \mathfrak{G} \to \mathfrak{G}/\mathfrak{H}$ be the natural homomorphism whose kernel is $\mathfrak{H}$. Then the mapping

$$\varphi f : \mathfrak{G} \to \mathfrak{G}/\mathfrak{H} \to \mathfrak{G}$$

is an endomorphism mapping $\mathfrak{G}$ onto $\mathfrak{G}$. But φf is not an automorphism because the mapping f is not one-one.

It is readily seen that a group which is isomorphic to the factor group with respect to one of its proper normal subgroups cannot be finite. As an example we consider the multiplicative group $\mathfrak{R}$ of all rational numbers (different from zero). If $\Pi = \{p_1, p_2, \ldots\}$ is the set of all prime numbers then $\mathfrak{R} = \langle\Pi\rangle$. Let $\Pi^* = \{p_{i_1}, p_{i_2}, \ldots\}$ be a proper infinite subset of Π and $\mathfrak{R}^* = \langle\Pi^*\rangle$; then $\mathfrak{R}^* < \mathfrak{R}$. Let $\langle\Pi\setminus\Pi^*\rangle = \mathfrak{H}$; then $\mathfrak{R}/\mathfrak{H} \simeq \mathfrak{R}^*$. On the other hand the mapping $\varphi : \Pi \to \Pi^*$ defined by $\varphi(p_\nu) = p_{i_\nu}$, $(\nu = 1, 2, \ldots)$ can easily be continued into an isomorphism $\mathfrak{R} \to \mathfrak{R}^*$ so that in fact there is an isomorphism $\mathfrak{R}/\mathfrak{H} \simeq \mathfrak{R}$.

2. The function $a^i \to a^{in}$ $(i = 0, \pm 1, \pm 2, \ldots)$ where n is a fixed integer, $n > 1$, represents a meromorphism of the infinite cyclic group $\langle a \rangle_\infty$.

3. Let $\mathfrak{Q}$ be the quaternion group (cf. Chap. I, §2, ex. 9 and Chap. II, §3, ex. 8). Its (normal) subgroup $\mathfrak{H}$ of order 2 consists of the two elements $e = 1$ and $e' = -1$. The factor group $\mathfrak{Q}/\mathfrak{H}$ is isomorphic to the non-cyclic group of order 4. But all subgroups of order 4 in $\mathfrak{Q}$ are cyclic, namely $\langle i \rangle, \langle j \rangle, \langle k \rangle$. Thus $\mathfrak{H}$ is not the kernel of an endomorphism of $\mathfrak{Q}$.

4. The non-cyclic group of order 4, namely

$$\mathfrak{G}_4 = \langle a, b \,|\, a^2 = b^2 = (ab)^2 = e \rangle$$

has an automorphism α interchanging a and b. The symbol α then is an operator for $\mathfrak{G}_4$. Find an α-admissible subgroup $\mathfrak{H}$ and verify that α is also an operator for $\mathfrak{H}$.

5. (a) Let $\mathfrak{V}_n$ be an n-dimensional vector space over a field $\mathbb{F}$. The set of all $n \times n$-matrices with elements from $\mathbb{F}$ may be taken as an operator domain for the additive group of $\mathfrak{V}_n$. The corresponding endomorphisms are the linear homogeneous transformations $\mathfrak{V}_n \to \mathfrak{V}_n$.

(b) The subgroup $\mathfrak{X}$ of all vectors $\begin{pmatrix} x \\ 0 \end{pmatrix}$ is an Ω-admissible subgroup of $\mathfrak{V}_2$ if Ω is the set of all matrices $\begin{pmatrix} a & b \\ 0 & c \end{pmatrix}$, $a, b, c, x \in \mathbb{F}$.

(c) Find the operator domain of all 2×2-matrices which induce in $\mathfrak{X}$ the identity endomorphism.

(d) For the vector space $\mathfrak{V}_n$ over the field $\mathbb{R}_p$ the integral domain $\mathbb{Z}$ is an operator domain. With all the operators $z = kp + r$, r fixed, $0 \leqslant r < p$, $k = 0, \pm 1, \pm 2, \ldots$ is associated the same endomorphism of $\mathfrak{V}_n$, namely the one effected by multiplication by r. Every subgroup of $\mathfrak{V}_n$ is $\mathbb{Z}$-admissible.

6. Prove the converse of theorem 3, that is: If the natural homomorphism $f_{\mathfrak{K}}$ is an Ω-homomorphism then its kernel $\mathfrak{K}$ is an Ω-admissible subgroup.

7. Determine the endomorphism ring $\mathrm{E}(\mathfrak{G})$
(a) if $\mathfrak{G}$ is the additive group of all integers;
(b) if $\mathfrak{G} = \mathbb{R}_m^{\,+}$, the additive group of the residues (mod m).

8. We define an endomorphism σ of $\mathfrak{G}$ to be *normal* if $\sigma\tau = \tau\sigma$ for all $\tau \in \Delta(\mathfrak{G})$.

Let σ be a normal endomorphism of $\mathfrak{G}$ and

$$\sigma(\mathfrak{G}) = \mathfrak{G}^* \leqslant \mathfrak{G}.$$

(a) Show that $t^{-1}\sigma(t) \in \mathfrak{C}_{\mathfrak{G}}(\mathfrak{G}^*)$, $t \in \mathfrak{G}$, (cf. §2, *e.*); if σ is an automorphism of $\mathfrak{G}$, then $t^{-1}\sigma(t) \in \mathfrak{Z}(\mathfrak{G})$.

(b) For every element c of the commutator subgroup of $\mathfrak{G}$

$$\sigma^2(c) = \sigma(c).$$

In particular if σ is a normal automorphism, $\sigma(c) = c$.

(c) All normal automorphisms form the subgroup $\mathfrak{C}_\Gamma(\Delta) \leqslant \Gamma(\mathfrak{G})$. All normal endomorphisms of $\mathfrak{G}$ form the sub-semigroup of the semigroup of all endomorphisms which "centralizes" the subgroup $\Delta(\mathfrak{G})$.

9. Let p be a fixed prime number and $\epsilon_n = e^{2\pi i/p^n}$ so that $\epsilon_{n+1}{}^p = \epsilon_n$ $(n = 1, 2, \ldots)$. We consider the multiplicative infinite abelian group $\mathfrak{G}_p = \langle \epsilon_1, \epsilon_2, \ldots \rangle = \langle \epsilon_m, \epsilon_{m+1}, \ldots \rangle$ (m an arbitrary natural number). Determine the endomorphism ring $\mathrm{E}(\mathfrak{G}_p)$.

(a) Every endomorphism α of $\mathfrak{G}_p$ is defined by its effect on the generators ϵ_n of $\mathfrak{G}_p$. Since ϵ_n is of order p^n one has $\alpha(\epsilon_n) = \epsilon_n{}^{h_n}$ where h_n is an integer and

$$(6) \qquad 0 \leqslant h_n < p^n.$$

Using $\epsilon_{n+1}{}^p = \epsilon_n$ it is readily shown that

$$(7) \qquad h_{n+1} \equiv h_n \pmod{p^n}.$$

Thus the endomorphism α of $\mathfrak{G}_p$ is represented by the infinite sequence $h = (h_1, h_2, \ldots)$ subject to the conditions (6), (7).

(b) Conversely every such sequence represents an endomorphism of $\mathfrak{G}_p$. Two such sequences h and $k = (k_1, k_2, \ldots)$ represent the same endomorphism if and only if $h_n \equiv k_n \pmod{p^n}$, i.e. if $h_n = k_n$ $(n = 1, 2, \ldots)$. Thus the relation $\alpha \to h$ is one-one onto.

(c) Using the definition (1) (suitably adapted to the multiplicative group $\mathfrak{G}_p$) show that

$$\alpha+\beta \to h+k = (h_1+k_1, h_2+k_2, \ldots),$$
$$\alpha\beta \to hk = (h_1k_1, h_2k_2, \ldots)$$

where the sums h_n+k_n and the products h_nk_n are to be replaced by their least non-negative residues modulo p^n ($n = 1, 2, \ldots$).

(d) Using (6) and (7) show that the ring $\mathrm{E}(\mathfrak{G}_p)$ contains no divisors of zero, that is, no non-zero elements α, β such that $\alpha\beta = o$, the zero endomorphism, represented by the sequence $(0, 0, \ldots)$.

(e) The h_n can be considered as the partial sums of an infinite series. Indeed by (6) and (7) there are integers a_n, $0 \leqslant a_n < p$, such that $h_{n+1} = h_n + a_np^n$ ($n = 1, 2, \ldots$), $h_0 = 0$; thus

$$h_n = a_0 + a_1p + a_2p^2 + \ldots + a_{n-1}p^{n-1}$$

is the n-th partial sum of the infinite series

$$\tilde{\alpha} = a_0 + a_1p + a_2p^2 + \ldots$$

which represents a p-adic integer. All p-adic integers form a ring $\mathbb{R}^{(p)}$ which is isomorphic to the sequence ring and therefore to the ring $\mathrm{E}(\mathfrak{G}_p)$. With regard to (*d*) it is possible to form the field of the quotients of $\mathbb{R}^{(p)}$ which is the *field of the p-adic numbers.*

Chapter IV

FINITE SERIES OF SUBGROUPS

§1 The fundamental concepts of lattice theory

In the greater part of the present chapter there will be little or no concern with group elements. We shall deal with questions on certain systems of subgroups of a given group $\mathfrak{G}$. It turns out that with regard to certain composition rules, specified systems of subgroups are "lattices" (in a technical sense to be explained) and that simple facts of lattice theory applied to such subgroup lattices lead to important theorems. For these the lattice viewpoint is not a necessity, but presents a solid framework for the development of this part of group theory.

a. *A partially ordered set* is a set **S**, whose elements we denote by capital Italic letters, together with a binary relation $\leqslant$ defined between certain couples of elements of **S** subject to the following three conditions:

(i) $A \leqslant A$ for all A in **S** (reflexivity);

(ii) If $C \leqslant B$ and $B \leqslant A$ then $C \leqslant A$ (transitivity);

(iii) If $A \leqslant B$ and $B \leqslant A$ then $A = B$ (antisymmetry).

A partially ordered set will be denoted by $\Sigma = (\mathbf{S}; \leqslant)$. The relation $\leqslant$ is called an *order relation* or an *inclusion*. We write $A < B$ if $A \leqslant B$, but $A \neq B$. The set Σ is an ordered set if for every A, B in S either $A \leqslant B$ or $B \leqslant A$.

An element N of a partially ordered set Σ is said to be *minimal* (*maximal*) in Σ if there is no X in **S** which satisfies the condition $X < N$ ($N < X$). A partially ordered set may have none, or one, or more than one minimal (maximal) element.

b. Let **S** be the element set of the partially ordered set $\Sigma = (\mathbf{S}; \leqslant)$. We impose on it the following two restrictions:

For every two elements A, B in S there exists in S

(a) an element D satisfying the conditions $D \leqslant A$, $D \leqslant B$ and if $X \leqslant A$, $X \leqslant B$ ($X \in$ S) then $X \leqslant D$;

(b) an element E satisfying the conditions $A \leqslant E$, $B \leqslant E$ and if $A \leqslant Y$, $B \leqslant Y$ ($Y \in$ S) then $E \leqslant Y$.

Thus D is maximal in the partially ordered set of all elements X in S for which $X \leqslant A$ and $X \leqslant B$. The element E is minimal in the set of all elements Y in S for which $A \leqslant Y$ and $B \leqslant Y$. A partially ordered set $\Sigma = (\mathrm{S}; \leqslant)$, subject to the conditions (a) and (b), is called a *lattice* (cf. Chap. I, §1, *b.*) or a *structure* (French: treillis, German: Verband). A lattice turns out to be an algebraic system $\Sigma = (\mathrm{S} \mid \wedge, \vee)$ with respect to the two composition laws

$$A \wedge B = D, \quad A \vee B = E,$$

to be read, e.g., "A meet B" and "A join B" respectively.

THEOREM 1. *The two composition laws* $\wedge$ *and* $\vee$ *of a lattice are well defined and satisfy the following conditions*

I. $A \wedge A = A$;

II. $A \wedge B = B \wedge A$;

III. $A \wedge (B \wedge C) = (A \wedge B) \wedge C$;

IV. $A \wedge (A \vee B) = A$;

I′. $A \vee A = A$;

II′. $A \vee B = B \vee A$;

III′. $A \vee (B \vee C) = (A \vee B) \vee C$;

IV′. $A \vee (A \wedge B) = A$

for all elements A, B, C *of the lattice.*

These laws are known as idempotence, commutativity, associativity, and absorption laws respectively.

Proof. Uniqueness, idempotence and commutativity are evident from the definition of the composition laws.

III. By (a) $(A \wedge B) \wedge C \leqslant A \wedge B \leqslant A$ and also $(A \wedge B) \wedge C \leqslant B$ and $(A \wedge B) \wedge C \leqslant C$, hence $(A \wedge B) \wedge C \leqslant B \wedge C$. Thus

$(A \wedge B) \wedge C \leqslant A \wedge (B \wedge C)$. In the same way it is shown that $A \wedge (B \wedge C) \leqslant (A \wedge B) \wedge C$. Thus by antisymmetry (iii) both sides are equal.

III'. By (b)

A and $B \leqslant A \vee B \leqslant (A \vee B) \vee C$, $C \leqslant (A \vee B) \vee C$

and therefore $B \vee C \leqslant (A \vee B) \vee C$. Thus again by (b)

$$A \vee (B \vee C) \leqslant (A \vee B) \vee C.$$

In the same way $(A \vee B) \vee C \leqslant A \vee (B \vee C)$; thus by (iii) both sides are equal.

IV. By (a) $A \wedge (A \vee B) \leqslant A$; by (b) $A \leqslant A \vee B$. Since $A \leqslant A$ we have $A \leqslant (A \vee B) \wedge A$; hence both sides are equal.

IV'. By (b) $A \leqslant A \vee (A \wedge B)$; by (a) $A \wedge B \leqslant A$. Since $A \leqslant A$ we have $A \vee (A \wedge B) \leqslant A$; hence both sides are equal.

c. LEMMA. *Let* $\Lambda = (\mathbf{S} \mid \wedge, \vee)$ *be an algebraic system where the composition laws* $\wedge, \vee$ *satisfy conditions I–IV and I'–IV'; then the two statements*

$$A \wedge B = B \quad \textit{and} \quad A \vee B = A$$

are equivalent.

Proof. If $A \wedge B = B$ it follows that $A \vee (A \wedge B) = A \vee B$; by IV' $A \vee (A \wedge B) = A$, hence $A \vee B = A$.

If $A \vee B = A$ it follows that $(A \vee B) \wedge B = A \wedge B$; by IV $(A \vee B) \wedge B = B$, hence $A \wedge B = B$.

In every system $\Lambda = (\mathbf{S} \mid \wedge, \vee)$ (as assumed in the lemma) one can define the binary relation $\lesssim$:

(1) $B \lesssim A$ if $A \wedge B = B$ or equivalently $A \vee B = A$.

THEOREM 2. *The relation* $\lesssim$ *is the "natural" inclusion: It satisfies the conditions* (*i*), (*ii*), (*iii*) *of subsection a.*

Proof. (i) $A \lesssim A$ because $A \wedge A = A$.

(ii) Suppose that $C \lesssim B$ and $B \lesssim A$, i.e. $B \wedge C = C$ and $A \wedge B = B$; then $A \wedge C = A \wedge (B \wedge C) = (A \wedge B) \wedge C = B \wedge C = C$, that is $C \lesssim A$.

(iii) By supposition $A \lesssim B$ and $B \lesssim A$; thus $A \wedge B = A$ and $A \wedge B = B$. Hence $A = B$.

COROLLARY. *The element set* **S** *of an algebraic system* $(\mathbf{S} \mid \wedge, \vee)$ *subject to the conditions* I–IV *and* I′–IV′ *is partially ordered by the inclusion defined by* (1) *and* $(\mathbf{S} \mid \wedge, \vee)$ *is a lattice.*

Resumé. A partially ordered set, subject to the conditions (a) and (b) is (by definition) a lattice. Every lattice is an algebraic system with two composition laws $\wedge$, $\vee$ subject to the conditions I–IV, I′–IV′. Every algebraic system with the composition laws $\wedge$, $\vee$ subject to I–IV, I′–IV′ is a lattice.

Remark 1. If in the conditions I–IV the symbols $\wedge$ and $\vee$ are interchanged we obtain the laws I′–IV′ and conversely. This fact is known as the *Principle of Duality.* It has an immediate practical consequence: If in a correct statement involving $\wedge$ and $\vee$ and elements of a lattice we interchange $\wedge$ and $\vee$ then we obtain another (or the same) correct statement.

Remark 2. A lattice Λ may contain a *zero element*, i.e. an element N which satisfies the condition $N \leqslant A$ for all A in Λ, i.e. $A \wedge N = N$, $A \vee N = A$. The zero element N is the unique *minimal element* in Λ. There may also exist a "unit element" in Λ, i.e. an element M such that $A \leqslant M$ for all A in Λ, i.e. $A \wedge M = A$, $A \vee M = M$. If it exists, this M is the unique maximal element of Λ (cf. *a.*).

d. Two special types of lattices will occur in our discussion:

(α) A lattice $\Lambda = (\mathbf{S} \mid \wedge, \vee)$ is called a *modular lattice* or also a *Dedekind lattice* if (in addition to conditions I–IV and I′–IV′) it has the following property: For each triple A, B, C in Λ the condition

V. $(A \vee B) \wedge C = A \vee (B \wedge C)$ if $A \leqslant C$, i.e. $A \wedge C = A$

is satisfied.

This axiom is "self-dual", that is, it remains unchanged if in it (incl. the side condition $A \wedge C = A$) we replace $\wedge$ by $\vee$ and $\vee$ by $\wedge$. Indeed this process leads to

V′. $(A \wedge B) \vee C = A \wedge (B \vee C)$ if $C \leqslant A$, i.e. $A \vee C = A$,

which after an interchange of A and C is equivalent with V (left and right side interchanged). Thus in a modular lattice the principle of duality is valid.

(β) *A distributive lattice* is a lattice Λ where the two laws $\wedge$, $\vee$ are connected by a distributive law

VI. $A \vee (B \wedge C) = (A \vee B) \wedge (A \vee C), \quad A, B, C \in \Lambda.$

Every distributive lattice is modular. Indeed if $A \leqslant C$, i.e. $A \vee C = C$, we have from VI

$$A \vee (B \wedge C) = (A \vee B) \wedge (A \vee C) = (A \vee B) \wedge C,$$

that is, V. Hence the duality principle remains applicable in a distributive lattice Λ and so from VI follows the second distributive law:

VI′. $A \wedge (B \vee C) = (A \wedge B) \vee (A \wedge C)$ (cf. ex. 6).

Examples and exercises

1. Let **S** be the set of all non-negative integers. Let $m, n \in \mathbf{S}$ and let inclusion be defined by $m \mid n$.

(a) Show that $(\mathbf{S}; \mid)$ is a partially ordered set, but not an ordered set.

(b) Define the composition laws $\wedge$, $\vee$ by which $(\mathbf{S}; \mid)$ will become a lattice $\Lambda = (\mathbf{S} \mid \wedge, \vee)$.

(c) Is Λ a modular lattice? (Cf. Chap. I, §1, ex. 5.)

(d) Does the lattice Λ have a minimal or a maximal element?

2. Let **M** be a finite set and **S** the set of all subsets of **M**. Let $A, B \subseteq \mathbf{M}$, i.e. $A, B \in \mathbf{S}$. Define

$$A \wedge B = A \cap B, \quad A \vee B = A \cup B.$$

(a) Show that $(\mathbf{S} \mid \cap, \cup)$ is a distributive lattice.

(b) The partially ordered system $\Sigma = (\mathbf{S}; \subseteq)$ of all subsets of **M** may be realized by a diagram, called a *graph*, consisting of vertices and edges joining certain vertices. Every vertex of the graph is a representative of a subset A of **M** and A is joined by an edge with each of those vertices which represent subsets containing A and exactly one more element of **M**. This graph, beginning at the vertex representing the empty set (which is a unique subset of **M** contained in every subset of **M**) and ending at the vertex representing the whole set **M**, clearly indicates all inclusions of Σ.

Draw the graph of Σ in the cases that **M** is a set of 2, of 3, and of 4 elements.*

3. Let $\Lambda = (\mathbf{S} \mid \wedge, \vee)$ be a lattice and **T** a subset of **S** for which $\mathrm{M} = (\mathbf{T} \mid \wedge, \vee)$ is a lattice (with the same composition laws $\wedge, \vee$ as Λ). Then M is called a *sublattice* of Λ.

As in ex. 1 assume that **S** is the set of all non-negative integers and **T** the set of all non-negative even numbers. Does the symbol $(\mathbf{T} \mid \wedge, \vee)$ represent a lattice (and therefore a sublattice of $(\mathbf{S} \mid \wedge, \vee)$)?

4. Two lattices $(\mathbf{S} \mid \wedge, \vee)$ and $(\mathbf{S}' \mid \wedge', \vee')$ are said to be isomorphic if there is a bijection $f: \mathbf{S} \to \mathbf{S}'$ such that for all $A, B \in \mathbf{S}$ we have

$$f(A \wedge B) = f(A) \wedge' f(B), \quad f(A \vee B) = f(A) \vee' f(B).$$

If $\leqslant$ denotes the inclusion relation in both isomorphic lattices $(\mathbf{S} \mid \wedge, \vee)$ and $(\mathbf{S}' \mid \wedge', \vee')$, i.e.

$$A \leqslant B \text{ if } A \wedge B = A, \quad A' \leqslant B' \text{ if } A' \wedge' B' = A',$$

show that if $A \leqslant B$ then $f(A) \leqslant f(B)$.

5. A lattice $(\mathbf{S} \mid \wedge, \vee)$ is modular if and only if for all triples $A, B, C \in \mathbf{S}$ one has $((A \wedge C) \vee B) \wedge C = (A \wedge C) \vee (B \wedge C)$.

To prove that the condition is necessary, notice that $A \wedge C \leqslant C$; substitute A for $A \wedge C$. To prove sufficiency let $A \leqslant C$.

6. Here is a formal proof for the fact that in a distributive lattice the two distributive laws VI and VI′ are equivalent. Assume VI; then, using V′, we obtain

$$A \wedge (B \vee C) = (A \wedge (A \vee C)) \wedge (B \vee C)$$
$$= A \wedge ((A \vee C) \wedge (B \vee C))$$

by VI:

$$= A \wedge ((A \wedge B) \vee C)$$
$$= ((A \wedge B) \vee A) \wedge ((A \wedge B) \vee C)$$

* It appears that the graph of Σ for $|\mathbf{M}| = 2$ is a square. For $|\mathbf{M}| = 3$ it can be considered as a projection of vertices and edges of a cube into a plane. Similarly if $|\mathbf{M}| = 4$ the graph of Σ is a projection of a four-dimensional cube into a plane, etc.

by VI:

$$= (A \wedge B) \vee (A \wedge C), \quad \text{q.e.d.}$$

Similarly VI follows from VI′.

§2 Lattices of subgroups

a. Let S be the system of all subgroups of a group $\mathfrak{G}$. This set is partially ordered by the ordinary inclusion of subgroups. Moreover we can introduce in S two composition laws: For every two subgroups $\mathfrak{A}, \mathfrak{B} \leqslant \mathfrak{G}$ there is in S

(a) a subgroup $\mathfrak{D}$ such that $\mathfrak{D} \leqslant \mathfrak{A}$ and $\mathfrak{D} \leqslant \mathfrak{B}$ and if for some $\mathfrak{X} \in \mathbf{S}$ $\mathfrak{X} \leqslant \mathfrak{A}$ and $\mathfrak{X} \leqslant \mathfrak{B}$, then $\mathfrak{X} \leqslant \mathfrak{D}$. Thus $\mathfrak{D}$ is "the greatest subgroup" of $\mathfrak{G}$ which is a subgroup of $\mathfrak{A}$ and a subgroup of $\mathfrak{B}$; indeed $\mathfrak{D} = \mathfrak{A} \wedge \mathfrak{B} \overset{\text{def}}{=} \mathfrak{A} \cap \mathfrak{B}$.

(b) a subgroup $\mathfrak{V}$ such that $\mathfrak{A} \leqslant \mathfrak{V}$ and $\mathfrak{B} \leqslant \mathfrak{V}$ and if for some $\mathfrak{Y} \in \mathbf{S}$ $\mathfrak{A} \leqslant \mathfrak{Y}$ and $\mathfrak{B} \leqslant \mathfrak{Y}$, then $\mathfrak{V} \leqslant \mathfrak{Y}$, or, expressed intuitively: $\mathfrak{V}$ is the smallest subgroup of $\mathfrak{G}$ which contains both $\mathfrak{A}$ and $\mathfrak{B}$. Thus $\mathfrak{V} = \mathfrak{A} \vee \mathfrak{B} \overset{\text{def}}{=} \langle \mathfrak{A}, \mathfrak{B} \rangle$. (This symbol has been introduced in Chap. II, §1, *d.*; cf. ex. 1.)

Hence we have the following theorem:

Theorem 1. *With regard to the operations* $\mathfrak{A} \wedge \mathfrak{B} = \mathfrak{A} \cap \mathfrak{B}$ *and* $\mathfrak{A} \vee \mathfrak{B} = \langle \mathfrak{A}, \mathfrak{B} \rangle$, $\mathfrak{A}, \mathfrak{B} \leqslant \mathfrak{G}$, *the system* S *of all subgroups of a group* $\mathfrak{G}$ *represents a lattice*

$$\Lambda = (\mathbf{S} \mid \wedge, \vee) = \Lambda(\mathfrak{G}).$$

Remark 1. The intersection of two subgroups is a subgroup (cf. Chap. II, §1, ex. 1). The union of two subgroups, in general, is not a subgroup (cf. Chap. II, §1, Theorem 3); thus the ordinary set union could not be taken as the second composition law in $\Lambda(\mathfrak{G})$. With regard to Chap. II, §1, Theorem 2, Corollary, also $\mathfrak{A}\mathfrak{B}$ is not always a subgroup; it is so only when $\mathfrak{A}\mathfrak{B} = \mathfrak{B}\mathfrak{A}$. So we have

$$\mathfrak{A} \vee \mathfrak{B} = \mathfrak{A}\mathfrak{B} \quad \text{if and only if} \quad \mathfrak{A}\mathfrak{B} = \mathfrak{B}\mathfrak{A}.$$

Remark 2. In the lattice $\Lambda(\mathfrak{G})$ inclusion can be defined by means of the operation $\wedge$ or $\vee$:

$$\mathfrak{A} \leqslant \mathfrak{B} \text{ if and only if } \mathfrak{A} \wedge \mathfrak{B} = \mathfrak{A}, \text{ or equivalently, } \mathfrak{A} \vee \mathfrak{B} = \mathfrak{B},$$

which evidently is the usual inclusion within the set **S** of all subgroups of $\mathfrak{G}$.

b. Let Ω be a system of endomorphisms of the group $\mathfrak{G}$ and consider the set $\mathbf{S}_\Omega$ of all Ω-admissible subgroups of $\mathfrak{G}$ (cf. III, §5, *c.*).

THEOREM 2. *All Ω-admissible subgroups of a group $\mathfrak{G}$ form a sublattice*

$$(\mathbf{S}_\Omega | \wedge, \vee) = \Lambda_\Omega(\mathfrak{G}) \subseteq \Lambda(\mathfrak{G}) = (\mathbf{S} | \wedge, \vee).$$

Proof. Let $\mathfrak{A}$, $\mathfrak{B}$ be two Ω-admissible subgroups of $\mathfrak{G}$ and $\omega \in \Omega$ then $\omega(\mathfrak{A} \wedge \mathfrak{B}) \leqslant \mathfrak{A} \wedge \mathfrak{B}$. If $a_1, a_2, \ldots \in \mathfrak{A}$, $b_1, b_2, \ldots \in \mathfrak{B}$, then

$$\omega(a_1 b_1 a_2 b_2 \ldots) \in \langle \mathfrak{A}, \mathfrak{B} \rangle = \mathfrak{A} \vee \mathfrak{B}.$$

Hence $\mathfrak{A} \wedge \mathfrak{B}$ and $\mathfrak{A} \vee \mathfrak{B}$ are Ω-admissible subgroups of $\mathfrak{G}$.

This is so in particular if $\Omega = \mathrm{E}$, the semigroup of all endomorphisms of $\mathfrak{G}$, if $\Omega = \Gamma$, and if $\Omega = \Delta$ (the automorphism group and the inner automorphism group of $\mathfrak{G}$ respectively). Thus Λ_Δ is the lattice of all normal subgroups, Λ_Γ the lattice of all characteristic subgroups of $\mathfrak{G}$, and

$$\Lambda_\mathrm{E} \subseteq \Lambda_\Gamma \subseteq \Lambda_\Delta \subseteq \Lambda(\mathfrak{G}).$$

c. In particular we consider the lattice Λ_Δ and prove the following important result:

THEOREM 3. *The lattice $\Lambda_\Delta(\mathfrak{G})$ of all normal subgroups of $\mathfrak{G}$ is a modular lattice.*

Instead of this we shall in fact prove the following more general theorem:

THEOREM 3′. *If $\mathfrak{A}, \mathfrak{B}, \mathfrak{C} \leqslant \mathfrak{G}$, $\mathfrak{A}\mathfrak{B} = \mathfrak{B}\mathfrak{A}$ and $\mathfrak{A} \leqslant \mathfrak{C}$, then*

$$(\mathfrak{A} \vee \mathfrak{B}) \wedge \mathfrak{C} = \mathfrak{A} \vee (\mathfrak{B} \wedge \mathfrak{C}).$$

Remark. Since normal subgroups commute, this theorem implies theorem 3.

Proof of Theorem 3′. Put $(\mathfrak{A} \vee \mathfrak{B}) \wedge \mathfrak{C} = \mathfrak{H}$. Then

$$\mathfrak{B} \wedge \mathfrak{H} = \mathfrak{B} \wedge (\mathfrak{A} \vee \mathfrak{B}) \wedge \mathfrak{C} = \mathfrak{B} \wedge \mathfrak{C}$$

so that $\mathfrak{A} \vee (\mathfrak{B} \wedge \mathfrak{C}) = \mathfrak{A} \vee (\mathfrak{B} \wedge \mathfrak{H})$. Thus we shall have proved the theorem when we have shown that

$$\text{(a)} \qquad \mathfrak{A} \vee (\mathfrak{B} \wedge \mathfrak{C}) \leqslant \mathfrak{H} \quad \text{and} \quad \text{(b)}\ \mathfrak{H} \leqslant \mathfrak{A} \vee (\mathfrak{B} \wedge \mathfrak{H}).$$

To prove (a):

(i) $\mathfrak{A} \leqslant \mathfrak{H}$ because

$$\mathfrak{A} \wedge \mathfrak{H} = \mathfrak{A} \wedge (\mathfrak{A} \vee \mathfrak{B}) \wedge \mathfrak{C} = \mathfrak{A} \wedge \mathfrak{C} = \mathfrak{A}$$

(ii) $\mathfrak{B} \wedge \mathfrak{C} \leqslant \mathfrak{H}$ because

$$(\mathfrak{B} \wedge \mathfrak{C}) \wedge \mathfrak{H} = \mathfrak{B} \wedge (\mathfrak{A} \vee \mathfrak{B}) \wedge \mathfrak{C} = \mathfrak{B} \wedge \mathfrak{C}$$

whence (a) follows immediately.

(b) Notice that by the definition of $\mathfrak{H}$ we have $\mathfrak{H} \leqslant \mathfrak{A} \vee \mathfrak{B}$. Because $\mathfrak{A}\mathfrak{B} = \mathfrak{B}\mathfrak{A}$ we have $\mathfrak{A} \vee \mathfrak{B} = \mathfrak{A}\mathfrak{B}$. Hence if $h \in \mathfrak{H}$ we can write $h = ab$, $a \in \mathfrak{A}$, $b \in \mathfrak{B}$. Thus $b = a^{-1}h$. Now by (a), (i) $\mathfrak{A} \leqslant \mathfrak{H}$ so that $b \in \mathfrak{H}$ and therefore $b \in \mathfrak{B} \wedge \mathfrak{H}$ and $h \in \mathfrak{A} \vee (\mathfrak{B} \wedge \mathfrak{H})$ so that $\mathfrak{H} \leqslant \mathfrak{A} \vee (\mathfrak{B} \wedge \mathfrak{H})$ q.e.d.

Corollary 1. *If* $\mathfrak{A}\mathfrak{B} = \mathfrak{B}\mathfrak{A}$ *and* $\mathfrak{A} \leqslant \mathfrak{C}$ *then* $(\mathfrak{A}\mathfrak{B}) \cap \mathfrak{C} = \mathfrak{A}(\mathfrak{B} \cap \mathfrak{C}) = (\mathfrak{B} \cap \mathfrak{C})\mathfrak{A} = \mathfrak{A} \vee (\mathfrak{B} \wedge \mathfrak{C})$ *is a subgroup of* $\mathfrak{G}$.

Corollary 2. *The lattice* $\Lambda_\Gamma(\mathfrak{G})$ *of all characteristic subgroups of* $\mathfrak{G}$ *is a modular lattice.*

d. The Lemma of Zassenhaus. It will not be attempted to give a motivation for the subject of the present subsection except to say that its appearance will be understood and fully justified in §3.

Theorem 4. (Lemma of Zassenhaus, 1934.) Let $\mathfrak{A}' \trianglelefteq \mathfrak{A} \leqslant \mathfrak{G}$ *and* $\mathfrak{B}' \trianglelefteq \mathfrak{B} \leqslant \mathfrak{G}$. *Then*

(1) $\quad \mathfrak{A}' \vee (\mathfrak{A} \wedge \mathfrak{B}') \trianglelefteq \mathfrak{A}' \vee (\mathfrak{A} \wedge \mathfrak{B})$,

i.e. $\mathfrak{A}'(\mathfrak{A} \cap \mathfrak{B}') \trianglelefteq \mathfrak{A}'(\mathfrak{A} \cap \mathfrak{B})$

$$(2) \quad \mathfrak{B}' \vee (\mathfrak{B} \wedge \mathfrak{A}') \trianglelefteq \mathfrak{B}' \vee (\mathfrak{B} \wedge \mathfrak{A}),$$

$$\text{i.e.} \quad \mathfrak{B}'(\mathfrak{B} \cap \mathfrak{A}') \trianglelefteq \mathfrak{B}'(\mathfrak{B} \cap \mathfrak{A})$$

$$(3) \quad \frac{\mathfrak{A}' \vee (\mathfrak{A} \wedge \mathfrak{B})}{\mathfrak{A}' \vee (\mathfrak{A} \wedge \mathfrak{B}')} \simeq \frac{\mathfrak{B}' \vee (\mathfrak{B} \wedge \mathfrak{A})}{\mathfrak{B}' \vee (\mathfrak{B} \wedge \mathfrak{A}')},$$

$$\text{i.e.} \quad \frac{\mathfrak{A}'(\mathfrak{A} \cap \mathfrak{B})}{\mathfrak{A}'(\mathfrak{A} \cap \mathfrak{B}')} \simeq \frac{\mathfrak{B}'(\mathfrak{B} \cap \mathfrak{A})}{\mathfrak{B}'(\mathfrak{B} \cap \mathfrak{A}')}.$$

Remark. The lattice notation $\wedge$ and $\vee$ is of course avoidable, but it gives the relations of the theorem a symmetric form.

In preparation for the proof we shall state two primitive facts which are easy to verify:

Let $\mathfrak{A}, \mathfrak{B}, \mathfrak{C} \leqslant \mathfrak{G}$ and $a \in \mathfrak{A}, b \in \mathfrak{B}, c \in \mathfrak{C}$.

I. If $c\mathfrak{A}c^{-1} = \mathfrak{A}$ and $c\mathfrak{B}c^{-1} = \mathfrak{B}$, that is: $c \in \mathfrak{N}(\mathfrak{A}) \wedge \mathfrak{N}(\mathfrak{B})$, then

(α) $c(\mathfrak{A} \wedge \mathfrak{B})c^{-1} = \mathfrak{A} \wedge \mathfrak{B}$, i.e. $c \in \mathfrak{N}(\mathfrak{A} \wedge \mathfrak{B})$

(β) $c(\mathfrak{A} \vee \mathfrak{B})c^{-1} = \mathfrak{A} \vee \mathfrak{B}$, i.e. $c \in \mathfrak{N}(\mathfrak{A} \vee \mathfrak{B})$.

II. If $\mathfrak{A} \leqslant \mathfrak{N}(\mathfrak{C})$, $\mathfrak{B} \leqslant \mathfrak{N}(\mathfrak{C})$, then
$\mathfrak{A} \wedge \mathfrak{B} \leqslant \mathfrak{N}(\mathfrak{C})$ and $\mathfrak{A} \vee \mathfrak{B} \leqslant \mathfrak{N}(\mathfrak{C})$.

Proof of Theorem 4. (1). The two facts I, II will enable us to verify the invariance of the left side of (1) under transformation with elements of parts, and then of the whole, of the right side:

Every element of $\mathfrak{A} \wedge \mathfrak{B}$ transforms

(i) $\mathfrak{A}$ into itself because $\mathfrak{A} \wedge \mathfrak{B} \leqslant \mathfrak{A}$

(ii) $\mathfrak{B}'$ into itself because $\mathfrak{B}' \trianglelefteq \mathfrak{B}$ and $\mathfrak{A} \wedge \mathfrak{B} \leqslant \mathfrak{B}$

(iii) $\mathfrak{A} \wedge \mathfrak{B}'$ into itself because of I (α)

(iv) $\mathfrak{A}'$ into itself because $\mathfrak{A}' \trianglelefteq \mathfrak{A}$

(v) $\mathfrak{A}' \vee (\mathfrak{A} \wedge \mathfrak{B}')$ into itself because of (iii), (iv) and I (β).

Further we observe that

(vi) $(\mathfrak{A}' \vee (\mathfrak{A} \wedge \mathfrak{B}')) \vee (\mathfrak{A} \wedge \mathfrak{B}) = \mathfrak{A}' \vee (\mathfrak{A} \wedge \mathfrak{B})$

because $\mathfrak{A} \wedge \mathfrak{B}' \leqslant \mathfrak{A} \wedge \mathfrak{B}$

and since obviously every element of the subgroup $\mathfrak{A}' \vee (\mathfrak{A} \wedge \mathfrak{B}')$ transforms $\mathfrak{A}' \vee (\mathfrak{A} \wedge \mathfrak{B}')$ into itself, we conclude with regard to (v), (vi) and II that every element of $\mathfrak{A}' \vee (\mathfrak{A} \wedge \mathfrak{B})$ transforms $\mathfrak{A}' \vee (\mathfrak{A} \wedge \mathfrak{B}')$ into itself. Thus (1) is proved.

Proof of (2). Interchange $\mathfrak{A}$ and $\mathfrak{B}$, $\mathfrak{A}'$ and $\mathfrak{B}'$ in the preceding proof.

Proof of (3). We recall that a normal subgroup commutes with every other subgroup. Thus, as $\mathfrak{A}' \leqslant \mathfrak{A}$ and $\mathfrak{A} \wedge \mathfrak{B}' \leqslant \mathfrak{A}$ we conclude that $\mathfrak{A}' \vee (\mathfrak{A} \wedge \mathfrak{B}') = \mathfrak{A}'(\mathfrak{A} \wedge \mathfrak{B}') = (\mathfrak{A} \wedge \mathfrak{B}')\mathfrak{A}'$. Now put

$$\mathfrak{D} = \mathfrak{A} \wedge \mathfrak{B}, \quad \mathfrak{K} = \mathfrak{A}' \vee (\mathfrak{A} \wedge \mathfrak{B}').$$

We shall derive a suitable expression for $\mathfrak{D} \wedge \mathfrak{K}$. For this purpose, in order to apply Theorem 3′, let us identify

$$\mathfrak{A} \wedge \mathfrak{B} \text{ with } \bar{\mathfrak{C}}, \quad \mathfrak{A}' \text{ with } \bar{\mathfrak{B}}, \quad \mathfrak{A} \wedge \mathfrak{B}' \text{ with } \bar{\mathfrak{A}}.$$

Then

$$\mathfrak{D} \wedge \mathfrak{K} = (\bar{\mathfrak{A}} \vee \bar{\mathfrak{B}}) \wedge \bar{\mathfrak{C}} = \bar{\mathfrak{A}} \vee (\bar{\mathfrak{B}} \wedge \bar{\mathfrak{C}});$$

therefore

$$\begin{aligned}(4) \qquad \mathfrak{D} \wedge \mathfrak{K} &= (\mathfrak{A} \wedge \mathfrak{B}') \vee (\mathfrak{A}' \vee (\mathfrak{A} \wedge \mathfrak{B})) \\ &= (\mathfrak{A} \wedge \mathfrak{B}') \vee (\mathfrak{B} \wedge \mathfrak{A}').\end{aligned}$$

Further we form

$$\begin{aligned}\mathfrak{D} \vee \mathfrak{K} &= (\mathfrak{A}' \vee (\mathfrak{A} \wedge \mathfrak{B}')) \vee (\mathfrak{A} \wedge \mathfrak{B}) = \mathfrak{A}' \vee (\mathfrak{A} \wedge \mathfrak{B}) \\ (5) \qquad &= \mathfrak{A}' \vee \mathfrak{D} = \mathfrak{A}'\mathfrak{D} \quad \text{because} \quad \mathfrak{A}' \trianglelefteq \mathfrak{A}, \mathfrak{D} \leqslant \mathfrak{A}.\end{aligned}$$

We shall show that $\mathfrak{K} \trianglelefteq \mathfrak{D} \vee \mathfrak{K}$ so that $\mathfrak{D} \vee \mathfrak{K} = \mathfrak{D}\mathfrak{K} = \mathfrak{K}\mathfrak{D}$. As in the proof of (1) we verify the invariance of $\mathfrak{K}$ under transformation with the elements of $\mathfrak{D} \vee \mathfrak{K}$:

Every element of $\mathfrak{A} \wedge \mathfrak{B} = \mathfrak{D}$ transforms

$\mathfrak{A}'$	into itself because	$\mathfrak{A}' \trianglelefteq \mathfrak{A}$ and $\mathfrak{D} \leqslant \mathfrak{A}$
$\mathfrak{A}$	into itself because	$\mathfrak{D} \leqslant \mathfrak{A}$
$\mathfrak{B}'$	into itself because	$\mathfrak{B}' \trianglelefteq \mathfrak{B}$ and $\mathfrak{D} \leqslant \mathfrak{B}$
$\mathfrak{K} = \mathfrak{A}' \vee (\mathfrak{A} \wedge \mathfrak{B}')$	into itself because	of I,

and by II every element of $\mathfrak{D} \vee \mathfrak{K}$ transforms $\mathfrak{K}$ into itself.

Now we apply the isomorphism theorem 5 of Chap. II, §5: With regard to (4) and (5)

$$\mathfrak{D}\mathfrak{K}/\mathfrak{K} \simeq \mathfrak{D}/(\mathfrak{D} \cap \mathfrak{K}),$$

that is

$$\frac{\mathfrak{A}' \vee (\mathfrak{A} \wedge \mathfrak{B})}{\mathfrak{A}' \vee (\mathfrak{A} \wedge \mathfrak{B}')} \simeq \frac{\mathfrak{A} \wedge \mathfrak{B}}{(\mathfrak{A} \wedge \mathfrak{B}') \vee (\mathfrak{B} \wedge \mathfrak{A}')}.$$

The symmetry in $\mathfrak{A}$, $\mathfrak{B}$ in the right-hand term of this isomorphism must exist also in the left-hand term; thus

$$\frac{\mathfrak{B}' \vee (\mathfrak{B} \wedge \mathfrak{A})}{\mathfrak{B}' \vee (\mathfrak{B} \wedge \mathfrak{A}')} \simeq \frac{\mathfrak{A} \wedge \mathfrak{B}}{(\mathfrak{A} \wedge \mathfrak{B}') \vee (\mathfrak{B} \wedge \mathfrak{A}')}.$$

So we have proved (3).

Examples and exercises

1. By verifying the axioms I–IV and I′–IV′ of §1, *b.*, Theorem 1, show that the system of all subgroups of $\mathfrak{G}$ represents a lattice if $\mathfrak{A} \wedge \mathfrak{B} = \mathfrak{A} \cap \mathfrak{B}$, $\mathfrak{A} \vee \mathfrak{B} = \langle \mathfrak{A}, \mathfrak{B} \rangle$.

2. Show that the lattice of all the subgroups of a cyclic group is distributive.

3. Give a detailed proof for Corollary 2 of Theorem 3.

4. Establish all possible significant special cases of the isomorphism (3) in Zassenhaus' Lemma. (Consider e.g. $\mathfrak{A} \leqslant \mathfrak{B}$, or $\mathfrak{A} \leqslant \mathfrak{B}$ and $\mathfrak{A}' = \mathfrak{e}$.)

5. Using the notion of Ω-isomorphism (Chap. III, §5, *c.*) formulate and prove the isomorphism theorem (Theorem 5 of Chap. II, §5) and Zassenhaus' Lemma for Ω-admissible subgroups of a group $\mathfrak{G}$ with the operator domain Ω.

6. Let $\Lambda_\Delta(\mathfrak{G})$ denote the modular lattice of all normal subgroups and $\Lambda_\Gamma(\mathfrak{G})$ that of the characteristic subgroups of a group $\mathfrak{G}$. Let

(i) $\mathfrak{H} < \mathfrak{G}$, (ii) $\mathfrak{H} \lhd \mathfrak{G}$, (iii) $\mathfrak{H} \,⋬\, \mathfrak{G}$.

In each case decide the following three questions:

(a) Is $\Lambda_\Delta(\mathfrak{H})$ a sublattice of $\Lambda_\Delta(\mathfrak{G})$?

(b) Is $\Lambda_\Gamma(\mathfrak{H})$ a sublattice of $\Lambda_\Delta(\mathfrak{G})$?

(c) Is $\Lambda_\Gamma(\mathfrak{H})$ a sublattice of $\Lambda_\Gamma(\mathfrak{G})$?

Give reasons or counterexample.

7. In §1, ex. 2(b) a graph has been associated with the partially ordered system of all subsets of a finite set. Similarly a graph can be associated with the lattice $\Lambda(\mathfrak{G})$ of all subgroups of a finite group $\mathfrak{G}$ in the following way: To every subgroup $\mathfrak{A}$ of $\mathfrak{G}$ corresponds a vertex. This vertex is joined by an edge to every subgroup $\mathfrak{B}$ of $\mathfrak{G}$ in which $\mathfrak{A}$ *is maximal*, that is, there is no subgroup $\mathfrak{H}$ of $\mathfrak{G}$ which satisfies the condition $\mathfrak{A} < \mathfrak{H} < \mathfrak{B}$. Thus the graph associated with $\Lambda(\mathfrak{G})$ begins with the vertex corresponding to the subgroup which consists of the unit element only and ends with the vertex corresponding to the whole group $\mathfrak{G}$.

Draw the graphs associated with some of the groups of small orders.

§3 The theory of O. Schreier

a. Chains and series of subgroups. A finite or infinite sequence of subgroups $\mathfrak{G}_i$ $(i = 0, 1, 2, \ldots)$ of a group $\mathfrak{G} = \mathfrak{G}_0$ will be called a *descending chain* of subgroups if each $\mathfrak{G}_i$ contains all the following ones: $\mathfrak{G}_{i+1} \leqslant \mathfrak{G}_i$; a *descending subnormal chain* if $\mathfrak{G}_{i+1} \trianglelefteq \mathfrak{G}_i$. We shall also write $\mathfrak{G}_i \geqslant \mathfrak{G}_{i+1}$ and $\mathfrak{G}_i \trianglerighteq \mathfrak{G}_{i+1}$ respectively.

If a subgroup chain has only a finite number of members and if its last member is the group consisting of the unit element e only, that is, if

$$(1) \qquad \mathfrak{G} = \mathfrak{G}_0 \geqslant \mathfrak{G}_1 \geqslant \mathfrak{G}_2 \geqslant \ldots \geqslant \mathfrak{G}_{s-1} \geqslant \mathfrak{G}_s = \mathfrak{e}$$

it will be called a *descending series* of subgroups. The number of proper inclusions (equality excluded) will be called the *length* of the series; thus the series

(2) $\mathfrak{G} = \mathfrak{G}_0 > \mathfrak{G}_1 > \mathfrak{G}_2 > \ldots > \mathfrak{G}_{l-1} > \mathfrak{G}_l = \mathfrak{e}$

is said to be a *proper* series of *length l.*

The series (1) is called

a *subnormal* series if $\mathfrak{G}_i \trianglelefteq \mathfrak{G}_{i-1}$

a *normal* series if $\mathfrak{G}_i \trianglelefteq \mathfrak{G}$ $(i = 1, 2, \ldots, s)$

a *characteristic* series if $\mathfrak{G}_i \,\underline{\lhd\!\!\lhd}\, \mathfrak{G}_{i-1}$.

At a first glance it would seem to be consistent to call the last case "subcharacteristic"; with regard to Chap. III, §1, Theorem 3(a), however, a subcharacteristic series is necessarily a characteristic series: $\mathfrak{G}_i \,\underline{\lhd\!\!\lhd}\, \mathfrak{G}$. A characteristic series is of course normal.

Remark. A subnormal series was formerly (and is still sometimes) called normal; a normal series was then called invariant.

In order to introduce another important kind of chain or series we need the following definition: Let $\mathfrak{H} < \mathfrak{G}$; then $\mathfrak{H}$ is a *maximal subgroup* of $\mathfrak{G}$, in symbols: $\mathfrak{H} \underset{\max}{<} \mathfrak{G}$, if there exists no proper subgroup $\mathfrak{L}$ of $\mathfrak{G}$ such that $\mathfrak{H} < \mathfrak{L} < \mathfrak{G}$. Likewise, let $\mathfrak{H} \lhd \mathfrak{G}$; then $\mathfrak{H}$ is a *maximal normal subgroup* of $\mathfrak{G}$, in symbols: $\mathfrak{H} \underset{\max}{\lhd} \mathfrak{G}$, if there exists no proper normal subgroup $\mathfrak{L}$ of $\mathfrak{G}$ such that $\mathfrak{H} \lhd \mathfrak{L} \lhd \mathfrak{G}$ (equality being excluded everywhere).

THEOREM 1. $\mathfrak{H} \underset{\max}{\lhd} \mathfrak{G}$ *if and only if* $\mathfrak{G}/\mathfrak{H}$ *is a simple group* (cf. Chap. II, §5, *c*.).

Proof. (a) If $\mathfrak{G}/\mathfrak{H}$ is not simple, then there is a proper normal subgroup $\mathfrak{L}/\mathfrak{H} < \mathfrak{G}/\mathfrak{H}$, $|\mathfrak{L}/\mathfrak{H}| > 1$, so that by the correspondence theorem 4 (Chap. II, §5, *b*.) $\mathfrak{H} \lhd \mathfrak{L} \lhd \mathfrak{G}$. Thus $\mathfrak{H}$ is not maximal normal in $\mathfrak{G}$.

(b) If $\mathfrak{H}$ is not a maximal normal subgroup of $\mathfrak{G}$, then there is an $\mathfrak{L}$ such that $\mathfrak{H} \lhd \mathfrak{L} \lhd \mathfrak{G}$, hence $|\mathfrak{L}/\mathfrak{H}| > 1$ and $\mathfrak{L}/\mathfrak{H} \lhd \mathfrak{G}/\mathfrak{H}$. Therefore $\mathfrak{G}/\mathfrak{H}$ is not simple.

Now we shall say that a subnormal series (2) is a *composition series* of length l in $\mathfrak{G}$ if

(3) $\mathfrak{G}_i \underset{\max}{\lhd} \mathfrak{G}_{i-1}$ $(i = 1, \ldots, l)$

or (according to Theorem 1) equivalently, if all the factor groups $\mathfrak{G}_{i-1}/\mathfrak{G}_i$, the so-called *composition factors* of the series, are simple groups.

In every finite group there is a composition series and there may be several different composition series, cf. ex. 3. It will be clear that an infinite group with an infinite descending subnormal chain with an infinite number of proper inclusions (as e.g. $\mathbb{Z}^+$) cannot have a composition series. Every simple group has exactly one composition series of length one.

b. Refinement. If for a series (1) where $\mathfrak{G}_{i+1} < \mathfrak{G}_i$ subgroups $\mathfrak{G}_{i,1}, \mathfrak{G}_{i,2}, \ldots \mathfrak{G}_{i,k}$ exist so that

$$\mathfrak{G}_i > \mathfrak{G}_{i,1} > \ldots > \mathfrak{G}_{i,k} = \mathfrak{G}_{i+1},$$

we shall say that the series (1) is refinable or "refined by" these subgroups $\mathfrak{G}_{i,k}$. The series

$$\mathfrak{G} \geqslant \mathfrak{H}_1 \geqslant \mathfrak{H}_2 \geqslant \ldots \geqslant \mathfrak{H}_m$$

is called a *refinement* of the series (1) if every $\mathfrak{G}_i$ is an $\mathfrak{H}_j$ but not necessarily every $\mathfrak{H}_j$ is a $\mathfrak{G}_i$.

Usually it is understood that an essential property of a given series should not be lost by a refinement. If e.g. the given series is subnormal then we take into consideration only subnormal refinements of this series.

A composition series of $\mathfrak{G}$ clearly is one which, as a subnormal series, cannot be refined. Conversely, every (in this sense) non-refinable subnormal series is a composition series.

DEFINITION. *Two proper subnormal series of* $\mathfrak{G}$

$$(4) \qquad \mathfrak{G} = \mathfrak{G}_0 \triangleright \mathfrak{G}_1 \triangleright \mathfrak{G}_2 \triangleright \ldots \triangleright \mathfrak{G}_l = \mathfrak{e}$$

$$(5) \qquad \mathfrak{G} = \mathfrak{H}_0 \triangleright \mathfrak{H}_1 \triangleright \mathfrak{H}_2 \triangleright \ldots \triangleright \mathfrak{H}_m = \mathfrak{e}$$

are said to be ***equivalent*** *or* ***isomorphic*** *if*

(a) *Both have the same length*: $m = l$, *and*

(b) *There is a permutation* $P = \binom{i}{j}$ $(i = 0, 1, \ldots, l-1)$ *of degree* l *such that* $\mathfrak{G}_i/\mathfrak{G}_{i+1} \simeq \mathfrak{H}_j/\mathfrak{H}_{j+1}$.

Remark. The permutation P is in general not uniquely defined because several of the $\mathfrak{G}_i/\mathfrak{G}_{i+1}$ may be isomorphic to each other.

The preceding definition is the basis for the following fundamental theorem:

THEOREM 2. (O. Schreier, 1928). *Two distinct proper subnormal series of a group $\mathfrak{G}$ can be so refined that the two resulting proper subnormal series are isomorphic.*

Proof. (H. Zassenhaus, 1934). Let us assume that the two distinct proper subnormal series of $\mathfrak{G}$ are given by (4) and (5). For a fixed i $(0 \leqslant i < l)$ consider the gap between $\mathfrak{G}_i$ and $\mathfrak{G}_{i+1}$. In this we insert the subgroups $\mathfrak{G}_{i+1} \vee (\mathfrak{G}_i \wedge \mathfrak{H}_j) = \mathfrak{G}_{i+1}(\mathfrak{G}_i \cap \mathfrak{H}_j)$ $(j = 0, 1, \ldots, m)$ which by §2, Theorem 4 (Zassenhaus' Lemma) are normal:

$$\mathfrak{G}_{i+1}(\mathfrak{G}_i \cap \mathfrak{H}_j) \trianglelefteq \mathfrak{G}_{i+1}(\mathfrak{G}_i \cap \mathfrak{H}_{j-1}).$$

In this way we obtain a refinement of the series (4)

$$\mathfrak{G}_i = \mathfrak{G}_{i+1}\mathfrak{G}_i = \mathfrak{G}_{i+1}(\mathfrak{G}_i \cap \mathfrak{H}_0) \trianglerighteq \mathfrak{G}_{i+1}(\mathfrak{G}_i \cap \mathfrak{H}_1) \trianglerighteq \cdots$$
$$\cdots \trianglerighteq \mathfrak{G}_{i+1}(\mathfrak{G}_i \cap \mathfrak{H}_m) = \mathfrak{G}_{i+1}\mathfrak{e} = \mathfrak{G}_{i+1}.$$

Likewise for a fixed j $(0 \leqslant j < m)$ we fill the gap between $\mathfrak{H}_j$ and $\mathfrak{H}_{j+1}$ by inserting the normal subgroups $\mathfrak{H}_{j+1}(\mathfrak{H}_j \cap \mathfrak{G}_i)$ $(i = 0, 1, \ldots, l)$:

$$\mathfrak{H}_j = \mathfrak{H}_{j+1}\mathfrak{H}_j = \mathfrak{H}_{j+1}(\mathfrak{H}_j \cap \mathfrak{G}_0) \trianglerighteq \mathfrak{H}_{j+1}(\mathfrak{H}_j \cap \mathfrak{G}_1) \trianglerighteq \cdots$$
$$\cdots \trianglerighteq \mathfrak{H}_{j+1}(\mathfrak{H}_j \cap \mathfrak{G}_l) = \mathfrak{H}_{j+1}\mathfrak{e} = \mathfrak{H}_{j+1}.$$

In order to apply part (3) of Zassenhaus' lemma we identify $\mathfrak{G}_i$ with $\mathfrak{A}$, $\mathfrak{G}_{i+1}$ with $\mathfrak{A}'$, $\mathfrak{H}_j$ with $\mathfrak{B}$, $\mathfrak{H}_{j+1}$ with $\mathfrak{B}'$. Then for the fixed i and the fixed j which we had chosen we have

$$(6) \qquad \frac{\mathfrak{G}_{i+1}(\mathfrak{G}_i \cap \mathfrak{H}_j)}{\mathfrak{G}_{i+1}(\mathfrak{G}_i \cap \mathfrak{H}_{j+1})} \simeq \frac{\mathfrak{H}_{j+1}(\mathfrak{H}_j \cap \mathfrak{G}_i)}{\mathfrak{H}_{j+1}(\mathfrak{H}_j \cap \mathfrak{G}_{i+1})}$$

or using the abbreviated notation

$$\mathfrak{G}_{i,j} = \mathfrak{G}_{i+1}(\mathfrak{G}_i \cap \mathfrak{H}_j), \quad \mathfrak{H}_{j,i} = \mathfrak{H}_{j+1}(\mathfrak{H}_j \cap \mathfrak{G}_i)$$

we may write instead of (6)

$$(6') \qquad \mathfrak{G}_{i,j}/\mathfrak{G}_{i,j+1} \simeq \mathfrak{H}_{j,i}/\mathfrak{H}_{j,i+1}.$$

If this refinement is carried out in all the intervals of the two series (4) and (5) we arrive at the two refined series

$$(4)' \qquad \mathfrak{G} = \mathfrak{G}_0 = \mathfrak{G}_{0,0} \trianglerighteq \mathfrak{G}_{0,1} \trianglerighteq \dots \trianglerighteq \mathfrak{G}_{0,m} = \mathfrak{G}_1$$
$$= \mathfrak{G}_{1,0} \trianglerighteq \mathfrak{G}_{1,1} \trianglerighteq \dots \trianglerighteq \mathfrak{G}_{1,m}$$
$$= \mathfrak{G}_2 = \mathfrak{G}_{2,0} \trianglerighteq \dots \trianglerighteq \mathfrak{G}_{l-1,m} = \mathfrak{G}_l = \mathfrak{e};$$

$$(5)' \qquad \mathfrak{G} = \mathfrak{H}_0 = \mathfrak{H}_{0,0} \trianglerighteq \mathfrak{H}_{0,1} \trianglerighteq \dots \trianglerighteq \mathfrak{H}_{0l} = \mathfrak{H}_1$$
$$= \mathfrak{H}_{1,0} \trianglerighteq \mathfrak{H}_{1,1} \trianglerighteq \dots \trianglerighteq \mathfrak{H}_{1,l}$$
$$= \mathfrak{H}_2 = \mathfrak{H}_{2,0} \trianglerighteq \dots \trianglerighteq \mathfrak{H}_{m-1,l} = \mathfrak{H}_m = \mathfrak{e}.$$

So the l gaps in (4) are filled each with m new terms and the m gaps in (5) are filled with l new terms and we have two new series (4′), (5)′, each with $l \cdot m$ terms, not necessarily all different. The isomorphism (6) between the factorgroups formed with successive members guarantees that in both series (4)′, (5)′ the same number of terms are actually distinct which means that after deletion of all repetitions both series (4)′ and (5)′ have the same length and there is a one-one correspondence between the remaining factor groups of (4)′ and those of (5)′ so that the corresponding factorgroups (6) are isomorphic. Thus Schreier's theorem is proved.

Remark. The proof of Schreier's theorem determines the correspondence between isomorphic factor groups from two equivalent subnormal series, obtained by a suitable refinement of two given subnormal series of a group. This correspondence may be illustrated by a graphical scheme where subgroups $\mathfrak{G}_{i,j}$ of a series are represented by dots on a vertical line, factor groups $\mathfrak{G}_{i,j}/\mathfrak{G}_{i,j+1}$ by small circles on the line between the dots associated with $\mathfrak{G}_{i,j}$ and $\mathfrak{G}_{i,j+1}$. Isomorphic factor groups of the two series are joined by cross lines. In the figure 8 it is assumed that $l = 3$ and $m = 4$.

c. An immediate consequence of Schreier's theorem is the classical *Theorem of Jordan and Hölder*:

Correspondence between isomorphic series

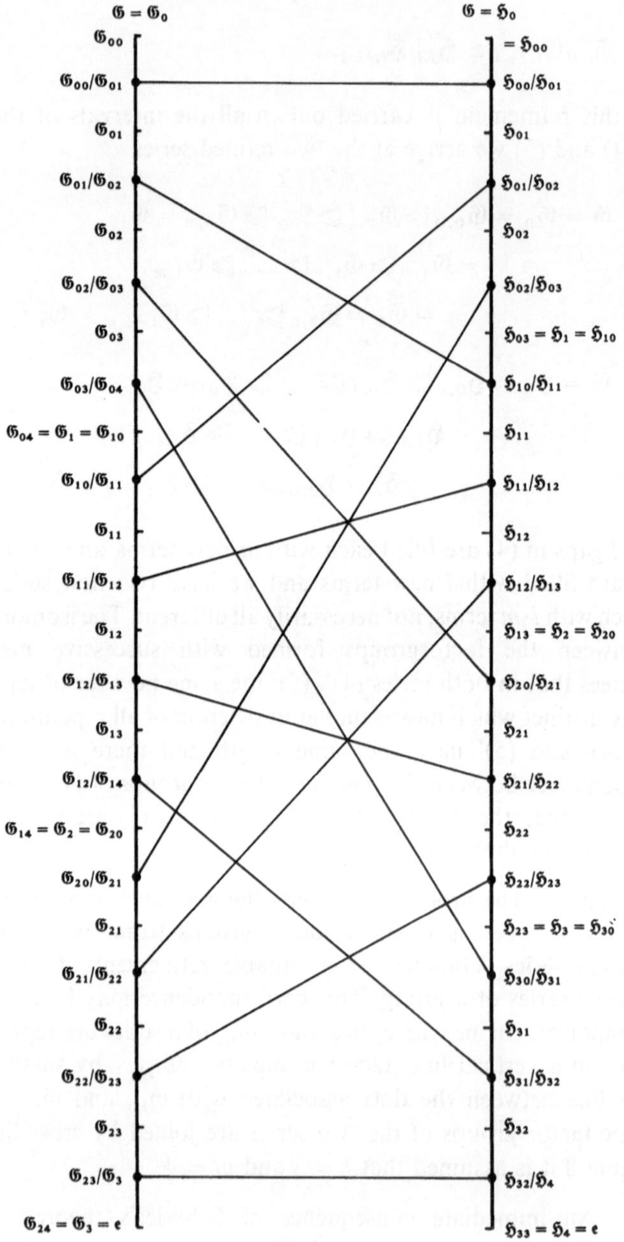

Figure 8

THEOREM 3. *Let $\mathfrak{G}$ be a group which possesses a composition series. Then every two composition series of $\mathfrak{G}$ have equal length and are isomorphic* (*equivalent*).

Remark. The French mathematician C. Jordan proved in 1869 that two composition series of a finite group have equal length and that there is a correspondence between the composition factors so that corresponding factors have equal order. The German mathematician O. Hölder completed the theorem in 1889 by showing that corresponding composition factors are isomorphic (cf. ex. 2).

Proof. A composition series does not admit refinement as a subnormal series; thus it is reproduced by the refinement process described in the proof of Schreier's theorem which intends to put normal subgroups in the gaps of the series: A normal subgroup of $\mathfrak{G}_i$ inserted between $\mathfrak{G}_i$ and $\mathfrak{G}_{i+1}$ will coincide either with $\mathfrak{G}_i$ or with $\mathfrak{G}_{i+1}$. Hence a composition series being the final result of any subnormal refinement process is isomorphic to any other composition series of $\mathfrak{G}$.

A parallel development ensues when we start with normal series instead of subnormal series. Indeed let (4) and (5) now represent two normal series in the group $\mathfrak{G}$. Refinement by means of the process of the proof of Schreier's theorem inserts the normal subgroups $\mathfrak{G}_{i+1}(\mathfrak{G}_i \cap \mathfrak{H}_j) = \mathfrak{G}_{i,j}$ $(j = 0, 1 \ldots, m)$ into the gap between $\mathfrak{G}_i$ and $\mathfrak{G}_{i+1}$, and since $\mathfrak{G}_i, \mathfrak{G}_{i+1}, \mathfrak{H}_j \trianglelefteq \mathfrak{G}$, also $\mathfrak{G}_{i,j} \trianglelefteq \mathfrak{G}$. This refinement of a normal series therefore is a normal series. Hence we have the result:

THEOREM 4. *If a group possesses a normal series then every two of its normal series admit isomorphic refinements.*

DEFINITION. *A normal series* (4) *which does not admit refinement into another normal series of $\mathfrak{G}$ is called a* **chief series** *or* **principal series** (*German*: *Hauptreihe*) *of $\mathfrak{G}$.*

An immediate consequence of Theorem 4 is then

COROLLARY. (Jordan and Hölder). *If $\mathfrak{G}$ possesses a chief series then two arbitrary chief series of $\mathfrak{G}$ are isomorphic.*

d. Applications. The theory of series of subgroups plays an important role in the classification of groups. In the present section we shall restrict ourselves to finite groups $\mathfrak{G}$.

LEMMA. *Given a subnormal series* (4) *in a finite group $\mathfrak{G}$, a composition series of $\mathfrak{G}$ can be constructed by refinement of the subnormal series.*

Proof. This refinement is effected as follows: If $\mathfrak{G}_1$ is not a maximum normal subgroup of $\mathfrak{G}_0$ one can find among all the normal subgroups of $\mathfrak{G}_0$ which contain $\mathfrak{G}_1$ a maximal one, say $\mathfrak{H}_1$. Then, however, $\mathfrak{H}_1$ is a maximum normal subgroup of $\mathfrak{G}_0$ in the absolute sense of the definition of subsection *a.*; for if this were not so, there would be a normal subgroup $\mathfrak{L}$ such that $\mathfrak{H}_1 \lhd \mathfrak{L} \lhd \mathfrak{G}_0$, implying that $\mathfrak{H}_1$ is not, as chosen, a maximum normal subgroup of $\mathfrak{G}_0$ containing $\mathfrak{G}_1$.

If $\mathfrak{G}_1 \lhd \mathfrak{H}_1$ it will be possible to find in $\mathfrak{H}_1$ a maximum normal subgroup $\mathfrak{H}_2$ such that $\mathfrak{G}_1 \leqslant \mathfrak{H}_2$, etc. In this way we fill the gap between $\mathfrak{G}_0$ and $\mathfrak{G}_1$ and obtain the first section of a composition series

$$\mathfrak{G}_0 \underset{\max}{\rhd} \mathfrak{H}_1 \underset{\max}{\rhd} \mathfrak{H}_2 \underset{\max}{\rhd} \cdots \underset{\max}{\rhd} \mathfrak{H}_k = \mathfrak{G}_1,$$

as a refinement of (4). Using the same argument we can also fill the other gaps. Thus the lemma is proved.

DEFINITION 1. *A finite group $\mathfrak{G}$ is said to be* ***solvable*** *if all its composition factors $\mathfrak{G}_i/\mathfrak{G}_{i+1}$ are simple abelian, i.e. cyclic of prime order.*

This definition cannot be extended satisfactorily to infinite groups. Therefore we introduce another definition which is based on the notion of the derived chain or commutator chain of $\mathfrak{G}$.

In Chap. III, §1, *c.* we had the commutator or derived subgroup $\mathfrak{K} = \mathfrak{K}(\mathfrak{G}) = \mathfrak{G}'$ as an example of a characteristic subgroup of $\mathfrak{G}$, namely $\mathfrak{G}' = \langle \mathbf{C} \rangle$ where $\mathbf{C}$ is the set of all commutators in $\mathfrak{G}$. Denoting by $\mathbf{C}'$ the set of all commutators in $\mathfrak{G}'$ we form $\mathfrak{G}'' = \langle \mathbf{C}' \rangle$, etc. In this way we obtain the so-called *derived chain*

$$\mathfrak{G} \trianglerighteq \mathfrak{G}' \trianglerighteq \mathfrak{G}'' \trianglerighteq \cdots \trianglerighteq \mathfrak{G}^{(r)} \trianglerighteq \cdots.$$

This is a characteristic series, called the *derived series* of $\mathfrak{G}$, if and only if there is a natural number r for which $\mathfrak{G}^{(r)} = \mathfrak{e}$.

Even if $\mathfrak{G}$ is finite the derived chain may not be a series: For a natural number r it may happen that $\mathfrak{G}^{(r+1)} = \mathfrak{G}^{(r)} > \mathfrak{e}$ and therefore all subsequent $\mathfrak{G}^{(r+n)} = \mathfrak{G}^{(r)}$. Indeed, if $\mathfrak{G}$ is non-abelian and simple or characteristically simple we have $\mathfrak{G}' = \mathfrak{G}$. Hence in a derived series repetitions cannot occur.

DEFINITION 2. *A group* $\mathfrak{G}$ *is said to be* ***solvable of length*** *r if its derived series has length r, that is if its r-th commutator group* $\mathfrak{G}^{(r)} = \mathfrak{e}$ *and r is the smallest natural number having this property.*

A group is solvable of length one if and only if it is abelian. A group $\mathfrak{G}$ which is solvable of length 2 is called *metabelian*; in this case $\mathfrak{G}'' = \mathfrak{e}$, i.e. $\mathfrak{G}'$ is abelian.

e. THEOREM 5. *Let* $\mathfrak{G}$ *be a finite group and l be the length of its composition series. There is an r such that* $\mathfrak{G}$ *is solvable of length r if and only if* $\mathfrak{G}$ *is solvable. Moreover,* $r \leqslant l$.

Proof. (a) Suppose that $\mathfrak{G}$ is solvable of length r; let

$$(7) \qquad \mathfrak{G} \trianglerighteq \mathfrak{G}' \trianglerighteq \mathfrak{G}'' \trianglerighteq \dots \trianglerighteq \mathfrak{G}^{(s-1)} \trianglerighteq \mathfrak{G}^{(s)} = \mathfrak{e}$$

be its derived series. By Chap. III, §1, Theorem 2, the factor groups $\mathfrak{G}/\mathfrak{G}', \mathfrak{G}'/\mathfrak{G}'', \dots, \mathfrak{G}^{(s-1)}/\mathfrak{G}^{(s)} \simeq \mathfrak{G}^{(s-1)}$ are abelian. We refine the series (7) according to the lemma in *d.* into a composition series whose first section may be

$$\mathfrak{G} \underset{\max}{\triangleright} \mathfrak{H}_1 \underset{\max}{\triangleright} \mathfrak{H}_2 \underset{\max}{\triangleright} \dots \underset{\max}{\triangleright} \mathfrak{H}_k = \mathfrak{G}'.$$

Because $\mathfrak{G}' \trianglelefteq \mathfrak{H}_1$ the factor group $\mathfrak{G}/\mathfrak{H}_1$ is abelian. Further $\mathfrak{H}_1' \trianglelefteq \mathfrak{G}'$, thus $\mathfrak{H}_1' \trianglelefteq \mathfrak{H}_2$ and $\mathfrak{H}_1/\mathfrak{H}_2$ is abelian; also $\mathfrak{H}_2' \trianglelefteq \mathfrak{G}'$ so that $\mathfrak{H}_2' \trianglelefteq \mathfrak{H}_3$ and $\mathfrak{H}_2/\mathfrak{H}_3$ is abelian, etc. As composition factors all these factor groups are simple. In the same way we deal with the other sections of the series obtained by the refinement of (7) which thus leads to a composition series with simple abelian composition factors; that is, $\mathfrak{G}$ is solvable in the sense of definition 1.

(b) Conversely, let $\mathfrak{G}$ be solvable in the sense of definition 1 and $\mathfrak{G} \triangleright \mathfrak{G}' \triangleright \mathfrak{G}'' \triangleright \dots \triangleright \mathfrak{G}^{(s)}$ its derived chain, $\mathfrak{G}^{(s)} = \mathfrak{G}^{(s+1)} \geqslant \mathfrak{e}$. If we add $\mathfrak{e}$ to the non-repetitive part of the chain we obtain a subnormal series $\mathfrak{G} \triangleright \mathfrak{G}' \triangleright \dots \triangleright \mathfrak{G}^{(s)} \trianglerighteq \mathfrak{e}$ where the factor groups $\mathfrak{G} / \mathfrak{G}', \dots, \mathfrak{G}^{(s-1)}/\mathfrak{G}^{(s)}$ are abelian. It remains to show that $\mathfrak{G}^{(s)}$ is abelian.

By the lemma we can determine a composition series "passing through" $\mathfrak{G}^{(s)}$ (indeed $\mathfrak{G} \triangleright \mathfrak{G}^{(s)} \trianglerighteq \mathfrak{e}$ is a subnormal series of $\mathfrak{G}$) and if $\mathfrak{G}^{(s)} \underset{\max}{\triangleright} \mathfrak{H}$ where $\mathfrak{H}$ is a member of this composition series then $\mathfrak{G}^{(s)}/\mathfrak{H}$ is abelian because $\mathfrak{G}$ is solvable. Hence $\mathfrak{G}^{(s)\prime} \trianglelefteq \mathfrak{H}$ and $\mathfrak{G}^{(s)\prime} =$

$\mathfrak{G}^{(s+1)} \underset{\max}{\leqslant} \mathfrak{H} < \mathfrak{G}$, thus contradicting our assumption that $\mathfrak{G}^{(s+1)} = \mathfrak{G}^{(s)}$. Therefore $\mathfrak{G}$ is solvable of length $s+1$.

COROLLARY. *A finite group $\mathfrak{G}$ which has a subnormal series with abelian factor groups is solvable.*

Indeed if (4) is such a subnormal series of $\mathfrak{G}$ then we can apply to it the same argument which, in part (a) of the proof, has been applied to the derived series. Thus we fill the gap between $\mathfrak{G}$ and $\mathfrak{G}_1$ with

$$\mathfrak{G} \underset{\max}{\triangleright} \mathfrak{H}_1 \underset{\max}{\triangleright} \mathfrak{H}_2 \underset{\max}{\triangleright} \cdots \underset{\max}{\triangleright} \mathfrak{H}_k = \mathfrak{G}_1.$$

Since $\mathfrak{G}/\mathfrak{G}_1$ is abelian it follows that $\mathfrak{G}' \trianglelefteq \mathfrak{G}_1$, hence $\mathfrak{H}_1 \trianglerighteq \mathfrak{G}'$ and $\mathfrak{G}/\mathfrak{H}_1$ is abelian and simple, etc.

The statement of the corollary can be used as another (indeed equivalent) definition of a finite solvable group. This definition, as well as Definition 2, have the advantage of being independent of the notion of a composition series. They are therefore extendable to a class of infinite groups.

Remark. The word "solvable" (or "soluble") originates in the Galois theory of algebraic equations. An algebraic equation is said to be solvable (German: auflösbar) if its roots can be represented by means of rational and root operations (radicals) applied to its coefficients. It is one of the principal results of the classical theory of Galois that an algebraic equation is solvable if and only if the group associated with the equation in the sense of Galois theory is solvable.

Examples and exercises

1. Let $\mathfrak{A}$, $\mathfrak{B}$ be two different maximum normal subgroups of $\mathfrak{G}$ and $\mathfrak{A} \cap \mathfrak{B} = \mathfrak{D}$. Then $|\mathfrak{A}/\mathfrak{D} \cap \mathfrak{B}/\mathfrak{D}| = 1$, $\mathfrak{A}/\mathfrak{D} \lhd \mathfrak{G}/\mathfrak{D}$, $\mathfrak{B}/\mathfrak{D} \lhd \mathfrak{G}/\mathfrak{D}$. Hence one can form the direct product

$$\mathfrak{H}/\mathfrak{D} = \mathfrak{A}/\mathfrak{D} \times \mathfrak{B}/\mathfrak{D} \trianglelefteq \mathfrak{G}/\mathfrak{D}.$$

Consider now the factor group

$$\mathfrak{B}/\mathfrak{D} \simeq \frac{\mathfrak{H}/\mathfrak{D}}{\mathfrak{A}/\mathfrak{D}} \trianglelefteq \frac{\mathfrak{G}/\mathfrak{D}}{\mathfrak{A}/\mathfrak{D}} \simeq \mathfrak{G}/\mathfrak{A} \qquad \text{(cf. Chap. II, §5, Theorem 3, Corollary)}$$

Since $\mathfrak{G}/\mathfrak{A}$ is simple by supposition, so is $(\mathfrak{G}/\mathfrak{D})/(\mathfrak{A}/\mathfrak{D})$; thus it cannot have a proper normal subgroup. Hence $\mathfrak{H}/\mathfrak{D} = \mathfrak{G}/\mathfrak{D}$ and

(8) $\quad \mathfrak{G}/\mathfrak{A} \simeq (\mathfrak{G}/\mathfrak{D})/(\mathfrak{A}/\mathfrak{D}) \simeq \mathfrak{B}/\mathfrak{D}.$

Further, with regard to Chap. II, §5, Theorem 5: $\mathfrak{A}\mathfrak{B}/\mathfrak{A} \simeq \mathfrak{B}/\mathfrak{D}$. Hence

$$\mathfrak{A}\mathfrak{B}/\mathfrak{A} \simeq \mathfrak{G}/\mathfrak{A}$$

so that if $\mathfrak{G}$ is finite:

$$\mathfrak{G} = \mathfrak{A}\mathfrak{B}.$$

THEOREM. *Every finite group is the product of any two of its maximum normal subgroups.*

2. On the formula (8) rests the classical proof of the Jordan–Hölder Theorem for finite groups $\mathfrak{G}$. Complete it using the following outline: Take two composition series (4), (5) of $\mathfrak{G}$ and assume the theorem to be correct for all groups of order $< |\mathfrak{G}|$. Show that if $\mathfrak{G}_1 \cap \mathfrak{H}_1 = \mathfrak{D}$ has the composition series

$$\mathfrak{D} \underset{\max}{\rhd} \mathfrak{D}_1 \underset{\max}{\rhd} \cdots \underset{\max}{\rhd} \mathfrak{D}_n = \mathfrak{e}$$

then $\mathfrak{G} \rhd \mathfrak{G}_1 \rhd \mathfrak{D} \rhd \mathfrak{D}_1 \rhd \ldots \rhd \mathfrak{D}_n = \mathfrak{e}$ is a composition series of $\mathfrak{G}$ which is isomorphic to the series (4) and also isomorphic to the composition series $\mathfrak{G} \rhd \mathfrak{H}_1 \rhd \mathfrak{D} \rhd \mathfrak{D}_1 \rhd \ldots \rhd \mathfrak{D}_n = \mathfrak{e}$ which is isomorphic to the series (5). Hence (4) and (5) are isomorphic.

3. Establish all composition series of the cyclic groups $\langle a \rangle_{30}$ and $\langle a \rangle_{p^3}$ where p is a prime number, and of the dihedral group $\mathfrak{D}_{20}$. In each case verify the Jordan–Hölder Theorem.

4. Every abelian group is solvable.

5. Show that the groups $\mathfrak{S}_3$ and $\mathfrak{S}_4$ are solvable and that $\mathfrak{A}_4$ is the only normal subgroup of index 2 in $\mathfrak{S}_4$.

6. Prove that $\mathfrak{A}_n$ is the only normal subgroup of $\mathfrak{S}_n$ if $n > 4$.

Let us assume that there were two: $\mathfrak{A}_n, \mathfrak{B} \lhd \mathfrak{S}_n$. Since $\mathfrak{S}_n \rhd \mathfrak{A}_n \rhd \mathfrak{I} = \{\mathfrak{I}\}$ is a composition series of $\mathfrak{S}_n$, every other one (if others exist) also has length 2; thus under our assumption $\mathfrak{S}_n \triangleright \mathfrak{B} \triangleright \mathfrak{I}$ is another composition series. The two could be isomorphic in two ways which implies the index relations

$$\text{(i)} \qquad |\mathfrak{S}_n/\mathfrak{B}| = |\mathfrak{S}_n/\mathfrak{A}_n| = 2, \quad |\mathfrak{B}| = |\mathfrak{A}_n| = \tfrac{1}{2}n!$$

$$\text{(ii)} \qquad |\mathfrak{S}_n/\mathfrak{B}| = |\mathfrak{A}_n/\mathfrak{I}| = \tfrac{1}{2}n!, \quad |\mathfrak{B}| = |\mathfrak{S}_n/\mathfrak{A}_n| = 2$$

Since $\mathfrak{S}_n$ has no normal subgroup of order 2 the case (ii) is ruled out. As to (i): Let $\mathfrak{B} \neq \mathfrak{A}_n$ and $\mathfrak{A}_n \cap \mathfrak{B} = \mathfrak{D}$; then $\mathfrak{D} \triangleleft \mathfrak{A}_n$ hence $\mathfrak{D} = \mathfrak{I}$ because $\mathfrak{A}_n$ is simple (Chap. III, §1, ex. 13). Since $\mathfrak{A}_n$ and $\mathfrak{B}$ are both maximum normal subgroups: $\mathfrak{S}_n = \mathfrak{A}_n \times \mathfrak{B}$ (cf. ex. 1) which is impossible because $|\mathfrak{A}_n \times \mathfrak{B}| = (\tfrac{1}{2}n!)^2 > n!$ $(n > 4)$*.

7. (a) Every subgroup of a finite solvable group is solvable.
(b) Every factor group of a finite solvable group is solvable.

8. Every metabelian group contains a proper abelian normal subgroup.

9. A finite group $\mathfrak{G}$ is solvable if and only if it contains a subnormal series $\mathfrak{G} \triangleright \mathfrak{G}_1 \triangleright \mathfrak{G}_2 \triangleright \ldots \triangleright \mathfrak{G}_s = \mathfrak{e}$ whose factor groups $\mathfrak{G}/\mathfrak{G}_1$, $\mathfrak{G}_1/\mathfrak{G}_2, \ldots, \mathfrak{G}_{s-1}/\mathfrak{G}_s$ are solvable.

10. Can a chief series always be obtained from a composition series by deleting certain of its members?

11. (a) Define the concept of maximum characteristic subgroup $\mathfrak{H} \underset{\max}{\triangleleft\!\!\!|} \mathfrak{G}$ (cf. subsection *a.*).
(b) Show that $\mathfrak{H} \underset{\max}{\triangleleft\!\!\!|} \mathfrak{G}$ if and only if the factor group $\mathfrak{G}/\mathfrak{H}$ is characteristically simple (elementary abelian). (Cf. Theorem 1 and Chap. III, §1, Theorem 6b).

12. Formulate and prove a Jordan–Hölder theorem for characteristic series.

§4 Central chains and series

The derived (or commutator) chain and series led in a natural manner to the definition of the class of the solvable groups which obviously

* Indeed, the arithmetical inequality is valid for $n > 2$.

contains the class of the abelian groups. In the present section we shall consider other chains and series which similarly enable us to characterize a more restricted class of groups containing the abelian groups, namely, the class of the *nilpotent groups.*

a. An ascending normal chain of subgroups of a group $\mathfrak{G}$, say

$$(1) \qquad \mathfrak{e} = \mathfrak{G}_0 \trianglelefteq \mathfrak{G}_1 \trianglelefteq \mathfrak{G}_2 \trianglelefteq \dots$$

is called *an ascending central chain* if the factor groups of successive members satisfy the conditions $\mathfrak{G}_{i+1}/\mathfrak{G}_i \leqslant \mathfrak{Z}(\mathfrak{G}/\mathfrak{G}_i)$ $(i = 0, 1, \dots)$, that is:

$$(2) \qquad \mathfrak{G}_1/\mathfrak{G}_0 \simeq \mathfrak{G}_1 \trianglelefteq \mathfrak{Z}(\mathfrak{G}), \quad \mathfrak{G}_2/\mathfrak{G}_1 \trianglelefteq \mathfrak{Z}(\mathfrak{G}/\mathfrak{G}_1),$$
$$\mathfrak{G}_3/\mathfrak{G}_2 \trianglelefteq \mathfrak{Z}(\mathfrak{G}/\mathfrak{G}_2), \dots.$$

A chain (1) which satisfies the conditions (2) is called a *central series* if it has only a finite number of distinct members and if its last member is the group $\mathfrak{G}$ itself (cf. ex. 1).

All the factor groups (2), as subgroups of centres, are abelian. Assuming that (1) is a central series, namely

$$(1)' \qquad \mathfrak{e} = \mathfrak{G}_0 \trianglelefteq \mathfrak{G}_1 \trianglelefteq \mathfrak{G}_2 \trianglelefteq \dots \trianglelefteq \mathfrak{G}_{s-1} \trianglelefteq \mathfrak{G}_s = \mathfrak{G},$$

we conclude from Chap. III, §1, Theorem 2 that because

$$(3) \qquad \begin{cases} \mathfrak{G}_s/\mathfrak{G}_{s-1} & \text{is abelian: } \mathfrak{G}' \trianglelefteq \mathfrak{G}_{s-1} \\ \mathfrak{G}_{s-1}/\mathfrak{G}_{s-2} & \text{is abelian: } \mathfrak{G}'' \trianglelefteq \mathfrak{G}_{s-1}' \trianglelefteq \mathfrak{G}_{s-2} \\ \mathfrak{G}_{s-2}/\mathfrak{G}_{s-3} & \text{is abelian: } \mathfrak{G}''' \trianglelefteq \mathfrak{G}_{s-2}' \trianglelefteq \mathfrak{G}_{s-3} \\ \dots\dots & \dots\dots \\ \mathfrak{G}_{i+1}/\mathfrak{G}_i & \text{is abelian: } \mathfrak{G}^{(s-i)} \trianglelefteq \mathfrak{G}_{i+1}' \trianglelefteq \mathfrak{G}_i \\ \dots\dots & \dots\dots \\ \mathfrak{G}_2/\mathfrak{G}_1 & \text{is abelian: } \mathfrak{G}^{(s-1)} \trianglelefteq \mathfrak{G}_2' \trianglelefteq \mathfrak{G}_1 \trianglelefteq \mathfrak{Z}(\mathfrak{G}). \end{cases}$$

Hence $\mathfrak{G}^{(s)} = \mathfrak{e}$, and if $\mathfrak{G}$ has an ascending central series, it also has a (descending) derived series.

A group $\mathfrak{G}$ is said to be *nilpotent* if it has an ascending central series. From the preceding it then follows that *every nilpotent group is solvable.*

Every abelian group $\mathfrak{G} > \mathfrak{e}$ has the central series $\mathfrak{G}_0 \lhd \mathfrak{G}$. Hence *every abelian group is nilpotent.* The nilpotent groups form a class between the abelian and the solvable groups. It will be seen that there are solvable groups which are not nilpotent (cf. ex. 2) and nilpotent groups which are not abelian (ex. 3).

b. Among the ascending central chains of $\mathfrak{G}$ there is one, called the *upper central chain*

$$\mathfrak{e} = \mathfrak{Z}_0 \trianglelefteq \mathfrak{Z}_1 \trianglelefteq \mathfrak{Z}_2 \trianglelefteq \dots$$

which is uniquely defined by the group $\mathfrak{G}$. The uniqueness is reached by substituting in (2) equalities for the inclusion conditions, i.e. the upper central chain is defined by

$$(4) \qquad \mathfrak{Z}_1/\mathfrak{Z}_0 \simeq \mathfrak{Z}_1 = \mathfrak{Z}(\mathfrak{G}), \quad \mathfrak{Z}_2/\mathfrak{Z}_1 = \mathfrak{Z}(\mathfrak{G}/\mathfrak{Z}_1), \quad \mathfrak{Z}_3/\mathfrak{Z}_2 = \mathfrak{Z}(\mathfrak{G}/\mathfrak{Z}_2), \dots .$$

Hence if (1) is a central chain of $\mathfrak{G}$, then $\mathfrak{G}_1 \trianglelefteq \mathfrak{Z}_1$. Further

$$\mathfrak{G}/\mathfrak{Z}_1 \simeq (\mathfrak{G}/\mathfrak{G}_1)/(\mathfrak{Z}_1/\mathfrak{G}_1);$$

thus there is a homomorphism $\mathfrak{G}/\mathfrak{G}_1 \to \mathfrak{G}/\mathfrak{Z}_1$ which induces the homomorphism of the subgroup $\mathfrak{G}_2/\mathfrak{G}_1 \to \mathfrak{G}_2/\mathfrak{Z}_1 \trianglelefteq \mathfrak{Z}(\mathfrak{G}/\mathfrak{Z}_1) = \mathfrak{Z}_2/\mathfrak{Z}_1$ whence we conclude that $\mathfrak{G}_2 \trianglelefteq \mathfrak{Z}_2$:

Let us assume that the inclusion

$$(5) \qquad \mathfrak{G}_i \trianglelefteq \mathfrak{Z}_i \quad (i \geqslant 1)$$

has already been established. Therefore $\mathfrak{G}/\mathfrak{Z}_i \simeq (\mathfrak{G}/\mathfrak{G}_i)/(\mathfrak{Z}_i/\mathfrak{G}_i)$ and we have the homomorphism $\mathfrak{G}/\mathfrak{G}_i \to \mathfrak{G}/\mathfrak{Z}_i$ under which $\mathfrak{G}_{i+1}/\mathfrak{G}_i \to \mathfrak{G}_{i+1}/\mathfrak{Z}_i \trianglelefteq \mathfrak{Z}(\mathfrak{G}/\mathfrak{Z}_i) = \mathfrak{Z}_{i+1}/\mathfrak{Z}_i$ so that $\mathfrak{G}_{i+1} \trianglelefteq \mathfrak{Z}_{i+1}$. Thus (5) is verified by induction for every ascending central chain (1).

If in $\mathfrak{G}$ there exists a central series, that is if for some natural number s one has $\mathfrak{G}_s = \mathfrak{G}$, then by (5) $\mathfrak{Z}_s = \mathfrak{G}$ which implies that then there exists an upper central series as well.

In the upper central chain there can occur only proper inclusions of consecutive terms:

$$(6) \qquad \mathfrak{Z}_i \lhd \mathfrak{Z}_{i+1} \quad (i = 0, 1, 2, \dots).$$

Indeed, suppose that $\mathfrak{Z}_j = \mathfrak{Z}_{j+1}$, then $\mathfrak{Z}_{j+1}/\mathfrak{Z}_j$ has the order 1 and the centre $\mathfrak{Z}(\mathfrak{G}/\mathfrak{Z}_j) = \mathfrak{Z}(\mathfrak{G}/\mathfrak{Z}_{j+1}) = \mathfrak{Z}_{j+2}/\mathfrak{Z}_{j+1}$ so that $\mathfrak{Z}_{j+2} = \mathfrak{Z}_{j+1}$. Continuing in this manner we see that if $\mathfrak{Z}_j = \mathfrak{Z}_{j+1}$ then all the subsequent $\mathfrak{Z}_{j+n} = \mathfrak{Z}_j$ $(n = 2, 3, \dots)$. Hence, as in the case of the derived series, in the upper central series of $\mathfrak{G}$ repetitions cannot occur, it must have the form

$$(6') \qquad \mathfrak{e} = \mathfrak{Z}_0 \lhd \mathfrak{Z}_1 \lhd \mathfrak{Z}_2 \lhd \dots \lhd \mathfrak{Z}_{r-1} \lhd \mathfrak{Z}_r = \mathfrak{G}.$$

The number r as well as the subgroups $\mathfrak{Z}_i$ are uniquely determined by the group $\mathfrak{G}$. Moreover the $\mathfrak{Z}_i$ are characteristic subgroups of $\mathfrak{G}$ so that every automorphism of $\mathfrak{G}$ will leave each of the $\mathfrak{Z}_i$ invariant and instead of (6′) we may write

$$(6'') \qquad \mathfrak{e} = \mathfrak{Z}_0 \blacktriangleleft \mathfrak{Z}_1 \blacktriangleleft \mathfrak{Z}_2 \blacktriangleleft \dots \blacktriangleleft \mathfrak{Z}_{r-1} \blacktriangleleft \mathfrak{Z}_r = \mathfrak{G}.$$

With regard to (5) it is evident that no ascending central series (1) can have more terms than the upper central series (6″) if repetitions in (1) are deleted.

A group $\mathfrak{G}$ having an upper central series of length r is said to be *nilpotent of class r*. A nilpotent group of class r is solvable of length $s \leqslant r$. At any rate, by (3) and (6′) the r-th derived group $\mathfrak{G}^{(r)} = \mathfrak{Z}_0 = \mathfrak{e}$; it is possible, however, that $\mathfrak{G}^{(s)} = \mathfrak{Z}_0$ for some $s < r$.

c. THEOREM 1. *Let $\mathfrak{G}$ be nilpotent, $\mathfrak{H} < \mathfrak{G}$, and $\mathfrak{Z}_{i-1}$ be the greatest member of the upper central series of $\mathfrak{G}$ which is contained in $\mathfrak{H}$. Then $\mathfrak{Z}_i \trianglelefteq \mathfrak{N}_{\mathfrak{G}}(\mathfrak{H})$, the normalizer of $\mathfrak{H}$ in $\mathfrak{G}$.*

The subgroup situation is illustrated in Fig. 9. The dots represent subgroups and subgroup $\mathfrak{X}$ is contained in (but not equal to) subgroup $\mathfrak{Y}$ when the $\mathfrak{X}$-dot lies below the $\mathfrak{Y}$-dot and both are joined by a possibly broken line (cf. §2, ex. 7).

Proof. Clearly $\mathfrak{H}/\mathfrak{Z}_{i-1} < \mathfrak{G}/\mathfrak{Z}_{i-1}$ and the elements of $\mathfrak{G}/\mathfrak{Z}_{i-1}$, in particular those of $\mathfrak{H}/\mathfrak{Z}_{i-1}$, commute with the elements of $\mathfrak{Z}(\mathfrak{G}/\mathfrak{Z}_{i-1}) = \mathfrak{Z}_i/\mathfrak{Z}_{i-1}$. Thus

$$\mathfrak{Z}_i/\mathfrak{Z}_{i-1} < \mathfrak{N}_{\mathfrak{G}/\mathfrak{Z}_{i-1}}(\mathfrak{H}/\mathfrak{Z}_{i-1}).$$

Hence, if $z_i \in \mathfrak{Z}_i$ and $h \in \mathfrak{H}$, then

$$z_i\mathfrak{Z}_{i-1}h\mathfrak{Z}_{i-1} = h\mathfrak{Z}_{i-1}z_i\mathfrak{Z}_{i-1}$$

and because $\mathfrak{Z}_{i-1} \lhd \mathfrak{G}$ we have

$$z_ih\mathfrak{Z}_{i-1} = hz_i\mathfrak{Z}_{i-1}.$$

Thus for each $z_i \in \mathfrak{Z}_i$ and $h \in \mathfrak{H}$ there is, for some $z_{i-1} \in \mathfrak{Z}_{i-1}$ an $h' = hz_{i-1}$ in $\mathfrak{H}$ such that $z_ih' = hz_i$ or $z_i^{-1}hz_i = h'$ and therefore $z_i \in \mathfrak{N}_{\mathfrak{G}}(\mathfrak{H})$, i.e. $\mathfrak{Z}_i \trianglelefteq \mathfrak{N}_{\mathfrak{G}}(\mathfrak{H})$.

If $\mathfrak{H} = \mathfrak{N}_{\mathfrak{G}}(\mathfrak{H}) = \mathfrak{N}(\mathfrak{H})$ we conclude that $\mathfrak{Z}_i \trianglelefteq \mathfrak{H}$ which contradicts the assumption of the theorem. Hence $\mathfrak{H} < \mathfrak{N}(\mathfrak{H})$ and by repeated application of this inclusion we have the result:

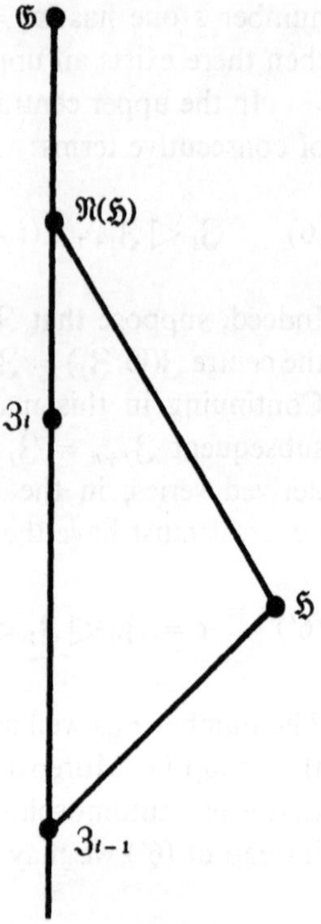

Figure 9

COROLLARY 1. *If $\mathfrak{G}$ is nilpotent of class r and $\mathfrak{H} < \mathfrak{G}$ then the normalizer chain*

$$\mathfrak{H} < \mathfrak{N}(\mathfrak{H}) < \mathfrak{N}[\mathfrak{N}(\mathfrak{H})] < \dots$$

is a reversed subnormal series which has at most r members and ends with $\mathfrak{G}$.

COROLLARY 2. *Every subgroup $\mathfrak{H}$ of a nilpotent group $\mathfrak{G}$ is a member of a subnormal series of $\mathfrak{G}$.*

d. Mixed commutator subgroups. Let **X** and **Y** be two subsets of elements of $\mathfrak{G}$. We define

$$[\mathbf{X}, \mathbf{Y}] = \langle [x, y] = xyx^{-1}y^{-1} \,|\, x \in \mathbf{X}, y \in \mathbf{Y} \rangle,$$

i.e. the subgroup of $\mathfrak{G}$ generated by all the commutators $[x, y]$ where

$x \in \mathbf{X}$ and $y \in \mathbf{Y}$. In particular $\mathfrak{G}' = \mathfrak{K}(\mathfrak{G}) = [\mathfrak{G}, \mathfrak{G}]$ is the ordinary commutator group; $\mathfrak{G}'' = [\mathfrak{G}', \mathfrak{G}']$, etc. Since $[x, y]^{-1} = yxy^{-1}x^{-1} \in [\mathbf{X}, \mathbf{Y}]$ we conclude that $[\mathbf{Y}, \mathbf{X}] \leqslant [\mathbf{X}, \mathbf{Y}]$, hence $[\mathbf{Y}, \mathbf{X}] = [\mathbf{X}, \mathbf{Y}]$.

For any $\mathfrak{H} \leqslant \mathfrak{G}$ one can form the mixed commutator subgroup $[\mathfrak{G}, \mathfrak{H}] \leqslant \mathfrak{G}$. However one has to take into account the following facts:

Lemma 1. $[\mathfrak{G}, \mathfrak{H}] \leqslant \mathfrak{H}$ *if and only if* $\mathfrak{H} \trianglelefteq \mathfrak{G}$.

Indeed if $\mathfrak{H} \trianglelefteq \mathfrak{G}$ we have $[x, y] = (xyx^{-1})y^{-1} \in \mathfrak{H}$ for all $x \in \mathfrak{G}$, $y \in \mathfrak{H}$. Conversely if $[x, y] \in \mathfrak{H}$ for all $x \in \mathfrak{G}$, $y \in \mathfrak{H}$ it follows that $(xyx^{-1})y^{-1} = h \in \mathfrak{H}$ and therefore $xyx^{-1} = hy \in \mathfrak{H}$, i.e. $\mathfrak{H} \trianglelefteq \mathfrak{G}$.

If $\mathfrak{H} \leqslant \mathfrak{Z}(\mathfrak{G})$ it follows that $[\mathfrak{G}, \mathfrak{H}] = \mathfrak{e}$.

Lemma 2. *Let* $\mathfrak{A} \triangleleft \mathfrak{G}$ *and* $\mathfrak{B} \triangleleft \mathfrak{G}$; *then* $[\mathfrak{A}, \mathfrak{B}] \triangleleft \mathfrak{G}$.

Proof. If $a \in \mathfrak{A}$ and $b \in \mathfrak{B}$ one has

$$[a, b] = (aba^{-1})b^{-1} \in \mathfrak{B}$$
$$= a(ba^{-1}b^{-1}) \in \mathfrak{A}.$$

Hence $[a, b] \in \mathfrak{D} = \mathfrak{A} \cap \mathfrak{B}$ and $[\mathfrak{A}, \mathfrak{B}] \trianglelefteq \mathfrak{D} \triangleleft \mathfrak{G}$. Moreover for all x in $\mathfrak{G}$ we have $x[a, b]x^{-1} = [a, b]^x = [a^x, b^x] \in [\mathfrak{A}, \mathfrak{B}]$, i.e. $[\mathfrak{A}, \mathfrak{B}] \triangleleft \mathfrak{G}$.

e. The lower (or descending) *central chain* of a group $\mathfrak{G}$ is a subnormal chain $\mathfrak{G} = \mathfrak{K}_0 \trianglerighteq \mathfrak{K}_1 \trianglerighteq \mathfrak{K}_2 \trianglerighteq \ldots$ defined as follows:

$$\mathfrak{K}_1 = \mathfrak{K}_1(\mathfrak{G}) = [\mathfrak{K}_0, \mathfrak{G}] = [\mathfrak{G}, \mathfrak{G}] = \mathfrak{G}', \quad \mathfrak{K}_2 = \mathfrak{K}_2(\mathfrak{G}) = [\mathfrak{K}_1, \mathfrak{G}] = [\mathfrak{G}', \mathfrak{G}], \ldots, \quad \mathfrak{K}_i = \mathfrak{K}_i(\mathfrak{G}) = [\mathfrak{K}_{i-1}, \mathfrak{G}] \quad (i = 2, 3, \ldots).$$

This chain is uniquely defined by the group $\mathfrak{G}$. Every endomorphism ω of $\mathfrak{G}$ maps a commutator into a commutator; hence $\mathfrak{K}_1 = \mathfrak{G}'$ is fully invariant and, thus, characteristic in $\mathfrak{G}$. Therefore ω maps every element of $\mathfrak{K}_2 = [\mathfrak{K}_1, \mathfrak{G}]$ into an element of $\mathfrak{K}_2$ and a simple induction shows that the lower central chain is a fully invariant, hence characteristic, chain of $\mathfrak{G}$.

If in the lower central chain an equality sign occurs for the first time at the term $\mathfrak{K}_r = \mathfrak{K}_{r+1} = [\mathfrak{K}_r, \mathfrak{G}]$, then equality exists between all the subsequent members of the chain, i.e. $\mathfrak{K}_r = \mathfrak{K}_{r+n}$ $(n = 1, 2, \ldots)$.

There are the following two possibilities. Either $\mathfrak{K}_{r+1} = \mathfrak{K}_r > \mathfrak{e}$; or $\mathfrak{K}_r = \mathfrak{e}$, that is $\mathfrak{K}_{r-1} \leqslant \mathfrak{Z}(\mathfrak{G}) = \mathfrak{Z}_1$ where $\mathfrak{Z}_1 > \mathfrak{e}$ so that $\mathfrak{K}_{r-2} \nleqslant \mathfrak{Z}(\mathfrak{G})$. In the second case the chain is a series, *the lower central series* of $\mathfrak{G}$.

THEOREM 2. *If the lower central chain of the group $\mathfrak{G}$ terminates with $\mathfrak{K}_r = \mathfrak{e}$ (that is, if it is the lower central series of $\mathfrak{G}$) then it is a reversed ascending central series and $\mathfrak{G}$ is nilpotent.*

Proof. For two consecutive terms in the lower central chain, $\mathfrak{K}_{i-1}$ and $\mathfrak{K}_i = [\mathfrak{K}_{i-1}, \mathfrak{G}]$ we consider the factor group $\mathfrak{K}_{i-1}/\mathfrak{K}_i$. Because $\mathfrak{K}_{i-1}' = [\mathfrak{K}_{i-1}, \mathfrak{K}_{i-1}] \leqslant \mathfrak{K}_i$, the factor group $\mathfrak{K}_{i-1}/\mathfrak{K}_i$ is abelian. Moreover it lies in the centre of $\mathfrak{G}/\mathfrak{K}_i$. Indeed

$$[\mathfrak{G}/\mathfrak{K}_i, \mathfrak{K}_{i-1}/\mathfrak{K}_i] = \langle [x\mathfrak{K}_i, x_{i-1}\mathfrak{K}_i] \mid x \in \mathfrak{G}, x_{i-1} \in \mathfrak{K}_{i-1} \rangle$$
$$= \langle [x, x_{i-1}]\mathfrak{K}_i \mid \ldots \rangle.$$

But $[x, x_{i-1}] \in \mathfrak{K}_i$ so that $[\mathfrak{G}/\mathfrak{K}_i, \mathfrak{K}_{i-1}/\mathfrak{K}_i] = \mathfrak{K}_i$, i.e. the unit element of the factor group $\mathfrak{G}/\mathfrak{K}_i$. All elements of this mixed commutator group being equal to the unit element implies that $\mathfrak{K}_{i-1}/\mathfrak{K}_i \leqslant \mathfrak{Z}(\mathfrak{G}/\mathfrak{K}_i)$ which proves the theorem.

The converse statement is likewise true:

THEOREM 3. *If the group $\mathfrak{G}$ is nilpotent then it has a lower central series.*

Proof. By supposition there exists in $\mathfrak{G}$ an ascending central series, say

$$\mathfrak{e} = \mathfrak{H}_0 \lhd \mathfrak{H}_1 \lhd \ldots \lhd \mathfrak{H}_s = \mathfrak{G}$$

so that $\mathfrak{H}_1 \trianglelefteq \mathfrak{Z}_1 = \mathfrak{Z}(\mathfrak{G})$, $\mathfrak{H}_2/\mathfrak{H}_1 \leqslant \mathfrak{Z}(\mathfrak{G}/\mathfrak{H}_1), \ldots, \mathfrak{H}_s/\mathfrak{H}_{s-1} = \mathfrak{G}/\mathfrak{H}_{s-1} \leqslant \mathfrak{Z}(\mathfrak{G}/\mathfrak{H}_{s-1})$. Therefore we can start the construction of a lower central series with $\mathfrak{K}_0 = \mathfrak{G} = \mathfrak{H}_s$. Further $\mathfrak{K}_1 = \mathfrak{G}' \trianglelefteq \mathfrak{H}_{s-1} \leqslant \mathfrak{Z}_{s-1}$ by (3) and (5) (note that $\mathfrak{Z}_{s-1} = \mathfrak{G}$ if $s > r$) and because $\mathfrak{H}_{s-1}/\mathfrak{H}_{s-2} \leqslant \mathfrak{Z}(\mathfrak{G}/\mathfrak{H}_{s-2})$ for $x \in \mathfrak{G}$ and $x_{s-1} \in \mathfrak{H}_{s-1}$ we have

$$[x_{s-1}\mathfrak{H}_{s-2}, x\mathfrak{H}_{s-2}] = [x_{s-1}, x]\mathfrak{H}_{s-2} = \mathfrak{H}_{s-2}.$$

Thus $[x_{s-1}, x] \in \mathfrak{H}_{s-2}$ which implies that $[\mathfrak{H}_{s-1}, \mathfrak{G}] \leqslant \mathfrak{H}_{s-2}$ whence

$$\mathfrak{K}_2 \leqslant \mathfrak{H}_{s-2} \leqslant \mathfrak{Z}_{s-2}.$$

By the same argument $\mathfrak{H}_{s-2}/\mathfrak{H}_{s-3} \leqslant \mathfrak{Z}(\mathfrak{G}/\mathfrak{H}_{s-3})$ so that

$$\mathfrak{K}_3 \leqslant \mathfrak{H}_{s-3} \leqslant \mathfrak{Z}_{s-3}, \text{ etc.}$$

Since the $\mathfrak{H}$-series starts with $\mathfrak{e}$, the $\mathfrak{K}$-series ends with $\mathfrak{e}$.

Examples and exercises

1. Each member of an ascending central chain, being normal in its successor, is automatically normal in the whole group. Thus an ascending central chain (series) is not only a subnormal, but a normal chain (series).

2. The symmetric groups $\mathfrak{S}_3$ and $\mathfrak{S}_4$ are solvable, but not nilpotent.

3. The quaternion group $\mathfrak{Q}$ is nilpotent, but not abelian.

4. If $\mathfrak{G}$ is nilpotent then so is every subgroup of $\mathfrak{G}$.
Proof. Let $\mathfrak{H} < \mathfrak{G}$. It is found that

$$\mathfrak{K}_i(\mathfrak{H}) = [\mathfrak{K}_{i-1}(\mathfrak{H}), \mathfrak{H}] < [\mathfrak{K}_{i-1}(\mathfrak{G}), \mathfrak{G}] = \mathfrak{K}_i(\mathfrak{G}).$$

5. If $\mathfrak{G}$ is nilpotent and $\mathfrak{H} \lhd \mathfrak{G}$, then $\mathfrak{G}/\mathfrak{H}$ is nilpotent.
Proof. By definition $\mathfrak{K}_i(\mathfrak{G}/\mathfrak{H}) = [\mathfrak{K}_{i-1}(\mathfrak{G}/\mathfrak{H}), \mathfrak{G}/\mathfrak{H}]$. Thus the lower central series begins with $\mathfrak{K}_0(\mathfrak{G}/\mathfrak{H}) = \mathfrak{G}/\mathfrak{H}$, $\mathfrak{K}_1(\mathfrak{G}/\mathfrak{H}) = (\mathfrak{G}/\mathfrak{H})'$ $= \langle [x\mathfrak{H}, y\mathfrak{H}] \,|\, x, y \in \mathfrak{G}\rangle = \langle [x, y]\mathfrak{H} \,|\, \ldots\rangle = \mathfrak{K}_1\mathfrak{H}$, this product to be understood as a system of cosets of $\mathfrak{H}$ with elements of $\mathfrak{K}_1$ as representatives. In the same manner:

$$\mathfrak{K}_2(\mathfrak{G}/\mathfrak{H}) = [\mathfrak{K}_1(\mathfrak{G}/\mathfrak{H}), \mathfrak{G}/\mathfrak{H}] = \langle [x_1\mathfrak{H}, x\mathfrak{H}] \,|\, x_1 \in \mathfrak{K}_1, x \in \mathfrak{G}\rangle$$
$$= \langle [x_1, x]\mathfrak{H}\rangle = \mathfrak{K}_2\,\mathfrak{H}, \text{ etc.}$$

and since the $\mathfrak{K}$-series of $\mathfrak{G}$ ends with $\mathfrak{e}$, the series $\{\mathfrak{K}_i(\mathfrak{G}/\mathfrak{H})\} = \{\mathfrak{K}_i(\mathfrak{G})\cdot\mathfrak{H}\}$ ends with $\mathfrak{H}$, the unit element of $\mathfrak{G}/\mathfrak{H}$.

6. Prove the statement of ex. 4 by referring to the definition of nilpotence, that is, using the ascendent central series only.

7. If $\mathfrak{G}$ is nilpotent and $\mathfrak{H} \underset{\max}{<} \mathfrak{G}$ then $\mathfrak{H} \underset{\max}{\lhd} \mathfrak{G}$. (Apply Theorem 1, Corollary 1.)

Chapter V

FINITE GROUPS AND PRIME NUMBERS

A primitive suggestion as to a connection between finite groups and prime numbers may be seen in the fact that groups of prime order are distinguished among all groups: They are the only groups which have no proper subgroups. Hence it seems natural to ask the following question: What influence has the occurrence of a prime divisor p of the order g of a group $\mathfrak{G}$ on the subgroup structure of $\mathfrak{G}$?

A first approach to this problem was made by Cauchy in 1844 with permutation groups in mind when he demonstrated that if p is a prime divisor of g then $\mathfrak{G}$ contains an element, and therefore a subgroup, of order p. Extending this result L. Sylow proved in 1872 that if $p^\lambda \mid g$ then $\mathfrak{G}$ contains a subgroup of order p^λ. In view of Cayley's theorem (cf. Chap. I, §2, *e.*), G. F. Frobenius realized (1887) that these theorems are facts of abstract group theory. Accordingly he proved them without using a representation of $\mathfrak{G}$ by permutations (cf. §2, *c.*).*

We shall prove Sylow's theorem in two ways; first, according to Wielandt, making use of some ideas of the theory of permutation groups in §1. The classical proof of Frobenius, followed by the other theorems of Sylow, will be given in §2; it can be read without previous knowledge of §1.

§1 Permutation groups

Elementary properties of permutations, defined as invertible mappings of a finite set, e.g. the set $\mathbf{N} = \{1, 2, \ldots, n\}$, onto itself, and groups of

*The title of Sylow's original paper "Théorèmes sur les groupes de substitutions" Mathemat. Annalen 5 (1872), p.584–594, as well as its content, suggest that he thought in terms of permutation groups only. The idea of an abstract group was not yet commonly accepted at his time.

permutations have been dealt with in earlier sections (Chap. I, §2; Chap. II, §4 *e*. and exx. 1–4, 11, 14; Chap. III, §1, ex. 13). In the present section we shall take up this subject in a different, slightly more general setting.

a. Action of a group on a set. Let **T** denote an arbitrary finite set. We shall say that a group $\mathfrak{G}$ acts on **T** as a group of permutations if for every element x of $\mathfrak{G}$ and for every element τ of **T** the formal product $x\tau$ represents a uniquely defined element of **T** and if the following conditions are satisfied:

(i) $e\tau = \tau$ for all τ in **T**;

(ii) $(yx)\tau = y(x\tau)$ for all x, y in $\mathfrak{G}$ and $\tau \in \mathbf{T}$.

This means that we consider the group elements as operations mapping the set **T** into itself; indeed, onto itself. For if there were an element τ_0 of **T** such that $x\tau \neq \tau_0$ whatever $\tau \in \mathbf{T}$, then for every element y in $\mathfrak{G}$ we would have $(yx)\tau \neq y\tau_0$; in particular for $y = x^{-1}$ it would follow that $e\tau \neq x^{-1}\tau_0 = \tau_1$ for all τ in **T**. But $e\tau_1 = \tau_1$.

The set of all elements x in $\mathfrak{G}$ which leave every element of **T** fixed, i.e. the set $\{x \mid x\tau = \tau, \tau \in \mathbf{T})$, is a *normal* subgroup of $\mathfrak{G}$. Indeed, let x and x_1 be elements of this set: Then clearly $x_1 x\tau = \tau$ and $x^{-1}\tau = \tau$. Thus this set forms a subgroup of $\mathfrak{G}$. Moreover, if $y \in \mathfrak{G}$ we have $yxy^{-1}\tau = y \cdot y^{-1}\tau = e\tau = \tau$. Hence this subgroup is normal in $\mathfrak{G}$.

The set of mappings of **T** onto itself, induced by the action of the elements of $\mathfrak{G}$ in **T** is a subset of $\mathfrak{S}(\mathbf{T})$, the symmetric group over **T**; it is by (ii) a homomorphic image of $\mathfrak{G}$ and therefore a subgroup $\tilde{\mathfrak{G}}$ of $\mathfrak{S}(\mathbf{T})$. The normal subgroup of those x in $\mathfrak{G}$ which leave all τ in **T** fixed is the kernel of this homomorphism $\varphi : \mathfrak{G} \to \tilde{\mathfrak{G}}$.

If $\mathbf{T} = \mathbf{G}$, the set of all elements of $\mathfrak{G}$, the formal product may be defined by the group product xt ($x \in \mathfrak{G}$, $t \in \mathbf{G}$). In this case $\tilde{\mathfrak{G}}$ is the Cayley representation of $\mathfrak{G}$ (cf. Chap. I, §2, *f*.). If **T** is the system of all cosets of some fixed subgroup $\mathfrak{H}$ of $\mathfrak{G}$, i.e. $\tau = t\mathfrak{H}$, $t \in \mathbf{G}$, $\mathfrak{H} < \mathfrak{G}$, we may set $x\tau = xt\mathfrak{H}$. This situation has been discussed in Chap. II, §4, *e*.

b. Transitivity. Let $\mathfrak{G}$ act as a group of permutations on the finite set $\mathbf{T} = \mathbf{N} = \{1, 2, \ldots, n\}$ so that $\tilde{\mathfrak{G}} \leqslant \mathfrak{S}_n$. Two symbols $i, j \in \mathbf{N}$ are said to be *related* with respect to $\mathfrak{G}$ if there is an element $x \in \mathfrak{G}$ which maps $i \to j$, i.e. $xi = j$. This relatedness evidently is an equivalence relation in $\mathbf{N}$. Thus $\mathbf{N}$ is, with respect to $\mathfrak{G}$, divided into disjoint subsets, called the *orbits* or the *transitivity systems* of $\mathfrak{G}$ in $\mathbf{N}$. Each orbit is a maximal set of pairwise related symbols in $\mathbf{N}$. An orbit $\mathbf{N}_1$ is defined by any one of its symbols; for, if $i \in \mathbf{N}_1$, then for every j, j' in $\mathbf{N}_1$ there are elements x, y of $\mathfrak{G}$ such that $xi = j$, $yi = j'$ and therefore $(yx^{-1})j = j'$

The group $\mathfrak{G}$ is said to act *transitively* or: to be *transitive*,* *on* $\mathbf{N}$ if there is one and only one orbit, namely $\mathbf{N}$ itself. The group $\mathfrak{G}$ acts *intransitively* on $\mathbf{N}$ if $\mathbf{N}$ splits into several orbits $\mathbf{N}_1, \ldots, \mathbf{N}_r$ $(r \geqslant 2)$ of mutually related symbols. The group $\mathfrak{G}$ acts transitively on each of the orbits.

If $\mathfrak{G}$ acts intransitively on $\mathbf{N}$, then for every x in $\mathfrak{G}$ its image $X = \varphi(x)$ in $\mathfrak{S}_n$ can be represented as a product $X = X_1X_2\ldots X_r$ where X_ρ, called the ρ-th component of X with respect to $\mathfrak{G}$, is a uniquely defined permutation operating only on the orbit $\mathbf{N}_\rho$ $(\rho = 1, 2, \ldots, r)$, leaving fixed the symbols in the other orbits $\mathbf{N}_\sigma$ $(\sigma \neq \rho)$. The permutation X_ρ is obtained by restricting the permutation X to the orbit $\mathbf{N}_\rho$. As an operation on $\mathbf{N}$ this X_ρ is an element of $\mathfrak{S}_n$, but in general it is not an element of the group $\tilde{\mathfrak{G}} = \varphi(\mathfrak{G})$. As an operation on $\mathbf{N}_\rho$ the permutation X_ρ has the degree $n_\rho = |\mathbf{N}_\rho|$. The factorization of $X = \varphi(x)$ into its components depends on the group $\mathfrak{G}$ of which x is an element (cf. ex. 1).

With the notations introduced above, we have the following theorem:

THEOREM 1. (a) *The set* $\mathbf{G}_\rho$ *of the* ρ*-th components* X_ρ *of all the elements* X *in* $\tilde{\mathfrak{G}}$ *represents a transitive group* $\mathfrak{G}_\rho$ *of permutations operating on the orbit* $\mathbf{N}_\rho$.

(b) *There is an epimorphism* $\omega_\rho : \tilde{\mathfrak{G}} \to \mathfrak{G}_\rho$ *whose kernel* $\tilde{\mathfrak{H}}_\rho$ *consists of all those elements of* $\tilde{\mathfrak{G}}$ *which have as* ρ*-th component the identity permutation* I_ρ *on* $\mathbf{N}_\rho$.

Proof. (a) Let $\varphi(x) = X = X_1X_2\ldots X_r$ and $\varphi(y) = Y = Y_1Y_2\ldots Y_r$; $x, y \in \mathfrak{G}$, $X_\rho \in \mathbf{G}_\rho$, $Y_\sigma \in \mathbf{G}_\sigma$. By Chap. I, §2, *c.*, Lemma, the permutations X_ρ and Y_σ commute if $\rho \neq \sigma$. Therefore

$$(1) \quad \varphi(xy) = (X_1X_2\ldots X_r)(Y_1Y_2\ldots Y_r) = (X_1Y_1)(X_2Y_2)\ldots(X_rY_r)$$

* If from the context it is clear which is the operational domain $\mathbf{T}$ of a group it may not be mentioned. Thus we say that $\mathfrak{S}_n$ is transitive, implying "transitive on $\mathbf{N}$".

has $X_\rho Y_\rho$ as its ρ-*th* component: $X_\rho Y_\rho \in \mathbf{G}_\rho$ if $X_\rho, Y_\rho \in \mathbf{G}_\rho$. Clearly $X^{-1} = X_r^{-1}\ldots X_1^{-1} = X_1^{-1}\ldots X_r^{-1}$; thus $X_\rho^{-1} \in \mathbf{G}_\rho$. Since $I = I_1 I_2 \ldots I_r$, also $I_\rho \in \mathbf{G}_\rho$. Thus the elements of $\mathbf{G}_\rho$ form a group $\mathfrak{G}_\rho$. If $\mathfrak{G}_\rho$ were not transitive on $\mathbf{N}_\rho$ then $\mathbf{N}_\rho$ would split into several orbits of $\mathfrak{G}_\rho$ which would be distinct orbits of $\mathfrak{G}$ in $\mathbf{N}_\rho$; hence $\mathbf{N}_\rho$ could not be an orbit of $\mathfrak{G}$.

(b) The mapping $\omega_\rho : \tilde{\mathfrak{G}} \to \mathfrak{G}_\rho$, defined by $\omega_\rho(X) = X_\rho$ for all X in $\tilde{\mathfrak{G}}$, with regard to (1), is seen to be a homomorphism. By definition $\ker \omega_\rho = \omega_\rho^{-1}(I_\rho) = \tilde{\mathfrak{H}}_\rho$ is the subgroup of $\tilde{\mathfrak{G}}$ consisting of all those X in $\tilde{\mathfrak{G}}$ which leave every symbol of the orbit $\mathbf{N}_\rho$ fixed.

In particular let $\rho = 1$ and $\mathbf{N}_1 = \{1, 2, \ldots, n_1\}$; then

$$X^{(1)} \in \tilde{\mathfrak{H}}_1 \Leftrightarrow X^{(1)} = I_1 X_2^{(1)} \ldots X_r^{(1)} = \begin{pmatrix} 1 \ldots n_1 & n_1+1 \ldots \\ 1 \ldots n_1 & p_{n_1+1} \ldots \end{pmatrix}$$

and the coset $X\tilde{\mathfrak{H}}_1 = \tilde{\mathfrak{H}}_1 X$ ($X \in \tilde{\mathfrak{H}}$) consists of all permutations in $\tilde{\mathfrak{G}}$ which have X_1, the first component of X, as their first component.

Remark. Denote by $\mathfrak{G}^*$ the direct product $(\mathfrak{G}_1, \mathfrak{G}_2, \ldots, \mathfrak{G}_r)$. Then $\tilde{\mathfrak{G}} \leqslant \mathfrak{G}^* \leqslant \mathfrak{S}_n$. If $\mathfrak{G}$ is transitive ($r = 1$) then $\tilde{\mathfrak{G}} = \mathfrak{G}_1 = \mathfrak{G}^*$.

c. Stabilizers. Throughout the rest of this section $\mathfrak{G}$ will denote a finite group acting as a permutation group on a finite set $\mathbf{T} = \{\tau_1, \tau_2, \ldots, \tau_n\}$. The elements y of $\mathfrak{G}$ which leave the symbol τ_i fixed form a subgroup $\mathfrak{G}^{(i)} < \mathfrak{G}$; this subgroup is called the *stabilizer* or the *stability subgroup* of τ_i in $\mathfrak{G}$. The homomorphic image $\tilde{\mathfrak{G}}^{(i)} = \varphi(\mathfrak{G}^{(i)})$ then is the stabilizer of i in the permutation group $\tilde{\mathfrak{G}}$.

Suppose now that $\mathbf{T}_1 = \{\tau_1, \ldots, \tau_{n_1}\}$ is an orbit of $\mathfrak{G}$ and therefore $\mathbf{N}_1 = \{1, \ldots, n_1\}$ an orbit of $\tilde{\mathfrak{G}}$. We shall show that *the stability subgroups* $\mathfrak{G}^{(1)}, \ldots, \mathfrak{G}^{(n_1)}$ of $\mathfrak{G}$ *form a complete set of conjugate subgroups of* $\mathfrak{G}$. Indeed if $y \in \mathfrak{G}^{(i)}$, i.e. $y\tau_i = \tau_i$, and $x\tau_i = \tau_j$, then $x\mathfrak{G}^{(i)}x^{-1}$ is the stability subgroup of τ_j. For since $xyx^{-1}\tau_j = xy\tau_i = x\tau_i = \tau_j$ we conclude that $x\mathfrak{G}^{(i)}x^{-1} \leqslant \mathfrak{G}^{(j)}$ and similarly $x^{-1}\mathfrak{G}^{(j)}x \leqslant \mathfrak{G}^{(i)}$, whence $\mathfrak{G}^{(j)} \leqslant x\mathfrak{G}^{(i)}x^{-1}$ and $\mathfrak{G}^{(j)} = x\mathfrak{G}^{(i)}x^{-1}$.

If $\mathbf{T} = \mathbf{G}$ (i.e. the set of the elements of $\mathfrak{G}$) and if the action is defined as multiplication in $\mathfrak{G}$, then the group $\mathfrak{G}$ acts transitively on $\mathbf{G}$ and the group $\tilde{\mathfrak{G}}$, i.e. Cayley's representation of $\mathfrak{G}$, is a transitive permutation group. The stabilizer of every element x of $\mathfrak{G}$ consists of the unit element alone.

Also if

(2) $\quad \mathbf{T}_{\mathfrak{H}} = \{t\mathfrak{H} \mid t \in \mathfrak{G}, \mathfrak{H} \leqslant \mathfrak{G}\}$

with the action defined by $x(t\mathfrak{H}) = (xt)\mathfrak{H}$, the group $\mathfrak{G}$ acts transitively on $\mathbf{T}_{\mathfrak{H}}$. The stabilizer of $\mathfrak{H}$ is the subgroup $\mathfrak{H}$ itself, and the stabilizer of $t\mathfrak{H}$ is the conjugate subgroup $t\mathfrak{H}t^{-1}$ of $\mathfrak{H}$.

Now suppose that $\mathfrak{G}$ acts on two sets, $\mathbf{T}$ and $\mathbf{T}'$. We shall say that $\mathbf{T}$ and $\mathbf{T}'$ are *action-equivalent with respect to* $\mathfrak{G}$ if there is an "action preserving" bijection $\psi : \mathbf{T} \to \mathbf{T}'$ such that for all $x \in \mathfrak{G}$ we have

(3) $\quad \psi(x\tau) = x\psi(\tau), \quad \tau \in \mathbf{T}.$

THEOREM 2. *If a group* $\mathfrak{G}$ *acts transitively on a set* $\mathbf{T}$, $|\mathbf{T}| = n$, *and if* $\mathfrak{H}$ *is the stabilizer of an element* τ *of* $\mathbf{T}$, *then the index* $(\mathfrak{G} : \mathfrak{H}) = n$ *and the two sets* $\mathbf{T}$ *and* $\mathbf{T}_{\mathfrak{H}}$ *[as defined by* (2)*] are action-equivalent with respect to* $\mathfrak{G}$.

Proof. Let $\mathbf{T} = \{\tau_1, \ldots, \tau_n\}$ and $\mathfrak{G}^{(1)}$ be the stabilizer of τ_1 in $\mathfrak{G}$. With regard to the transitivity of $\mathfrak{G}$ on $\mathbf{T}$ we can determine a subset $\{x_1, \ldots, x_n\}$ of $\mathfrak{G}$ such that $x_i\tau_1 = \tau_i$ $(i = 1, \ldots, n)$ are the n distinct elements of $\mathbf{T}$. Then $x_i\mathfrak{G}^{(1)} \neq x_j\mathfrak{G}^{(1)}$ if $i \neq j$. Indeed suppose that these two cosets of $\mathfrak{G}^{(1)}$ were equal; then $x_j^{-1}x_i \in \mathfrak{G}^{(1)}$ whence $x_j^{-1}x_i\tau_1 = \tau_1$ and $x_i\tau_1 = x_j\tau_1$ which implies that $x_i = x_j$. Hence $\mathfrak{G}^{(1)}$ has at least n distinct cosets in $\mathfrak{G}$.

Further, for every element x of $\mathfrak{G}$ there is an index i such that $x\tau_1 = x_i\tau_1 = \tau_i$. Hence $x\mathfrak{G}^{(1)} = x_i\mathfrak{G}^{(1)}$ and $\{x_1, \ldots, x_n\}$ is a transversal of the cosets of $\mathfrak{G}^{(1)}$ in $\mathfrak{G}$ and therefore $(\mathfrak{G} : \mathfrak{G}^{(1)}) = n$.

To complete the proof we define the mapping $\psi(\tau_i) = x_i\mathfrak{G}^{(1)}$ $(i = 1, \ldots, n)$. It satisfies the condition (3); for, if $x\tau_i = \tau_j$ then $xx_i\tau_1 = x_j\tau_1$ and

$$\psi(x\tau_i) = \psi(\tau_j) = x_j\mathfrak{G}^{(1)} = xx_i\mathfrak{G}^{(1)} = x\psi(\tau_i).$$

COROLLARY 1. *If the group* $\mathfrak{G}$ *acts transitively on a set* $\mathbf{T}$, *then the number* $n = |\mathbf{T}|$, *as the index of the stability subgroup* $\mathfrak{H}$ *of* $\mathfrak{G}$, *is a divisor of* $|\mathfrak{G}|$.

COROLLARY 2. *The degree of a transitive permutation group is a divisor of its order.*

COROLLARY 3. *If* $\mathbf{N}_1, \mathbf{N}_2, \cdots, \mathbf{N}_r$ *are the orbits of a permutation group* $\mathfrak{G}$, *then the numbers* $n_\rho = |\mathbf{N}_\rho|$ $(\rho = 1, \ldots, r)$ *are divisors of* $|\mathfrak{G}|$.

Indeed, $\mathfrak{G}$ acts transitively on each of the orbits $\mathbf{N}_\rho$ in $\mathbf{N}$.

d. Application. Let $|\mathfrak{G}| = g$, let h be a natural number, $h < g$ and $n = \binom{g}{h}$. Let $\mathbf{T}$ denote the system of the n distinct combinations $\mathbf{C}_1, \mathbf{C}_2, \ldots, \mathbf{C}_n$, each a combination of h different elements of $\mathfrak{G}$ without regard to order. For every element x of $\mathfrak{G}$ the sets $x\mathbf{C}_1, \ldots, x\mathbf{C}_n$ are defined; they represent again the n combinations $\mathbf{C}_1, \ldots, \mathbf{C}_n$ in a certain order of succession depending on x. Thus the group $\mathfrak{G}$ acts on the set $\mathbf{T}$. The corresponding group $\tilde{\mathfrak{G}}$ consists of the permutations

$$P[x] = \begin{pmatrix} \mathbf{C}_1 & \mathbf{C}_2 & \ldots & \mathbf{C}_n \\ x\mathbf{C}_1 & x\mathbf{C}_2 & \ldots & x\mathbf{C}_n \end{pmatrix}.$$

We shall show that the *homomorphism* $\mathfrak{G} \to \tilde{\mathfrak{G}}$, mapping $x \to P[x]$, *is an isomorphism.*

Proof. Let $x \neq e$. Since $h < g$ we can choose an i, $i \in \mathbf{N}$, such that $e \in \mathbf{C}_i$, but $x \notin \mathbf{C}_i$. Then clearly $x\mathbf{C}_i \neq \mathbf{C}_i$ and thus $P[x] \neq P[e] = I$. Hence the kernel of the homomorphism $x \to P[x]$ consists of the unit element e only.

The permutation group $\tilde{\mathfrak{G}}$ will in general be intransitive. Thus the system $\mathbf{T} = \{\mathbf{C}_1, \ldots, \mathbf{C}_n\}$ splits with respect to the group $\tilde{\mathfrak{G}}$ into r orbits $\mathbf{T}_1, \ldots, \mathbf{T}_r$, $|\mathbf{T}_\rho| = n_\rho$ $(\rho = 1, \ldots, r)$. If $\mathbf{C}_i \in \mathbf{T}$ and if $\mathfrak{G}^{(i)}$ is the stabilizer of $\mathbf{C}_i$ in $\tilde{\mathfrak{G}}$, then by Corollary 3 we have

$$(4) \qquad (\mathfrak{G} : \mathfrak{G}^{(i)}) = n_\rho \quad \text{and} \quad g = g_i\, n_\rho \quad \text{if} \quad g_i = |\mathfrak{G}^{(i)}|.$$

This result will be used in subsection *f*.

*e. Multiple transitivity. Imprimitivity.** A group $\mathfrak{G}$ of permutations of degree n is said to be *k-fold transitive* $(k = 1, 2, \ldots, n)$ if for every two sets of k different symbols from $\mathbf{N}$, say $\{i_1, \ldots, i_k\}$ and $\{j_1, \ldots, j_k\}$, there is a permutation P in $\mathfrak{G}$ which maps $i_1 \to j_1, \ldots, i_k \to j_k$ so that $Pi_\kappa = j_\kappa$ $(\kappa = 1, \ldots, k)$.

By its definition the symmetric group $\mathfrak{S}_n$ is n-fold transitive and no other permutation group of degree n has this property. Every k-fold transitive group is also $(k-1)$ fold transitive $(k \geqslant 2)$. A permutation group is said to be *strictly k-fold transitive* if it is k-fold, but not $(k+1)$-fold transitive.

* This subsection is not referred to in the subsequent text.

A *transitive* permutation group $\mathfrak{G}$ acting on $\mathbf{N} = \{1, \ldots, n\}$ is said to be *imprimitive* if $\mathbf{N}$ can be divided into s disjoint sets $\mathbf{M}_1, \mathbf{M}_2, \ldots, \mathbf{M}_s$, $s > 1$, so that for every permutation P in $\mathfrak{G}$ either $P\mathbf{M}_\lambda = \mathbf{M}_\lambda$ or $P\mathbf{M}_\lambda = \mathbf{M}_\mu$, $\mu \neq \lambda$; $\lambda, \mu = 1, 2, \ldots, s$. With regard to the transitivity of $\mathfrak{G}$ it is evident that $|\mathbf{M}_\lambda| = m$ for each λ and therefore $m|n$. If $1 < m < n$ the sets $\mathbf{M}_\lambda$ are called *blocks* or *systems of imprimitivity* for $\mathfrak{G}$.

If there is no division of $\mathbf{N}$ into blocks for $\mathfrak{G}$ then the transitive group $\mathfrak{G}$ is said to be primitive.*

A k-fold transitive group $\mathfrak{G}$ is primitive if $k \geqslant 2$. Indeed let $\mathbf{M}_1, \ldots, \mathbf{M}_s$ be a division of $\mathbf{N}$ into disjoint subsets. Let 1 and 2 lie in $\mathbf{M}_1$. Because $\mathfrak{G}$ is at least doubly transitive, there is in $\mathfrak{G}$ a permutation which maps $1 \to 1$ and $2 \to i \in \mathbf{M}_\mu$, $(\mu \geqslant 2)$. Thus the $\mathbf{M}_\lambda$ cannot be systems of imprimitivity for $\mathfrak{G}$.

A transitive group $\mathfrak{G}$ can be primitive: This is the case if the degree of $\mathfrak{G}$ is a prime number. More generally we have the following theorem:

THEOREM 3. *A transitive permutation group $\mathfrak{G}$ acting on* $\mathbf{N}$ *is primitive if and only if the stabilizer $\mathfrak{G}^{(i)}$ of every i in* $\mathbf{N}$ *is a maximal subgroup of $\mathfrak{G}$.*

Proof. Let $\mathfrak{G}$ be imprimitive and $\mathbf{M} = \{\mathbf{M}_1, \ldots, \mathbf{M}_s\}$ be a complete set of blocks $\mathbf{M}_\lambda$ for $\mathfrak{G}$ in $\mathbf{N}$. Let $\mathfrak{G}$ act on $\mathbf{M}$ and denote by $\mathfrak{H}$ the stability subgroup of $\mathbf{M}_1$ in $\mathfrak{G}$. Because of transitivity of $\mathfrak{G}$ it follows that $\mathfrak{H} < \mathfrak{G}$; but if $1 \in \mathbf{M}_1$ then $\mathfrak{G}^{(1)} < \mathfrak{H}$. Therefore $\mathfrak{G}^{(1)}$ is not a maximal subgroup of $\mathfrak{G}$. Thus if all $\mathfrak{G}^{(i)}$ are maximal subgroups of $\mathfrak{G}$ then $\mathfrak{G}$ is primitive.

To prove the converse let us suppose that there is a subgroup $\mathfrak{H}$ of $\mathfrak{G}$ such that $\mathfrak{G}^{(1)} < \mathfrak{H} < \mathfrak{G}$. We substitute for $\mathbf{N}$ the system $\mathbf{T}_{\mathfrak{G}^{(1)}}$ of the cosets of $\mathfrak{G}^{(1)}$ in $\mathfrak{G}$ which by theorem 2 is action-equivalent to $\mathbf{N}$ with respect to $\mathfrak{G}$. If $\mathfrak{G}$ acts on $\mathbf{T}_{\mathfrak{G}^{(1)}}$ then those cosets of $\mathfrak{G}^{(1)}$ which are contained in a given coset of $\mathfrak{H}$, form a block in $\mathbf{T}_{\mathfrak{G}^{(1)}}$ for $\mathfrak{G}$. Thus $\mathfrak{G}$ is imprimitive. Therefore we can say that if $\mathfrak{G}$ is primitive, a proper subgroup of $\mathfrak{G}$ cannot properly contain a stabilizer of an element of $\mathfrak{G}$.

Let $\mathfrak{G}$ be an imprimitive permutation group with the block system $\mathbf{M} = \{\mathbf{M}_1, \ldots, \mathbf{M}_s\}$. Consider the permutation group $\tilde{\mathfrak{G}}$ of degree s which results from the action of $\mathfrak{G}$ on $\mathbf{M}$. There is the homomorphism

* In this case $m = n$ or $m = 1$.

$\varphi : \mathfrak{G} \to \tilde{\mathfrak{G}}$ whose kernel is the normal subgroup of $\mathfrak{G}$ consisting of all those elements of $\mathfrak{G}$ which map each of the $\mathbf{M}_\lambda$ $(\lambda = 1, \ldots, s)$ onto itself. From the transitivity of $\mathfrak{G}$ follows the transitivity of $\mathfrak{G}/\ker\varphi$, acting on $\mathbf{M}$. This factor group is isomorphic to $\tilde{\mathfrak{G}}$. The subgroup $\ker\varphi$ itself is intransitive and each $\mathbf{M}_\lambda$ is a union of orbits of $\ker\varphi$.

Remark. The notion of imprimitivity can be extended to intransitive permutation groups $\mathfrak{G}$. Let $\mathfrak{G}$ have the component groups $\mathfrak{G}_1, \ldots, \mathfrak{G}_r$ (cf. *b.* theorem 1). This group may be called imprimitive if at least one of the transitive groups $\mathfrak{G}_\rho$, operating in $\mathbf{N}_\rho$, is imprimitive in the sense defined above. Blocks of $\mathfrak{G}$ are then the blocks of the groups $\mathfrak{G}_\rho$ in the orbits $\mathbf{N}_\rho$ respectively $(\rho = 1, \ldots, r)$.

f. Sylow's first theorem.

THEOREM 4. *Let $\mathfrak{G}$ be a finite group of order g and p a prime divisor of g. If $p^\lambda \mid g$ for $\lambda = 0, 1, \ldots, \alpha$ and $p^{\alpha+1} \nmid g$ then $\mathfrak{G}$ contains a subgroup of order p^λ.*

Proof. We begin with the adaptation of an elementary combinatorial formula. Let $g = p^\lambda m$ and assume that the group $\mathfrak{G}$ acts as a group of permutations on the set $\mathbf{T}$ of combinations $\mathbf{C}_\nu$ $(\nu = 1, \ldots, n)$ of p^λ elements of $\mathfrak{G}$ (as introduced in subsect. *d.*). The number of such combinations then equals

$$|\mathbf{T}| = n = \binom{g}{p^\lambda} = \frac{p^\lambda m(p^\lambda m - 1)(p^\lambda m - 2)\ldots(p^\lambda m - p^\lambda + 1)}{p^\lambda \cdot \quad 1 \quad \cdot \quad 2 \quad \cdot \ldots (p^\lambda - 1)}$$
$$= m \cdot k, \quad k = \binom{p^\lambda m - 1}{p^\lambda - 1}.$$

We show that the binomial coefficient k is not divisible by p. If $\lambda = 0$ we have $k = \binom{m-1}{0} = 1$. Thus let $\lambda > 0$ and $1 \leqslant \gamma \leqslant \lambda$. Then $p^\gamma \mid p^\lambda m - j$ if and only if $p^\gamma \mid j$. Therefore $p^\lambda m - j$ and j, corresponding terms in the numerator and denominator of k in its explicit form, have p as factor of the same multiplicity. Hence all factors p cancel and we have $(k, p) = 1$.

Now let $\lambda = \alpha - \beta$; then $p^\beta \mid m$, but $p^{\beta+1} \nmid m$. Therefore $p^\beta \mid n$ and $p^{\beta+1} \nmid n$. Again let $n_\rho = |\mathbf{T}_\rho|$, i.e. the order of the orbit $\mathbf{T}_\rho$ of $\mathfrak{G}$ in $\mathbf{T}$. Since $n = n_1 + \ldots + n_r$ not all n_ρ $(\rho = 1, \ldots, r)$ can be divisible by $p^{\beta+1}$. We may assume that $p^{\beta+1} \nmid n_1$, say $n_1 = p^{\beta'} q$,

$\beta' \leqslant \beta, (p, q) = 1$. By (4) $n_1 | g$. Further $p^\beta | m, m = p^\beta m', (m', p) = 1$. Thus $q | m'$ and $p^{\beta'} | m$. Hence $n_1 | m$ *and* $n_1 \leqslant m$.

Further let g_1 be the order of the stabilizer $\mathfrak{G}^{(1)}$ of $\mathbf{C}_1$, that is, the group of all $x \in \mathfrak{G}$ for which $x\mathbf{C}_1 = \mathbf{C}_1$. Then $g_1 \leqslant p^\lambda$. Indeed, let $a \in \mathbf{C}_1$ and suppose that $g_1 > p^\lambda$: Since there are at least g_1 elements xa $(x \in \mathfrak{G}^{(1)})$ in $x\mathbf{C}_1$ and only p^λ elements in $\mathbf{C}_1$ the coincidence of $\mathbf{C}_1$ and $x\mathbf{C}_1$ is impossible.

Using (4) and the inequalities established in the two preceding paragraphs we have

$$g = g_1 n_1 \leqslant g_1 m \leqslant p^\lambda m = g.$$

Therefore $g_1 = p^\lambda$ and the stabilizer $\mathfrak{G}^{(1)}$ is a subgroup of order p^λ. Thus the proof of the theorem is complete.

Examples and exercises

1. The permutation $P = (12)(34)$ is an element of the four group $\mathfrak{V}$ as well as an element of the group $\mathfrak{G}_4 = \langle(12), (34)\rangle$. They are isomorphic; but $\mathfrak{V}$ is transitive and $\mathfrak{G}_4$ is intransitive. The group $\mathfrak{G}_4$ has $\mathbf{N}_1 = \{1, 2\}$, $\mathbf{N}_2 = \{3, 4\}$ as orbits. The components of P with respect to both groups $\mathfrak{V}$ and $\mathfrak{G}_4$ are $P_1 = (12)$, $P_2 = (34)$. The groups $\langle P_1\rangle$, $\langle P_2\rangle$ are subgroups of $\mathfrak{G}_4$, but not of $\mathfrak{V}$. The group $\mathfrak{V}$ is imprimitive; $\mathbf{M} = \{\mathbf{N}_1, \mathbf{N}_2\}$ represents a system of blocks for $\mathfrak{V}$. There are two other systems of blocks for $\mathfrak{V}$.

2. Show that if $n > 2$ the alternating group $\mathfrak{A}_n$ is $(n-2)$-fold transitive.

3. THEOREM (Bertrand, 1845). *Let* $n \geqslant 5$. *The symmetric group* $\mathfrak{S}_n$ *contains no subgroup of index i if* $2 < i < n$.

Proof (N. G. Čebotarëv, 1913). Let $\mathfrak{H} < \mathfrak{S}_n$, $(\mathfrak{S}_n : \mathfrak{H}) = i$ and $2 < i \leqslant n$. By the action of $\mathfrak{S}_n$ on the system $\mathbf{T}$ of the cosets of $\mathfrak{H}$ in $\mathfrak{S}_n$ we obtain a permutation group $\tilde{\mathfrak{S}}_n$ of degree i and the mapping $f_{\mathfrak{H}} : \mathfrak{S}_n \to \tilde{\mathfrak{S}}_n$ is a homomorphism. The intersection $\mathfrak{D}$ of all conjugates of $\mathfrak{H}$ in $\mathfrak{S}_n$ is a normal subgroup of $\mathfrak{S}_n$. Since by Chap. IV, §3, ex. 6, $\mathfrak{A}_n$ is the only proper normal subgroup of $\mathfrak{S}_n$ we conclude that $\mathfrak{D} = \ker f_{\mathfrak{H}} = e$. Thus $f_{\mathfrak{H}}$ is an isomorphism and $i \geqslant n$. Therefore $i = n$.

COROLLARY 1. *If* $\mathfrak{H} < \mathfrak{S}_n$ *and* $(\mathfrak{S}_n : \mathfrak{H}) = n$, *i.e,* $|\mathfrak{H}| = (n-1)!$, *then* $\mathfrak{H} \simeq \mathfrak{S}_{n-1}$.

Proof. Consider the subgroup $\mathfrak{H} < \mathfrak{S}_n$ which under the isomorphism $f_{\mathfrak{H}}$ corresponds to the subgroup $\tilde{\mathfrak{H}}$ of $\tilde{\mathfrak{S}}_n$. For all elements X in $\tilde{\mathfrak{H}}$ and no other elements of $\tilde{\mathfrak{S}}_n$ we have $X\mathfrak{H} = \mathfrak{H}$. Hence $\tilde{\mathfrak{H}}$ is the stabilizer of $\mathfrak{H}$ (as an element of **T**) for $\mathfrak{S}_n$ and as such isomorphic to the symmetric group $\mathfrak{S}_{n-1}$.

COROLLARY 2. *Let $n \neq 6, n \geqslant 5$; every subgroup $\mathfrak{H} < \mathfrak{S}_n, (\mathfrak{S}_n : \mathfrak{H}) = n$, actually is the stabilizer of one of the symbols in* **N**.

Proof. Let us identify the set $\mathbf{N} = \{1, 2, \ldots, n\}$ with the system $\mathbf{T} = \{\mathfrak{H}, P_2\mathfrak{H}, \ldots, P_n\mathfrak{H}\}$ of the cosets of $\mathfrak{H}$ in $\mathfrak{S}_n$ so that by theorem 2 **N** and **T** are action-equivalent with respect to $\mathfrak{S}_n$. Then the isomorphism $f_{\mathfrak{H}} : \mathfrak{S}_n \to \tilde{\mathfrak{S}}_n$ will be an automorphism of $\mathfrak{S}_n$ and therefore by Chap. III, §2, ex. 9 an inner automorphism of $\mathfrak{S}_n$. This implies that there is a permutation Q in $\mathfrak{S}_n$ for which $f_{\mathfrak{H}}(X) = QXQ^{-1}$, $X \in \mathfrak{S}_n$; in particular $f_{\mathfrak{H}}(\mathfrak{H}) = Q\mathfrak{H}Q^{-1} = \tilde{\mathfrak{H}}$ which was seen to be the stabilizer of $\mathfrak{H}$ in **T** and thus, after identifying **N** and **T**, is the stabilizer of the symbol 1 in **N**. Thus $\mathfrak{H}$ is the stabilizer of the symbol $Q^{-1}(1)$ in **N**.

Burnside has established the existence of six transitive subgroups of index 6 in $\mathfrak{S}_6$ which are not conjugate to the six stabilizers, but isomorphic according to Corollary 1. (Cf. W. Burnside, *Theory of Groups of Finite Order*, 2nd ed. 1911, p. 208–209; also Chap. III, §2, ex. 9.)

4. DEFINITION. A permutation group is said to be regular if it is transitive and if the identity subgroup is a stabilizer. In particular every transitive abelian permutation group is regular.

(a) Cayley's regular representation $\tilde{\mathfrak{G}}$ of a finite group $\mathfrak{G}$ is a regular permutation group (cf. Chap. I, §2, ex. 10).

(b) All elements of a regular permutation group are regular permutations (cf. Chap. I, §2, ex. 10).

5. (a) A transitive permutation group is regular if all its subgroups are normal.

(b) The order of a regular permutation group is equal to its degree (cf. theorem 2).

6. Let $\tilde{\mathfrak{G}}$ be the regular (Cayley) representation of a finite group $\mathfrak{G}$. If $\mathfrak{H} < \mathfrak{G}$ show that the left cosets $a\mathfrak{H}$ of $\mathfrak{H}$ in $\mathfrak{G}$ form a set of blocks for $\tilde{\mathfrak{G}}$.

7. Discuss the statement: An imprimitive permutation group cannot be simple.

8. If a transitive permutation group contains a transposition then it is either a symmetric group or it is imprimitive.

9. Let $\mathfrak{G}$ be a finite group.

(a) The conjugate classes of $\mathfrak{G}$ represent a set of orbits for the group $\Delta(\mathfrak{G})$ of all inner automorphisms of $\mathfrak{G}$.

(b) The characteristic classes of $\mathfrak{G}$ represent a set of orbits for the group $\Gamma(\mathfrak{G})$ of all automorphisms of $\mathfrak{G}$ (cf. Chap. III, §1, e.).

10. Let N_1 be one of the orbits of the cyclic group $\langle P \rangle$, $P \in \mathfrak{S}_n$. The restriction of the permutation P to N_1 is then one of the disjoint cycle factors of P (cf. Chap. I, §2, *c.*).

§2 Sylow's theorems

In this section the fundamental theorems of L. Sylow will be proved without reference to permutation groups. This is preceded by a proof of the simplest forerunner of Sylow's first theorem (the lemma in *a.*) which is often referred to as Cauchy's theorem, and (in *b.*) by a discussion of the *class equation*, an essential element in the proof of Sylow's first theorem. The class equation is also an important device in other respects.

a. Preliminaries. According to Lagrange's theorem (Chap. II, §2, *a.*) the order h of a subgroup $\mathfrak{H}$ of a finite group $\mathfrak{G}$ is always a divisor of $g = |\mathfrak{G}|$. It is easy to give examples of groups $\mathfrak{G}$ which for every divisor h of g contain at least one subgroup $\mathfrak{H}$ of order h; yet this is not so in general. Indeed the alternating group $\mathfrak{A}_4$, $|\mathfrak{A}_4| = 12$, contains no subgroup of order 6. We have, however, the following fact:

LEMMA 1. *If $\mathfrak{G}$ is abelian and if p is a prime divisor of g then $\mathfrak{G}$ contains an element a of order p, i.e. $\langle a \rangle_p \leqslant \mathfrak{G}$.*

Proof. If $\mathfrak{G}$ is cyclic then it contains a subgroup of order p (cf. Chap. II, §2, Theorem 4). Thus we may assume that $\mathfrak{G}$ is not cyclic. We shall apply induction with respect to the group order g. Assuming the theorem to be correct for all groups of orders smaller than g we prove it for $\mathfrak{G}$.

Suppose first that $\mathfrak{G}$ contains a proper subgroup $\mathfrak{H}$ whose order h is divisible by p; then $\mathfrak{H}$, and therefore $\mathfrak{G}$, contains an element of order p.

Now suppose that $\mathfrak{H} \underset{\max}{<} \mathfrak{G}$ and $p \nmid h$. There is an element b in $\mathfrak{G}$, $b \notin \mathfrak{H}$. Consider the cyclic subgroup $\langle b \rangle = \mathfrak{B}$, $|\mathfrak{B}| = n$. Since $\mathfrak{G}$ is not cyclic, $\mathfrak{B} < \mathfrak{G}$. Clearly $\mathfrak{H} < \mathfrak{B}\mathfrak{H} \trianglelefteq \mathfrak{G}$. Since $\mathfrak{H}$ is a maximal normal subgroup in $\mathfrak{G}$, we conclude that $\mathfrak{B}\mathfrak{H} = \mathfrak{G}$. By Chap. II, §6, (1′), we have $g = |\mathfrak{B}\mathfrak{H}| = nh/d$ where $d = |\mathfrak{B} \cap \mathfrak{H}|$. Since $\mathfrak{B} \cap \mathfrak{H} < \mathfrak{H}$ clearly $d \mid h$ and thus $p \nmid d$ implies $p \mid n$. Thus the cyclic group $\mathfrak{B}$ contains an element of order p, which is of course an element of $\mathfrak{G}$.

b. The class equation. Let $x \in \mathfrak{G}$. By $\mathscr{C}(x)$ we denote (as in Chap. II, §4, *c.*) the class of all the conjugates $x^t = txt^{-1}$ $(t \in \mathfrak{G})$ of x in $\mathfrak{G}$. Let $|\mathscr{C}(x)| = \gamma_x$. Let $\mathfrak{Z} = \mathfrak{Z}(\mathfrak{G}) = \{x_1 = e, x_2, \ldots, x_z\}$, $|\mathfrak{Z}| = z$. Since $\mathfrak{G}$ is the union of its disjoint conjugacy classes, namely

$$\mathfrak{G} = x_1 \cup \ldots \cup x_z \cup \mathscr{C}(x_{z+1}) \cup \ldots \cup \mathscr{C}(x_k)$$

we obtain the so-called *class equation* of the group $\mathfrak{G}$, i.e.

$$(1) \qquad g = z + \gamma_{x_{z+1}} + \ldots + \gamma_{x_k}$$

where k is the number of different conjugacy classes. The set $x_1, \ldots, x_z, x_{z+1}, \ldots, x_k$ is a transversal of the k classes in $\mathfrak{G}$. For all $x \notin \mathfrak{Z}$ the number $\gamma_x \geqslant 2$ and by Chap. II, §4, *d.*, Corollary 1: $\gamma_x \mid g$, or more precisely

$$(2) \qquad g = \gamma_x \cdot n_x \quad \text{for all} \quad x \in \mathfrak{G}$$

where $n_x = |\mathfrak{N}(x)|$ if $\mathfrak{N}(x)$ denotes the normalizer of x in $\mathfrak{G}$.

Now let us write γ_κ instead of γ_{x_κ} $(\kappa = 1, 2, \ldots, k)$ and n_κ instead of n_{x_κ}. Taking the γ_κ in the order of their magnitude we have by (1)

$$(1)' \qquad g = \gamma_1 + \gamma_2 + \ldots + \gamma_k \quad \text{where} \quad \gamma_1 = 1 \leqslant \gamma_2 \leqslant \ldots \leqslant \gamma_k.$$

Division of this equation by g yields in virtue of (2)

$$(3) \qquad 1 = \frac{1}{g} + \frac{1}{n_2} + \ldots + \frac{1}{n_k} \quad \text{where} \quad n_1 = g \geqslant n_2 \geqslant \ldots \geqslant n_k, n_\kappa \mid g,$$

i.e. another form of the class equation which turns out to be useful (cf. ex. 3).

c. THEOREM 1. (L. Sylow, 1872). *Let $\mathfrak{G}$ be a finite group and p a prime divisor of $g = |\mathfrak{G}|$ so that $p^\lambda | g$ for $\lambda = 1, \ldots, \alpha$ and $p^{\alpha+1} \nmid g$. Then $\mathfrak{G}$ contains a subgroup $\mathfrak{P}_\lambda$ of order p^λ.*

Before the proof it is convenient to draw from this theorem a conclusion which suggests the introduction of an important concept. A group $\mathfrak{G}$, finite or infinite, is called a *p*-group if each of its elements has as order a power of the prime number p; so one speaks of 2-groups, 7-groups etc. We can then state:

A finite group $\mathfrak{G}$ is a *p*-group if and only if $|\mathfrak{G}| = p^\alpha$ where α is a non-negative integer.

By Lagrange's theorem the condition is sufficient. Conversely if we assume that every $x \in \mathfrak{G}$ has as order a power of p and that q is a prime divisor of g, it follows from theorem 1 that $\mathfrak{G}$ contains a subgroup of order q and thus an element of order q. This implies that $q = p$. Hence p is the only prime divisor of g.

DEFINITION. *A p-subgroup $\mathfrak{P}$ of $\mathfrak{G}$ of the highest possible order $|\mathfrak{P}| = p^\alpha$, $p^{\alpha+1} \nmid g$, is called a* ***Sylow p-subgroup*** *of $\mathfrak{G}$.*

COROLLARY 1. *If $p|g$ then $\mathfrak{G}$ contains a Sylow p-subgroup.*

Proof of Theorem 1 (G. Frobenius, 1887). Let p be a prime number, α a non-negative integer, and $\mathfrak{G}$ a group of order $g = p^\alpha m$, $p \nmid m$. It is to be established that for every non-negative integer $\lambda \leqslant \alpha$ there exists in $\mathfrak{G}$ a subgroup $\mathfrak{P}_\lambda$, $|\mathfrak{P}_\lambda| = p^\lambda$.

The statement is obvious for $\alpha = 0$ and we assume it to be true for all groups of order less than g.

Two distinct cases are to be considered:

(a) $p | z$ ($z = |\mathfrak{Z}(\mathfrak{G})|$). Since $\mathfrak{Z} = \mathfrak{Z}(\mathfrak{G})$ is abelian we know from lemma 1 that there is a subgroup $\mathfrak{P}_1 = \langle a \rangle_p \leqslant \mathfrak{Z}$ and therefore $\mathfrak{P}_1 \lhd \mathfrak{G}$. Consider the factor group $\mathfrak{G}_1 = \mathfrak{G}/\mathfrak{P}_1$ so that $|\mathfrak{G}_1| = g/p < g$. In view of the induction hypothesis the group $\mathfrak{G}_1$ contains for every $\lambda = 1, 2, \ldots, \alpha$ a subgroup $\mathfrak{H}_1$ of order $p^{\lambda-1}$. Under the homomorphism $\mathfrak{G} \to \mathfrak{G}_1$ the subgroup $\mathfrak{H}_1$ is the image of a subgroup $\mathfrak{H}$ of $\mathfrak{G}$ so that $\mathfrak{H}/\mathfrak{P}_1 = \mathfrak{H}_1$. Thus if $|\mathfrak{H}| = h$ we have $h/p = p^{\lambda-1}$ and $h = p^\lambda$. Therefore $\mathfrak{H} = \mathfrak{P}_\lambda$.

(b) $p \nmid z$. Since $p | g$ we conclude from the class equation (1) that p can *not* be a divisor of γ_x for every x in $\mathfrak{G} \setminus \mathfrak{Z}$. Hence for some element a in $\mathfrak{G}$, $a \notin \mathfrak{Z}$, one has $p \nmid \gamma_a$. By (2) $g = \gamma_a \cdot n_a$, $n_a = |\mathfrak{N}(a)|$ whence $p^\alpha | n_a$. Because $a \notin \mathfrak{Z}$ we have $\mathfrak{N}(a) < \mathfrak{G}$. By induction $\mathfrak{N}(a)$ contains a subgroup $\mathfrak{P}_\lambda$ for each $\lambda = 0, 1, \ldots, \alpha$. Hence $\mathfrak{P}_\lambda < \mathfrak{G}$.

d. Obviously every conjugate of a Sylow p-subgroup is a Sylow p-subgroup. Conversely, every Sylow p-subgroup of a finite group $\mathfrak{G}$ is conjugate to one of the Sylow p-subgroups of $\mathfrak{G}$. This follows from the following theorem.

THEOREM 2. *If $\mathfrak{A}$ is a Sylow p-subgroup of a finite group $\mathfrak{G}$ and $\mathfrak{B}$ is an arbitrary p-subgroup of $\mathfrak{G}$ then $\mathfrak{B}$ is contained in a conjugate of $\mathfrak{A}$.*

Proof. Let $|\mathfrak{A}| = p^\alpha$, $|\mathfrak{B}| = p^\lambda$, $\lambda \leqslant \alpha$. For an arbitrary $x \in \mathfrak{G}$ let $x^{-1}\mathfrak{A}x \cap \mathfrak{B} = \mathfrak{D}_x$ and $|\mathfrak{D}_x| = p^{\delta_x}$. By Chap. II, §6, (6)

$$(\mathfrak{G} : \mathfrak{A}) = g/p^\alpha = m = \sum_{x \in \mathbf{R}} (\mathfrak{B} : \mathfrak{D}_x) = \sum_{x \in \mathbf{R}} p^{\lambda - \delta_x}$$

if $\mathbf{R}$ represents a transversal of the double cosets mod $(\mathfrak{A}, \mathfrak{B})$ in $\mathfrak{G}$.

Suppose now that for all $x \in \mathbf{R}$ we had $\lambda > \delta_x$. It would then follow that p is a divisor of the sum $\Sigma p^{\lambda - \delta_x}$ and therefore $p | m$ which by supposition is not the case. Hence there is an element c in $\mathfrak{G}$ for which $\delta_c = \lambda$. This implies that $(\mathfrak{B} : \mathfrak{D}_c) = 1$, thus $|\mathfrak{B}| = |\mathfrak{D}_c|$ and since $\mathfrak{D}_c \leqslant \mathfrak{B}$ it follows that $\mathfrak{D}_c = \mathfrak{B} \leqslant c^{-1}\mathfrak{A}c$.

For $\lambda = \alpha$ we have the result:

COROLLARY 2. *Every Sylow p-subgroup $\mathfrak{P}_\alpha$ of $\mathfrak{G}$ is a conjugate of $\mathfrak{A}$.*

COROLLARY 3. *The group $\mathfrak{G}$ has a unique Sylow p-subgroup $\mathfrak{A}$ if and only if $\mathfrak{A}$ is a normal subgroup of $\mathfrak{G}$.*

COROLLARY 4. *Every p-subgroup of a finite group $\mathfrak{G}$ is contained in a Sylow p-subgroup of $\mathfrak{G}$.*

e. The preceding theorem is completed by the following often very useful theorem:

THEOREM 3. *If s_p represents the number of distinct Sylow p-subgroups of the group $\mathfrak{G}$ then $s_p \equiv 1 \ (mod\, p)$.*

Proof. Let $\mathfrak{A}$ be one of the Sylow p-subgroups of $\mathfrak{G}$. By Theorem 2 and by Chap. II, §4, Theorem 5 it follows that

$$(4) \qquad s_p = (\mathfrak{G} : \mathfrak{N}(\mathfrak{A})) = g/n$$

where $g = p^\alpha m$, $p \nmid m$ and $n = |\mathfrak{N}(\mathfrak{A})| = p^\alpha m'$, $m' | m$.

Now we apply the double coset expansion of Chap. II, §6, (3) to the case $\mathfrak{B} = \mathfrak{N}(\mathfrak{A})$ so that $\mathfrak{A}\mathfrak{B} = \mathfrak{N}(\mathfrak{A})$. Denoting by $\mathbf{R}$ a transversal of the double cosets mod$(\mathfrak{A}, \mathfrak{B})$ and putting $\mathbf{R}' = \mathbf{R}\setminus\{e\}$ we have

$$\mathfrak{G} = \mathfrak{N}(\mathfrak{A}) \cup \bigcup_{x\in\mathbf{R}'} \mathfrak{A}x\mathfrak{N}(\mathfrak{A})$$

and if $\mathfrak{D}_x = \mathfrak{A}^{x^{-1}} \cap \mathfrak{N}(\mathfrak{A})$, then

$$g = n + \sum_{x\in\mathbf{R}'} \frac{|\mathfrak{A}|\,|\mathfrak{N}(\mathfrak{A})|}{|\mathfrak{D}_x|} = n + \sum_{x\in\mathbf{R}'} p^\alpha n \Big/ |\mathfrak{D}_x|.$$

Clearly $\mathfrak{D}_x$, as a subgroup of $\mathfrak{A}$, is a p-group; hence $|\mathfrak{D}_x| = p^{\delta_x}$, $0 \leqslant \delta_x \leqslant \alpha$. Thus the left and the right side of the last equation can be divided by n, and using (4) we obtain

$$s_p = g/n = 1 + \sum_{x\in\mathbf{R}'} p^{\alpha-\delta_x}.$$

The theorem will be established once it has been shown that $\alpha > \delta_x$ for all $x \in \mathbf{R}'$, in which case the sum $\Sigma p^{\alpha-\delta_x}$ is indeed divisible by p.

Accordingly let $x \in \mathbf{R}'$; then $x \notin \mathfrak{N}(\mathfrak{A})$, i.e. $x^{-1}\mathfrak{A}x \neq \mathfrak{A}$. Now assuming that $\delta_x = \alpha$ for some $x \in \mathbf{R}'$ we conclude that $|\mathfrak{D}_x| = p^\alpha$ whence $x^{-1}\mathfrak{A}x \leqslant \mathfrak{N}(\mathfrak{A})$. But obviously $\mathfrak{A} \trianglelefteq \mathfrak{N}(\mathfrak{A})$ and since $\mathfrak{A}$ is a normal Sylow p-group of $\mathfrak{N}(\mathfrak{A})$, it is, by Corollary 3, the only Sylow p-group of $\mathfrak{N}(\mathfrak{A})$, and therefore $x^{-1}\mathfrak{A}x = \mathfrak{A}$, a contradiction. Thus $\delta_x < \alpha$ and $s_p = 1 + pk$.

The preceding three theorems of this section are often combined into one single theorem which then is called "*Sylow's theorem*".

f. We conclude this section with one more theorem on Sylow subgroups which will be useful at a later stage.

THEOREM 4. *Let $\mathfrak{P}$ be a Sylow p-subgroup of the finite group $\mathfrak{G}$ and let $\mathfrak{H} \lhd \mathfrak{G}$. Then*

(a) *$\mathfrak{H}\mathfrak{P}/\mathfrak{H}$ is a Sylow p-subgroup of $\mathfrak{G}/\mathfrak{H}$.*

(b) *$\mathfrak{D} = \mathfrak{H} \cap \mathfrak{P}$ is a Sylow p-subgroup of $\mathfrak{H}$.*

Proof. (a) With regard to Chap. II, §5, *b*., Theorem 5, we have $\mathfrak{H}\mathfrak{P}/\mathfrak{H} \simeq \mathfrak{P}/\mathfrak{D}$. Because subgroups of a p-group are p-groups,

$|\mathfrak{D}| = p^\delta$. Likewise $\mathfrak{P}/\mathfrak{D}$ is a p-group and therefore $\mathfrak{H}\mathfrak{P}/\mathfrak{H}$ is a p-subgroup of $\mathfrak{G}/\mathfrak{H}$. By the index law

$$(\mathfrak{G}/\mathfrak{H} : \mathfrak{H}\mathfrak{P}/\mathfrak{H}) = (\mathfrak{G} : \mathfrak{H}\mathfrak{P}) = p^\alpha m \Big/ \frac{p^\alpha h}{p^\delta}, \quad h = |\mathfrak{H}|,$$

and this expression is not divisible by p because $p^\delta | h$. Thus $\mathfrak{H}\mathfrak{P}/\mathfrak{H}$ is a Sylow p-subgroup of $\mathfrak{G}/\mathfrak{H}$.

(b) By the index law

$$(\mathfrak{H}\mathfrak{P} : \mathfrak{P}) = (\mathfrak{H} : \mathfrak{D}) = h/p^\delta$$

and h/p^δ is not divisible by p because $\mathfrak{P}$, as a Sylow p-subgroup of $\mathfrak{G}$, must be a Sylow p-subgroup of $\mathfrak{H}\mathfrak{P}$. Hence $\mathfrak{D}$ is a Sylow p-subgroup of $\mathfrak{H}$.

Examples and exercises

1. Show that the alternating group $\mathfrak{A}_n$, $|\mathfrak{A}_n| = \frac{1}{2}n!$, has no subgroup of the order $\frac{1}{4}n!$

2. If $\mathfrak{G}$ is a finite abelian group, $|\mathfrak{G}| = g$, and $h|g$, $h < g$, then there is a subgroup $\mathfrak{H} < \mathfrak{G}$, $|\mathfrak{H}| = h$.

3. In order to demonstrate the power of the class equation [cf. (1) or (3)] we shall prove the following

THEOREM (E. Landau, 1903). *Let k be a given positive integer. There exists only a finite number of finite groups $\mathfrak{G}$ which have exactly k conjugacy classes.*

In order to prove the theorem one has to show that the class equation (3), considered as a diophantine equation in the unknowns $g = n_1, n_2, \ldots, n_k$, that is an equation where the unknowns are integers, with the order- and divisibility-conditions indicated in (3), has only a finite number of solutions $\{n_1, n_2, \ldots, n_k\}$ to which may, or may not, correspond groups having $\gamma_\kappa = g/n_\kappa$ ($\kappa = 1, \ldots, k$) as class orders.

If $k = 1$ we have $g = 1$ and the group $\mathfrak{e}$ is unique. If $k > 1$, the value $n_k = 1$ is impossible; thus $n_k \geqslant 2$.

If in (3) we replace $g, n_2, \ldots, n_{k-1}$ each by n_k (which by supposition is not greater than any one of them) we obtain the inequality

$$1 \leqslant \frac{1}{n_k}+\frac{1}{n_k}+\ldots+\frac{1}{n_k} = \frac{k}{n_k} \quad \text{whence} \quad n_k \leqslant k.$$

Therefore

(5) $\quad 2 \leqslant n_k \leqslant k.$

If $n_k = k$ it follows that all $n_\kappa = k$; thus $\{k, k, \ldots, k\}$ is a solution of (3) (satisfying the order- and divisibility-conditions) to which corresponds every abelian group of order k.

Interrupting the proof let us consider some special cases: $k = 2$: $1 = 1/g + 1/n_2$, $n_2 = g/(g-1)$ which will be an integer only for $g = 2$. The only solution of the diophantine problem therefore is $\{2, 2\}$ and the corresponding group is the group of order 2.

$k = 3$: $1 = 1/g + 1/n_2 + 1/n_3$; excluding the (trivial) case $n_3 = 3$ there remains $n_3 = 2$ so that $n_2 = 2g/(g-2)$. This will be an integer if and only if $(g-2)|2g$; but $2g = 2(g-2)+4$ so that $g-2|4$ and thus $2 < g \leqslant 6$. Only $g = 4$ and $g = 6$ yield solutions, namely $\{4, 4, 2\}$ and $\{6, 3, 2\}$. No group corresponds to the first solution; for $g = 6$ we have the class equation (1) $6 = 1+2+3$ to which corresponds the group $\mathfrak{G} = \mathfrak{S}_3$.

We continue the proof and consider the function

(6) $$1-\frac{1}{n_k}-\frac{1}{n_{k-1}}-\ldots-\frac{1}{n_{k-\lambda}} \quad (\lambda = 0, 1, \ldots, k-2)$$

of the positive integral variables $n_k, \ldots, n_{k-\lambda}$. Subject to the order- and divisibility-conditions in (3) there is a value system for these variables for which the function (6) assumes its least positive value which we denote by $\mu_{k-\lambda}$. So we have

$$\mu_k = \text{Min}\left(1-\frac{1}{n_k}\right) = \frac{1}{2} \quad \text{assumed for} \quad n_k = 2,$$

$$\mu_{k-1} = \text{Min}\left(1-\frac{1}{n_k}-\frac{1}{n_{k-1}}\right) = \frac{1}{6}$$

assumed for $n_k = 2, n_{k-1} = 3$;

$$\mu_{k-2} = \text{Min}\left(1-\frac{1}{n_k}-\frac{1}{n_{k-1}}-\frac{1}{n_{k-2}}\right) = \frac{1}{42}$$

assumed for $n_k = 2, n_{k-1} = 3, n_{k-2} = 7$; etc.

Now we can successively derive upper bounds for the $n_k, n_{k-1}, \ldots, n_1 = g$ which constitute a solution of (3). In the first place by (5) we have $n_k \leqslant k$. Further

$$1-\frac{1}{n_k} = \frac{1}{g}+\frac{1}{n_2}+\ldots+\frac{1}{n_{k-1}} \leqslant \frac{1}{n_{k-1}}+\ldots+\frac{1}{n_{k-1}} = \frac{k-1}{n_{k-1}},$$

hence

$$n_{k-1} \leqslant \frac{k-1}{1-\dfrac{1}{n_k}} \leqslant \frac{k-1}{\mu_k} = 2(k-1).$$

Similarly

$$1-\frac{1}{n_k}-\frac{1}{n_{k-1}} \leqslant \frac{k-2}{n_{k-2}} \quad \text{so that} \quad n_{k-2} \leqslant \frac{k-2}{\mu_{k-1}} = 6(k-2).$$

By the same argument:

$$n_{k-3} \leqslant \frac{k-3}{\mu_{k-2}} = 42(k-3), \ldots,$$

$$n_{k-\lambda-1} \leqslant \frac{k-\lambda-1}{\mu_{k-\lambda}}, \ldots, n_1 = g \leqslant \frac{1}{\mu_2}.$$

This shows that the order g of a group having k conjugacy classes is bounded.

Remark. To each of the groups having k classes corresponds a solution $\{n_1, \ldots, n_k\}$ of the diophantine problem (3) and it has not been excluded that to several groups corresponds the same solution. On the other hand, the case $k = 3$ shows that a solution may be spurious in so far as it does not correspond to a group.

In the following examples 4–8 we develop briefly the basic theorems on finite p-groups (Frobenius 1895).

4. Every p-group has a non-trivial center. (Hint: From the class equation derive the fact that the order of the center of a p-group is divisible by p.)

5. Every maximal subgroup of the p-group $\mathfrak{P}_\alpha$, $|\mathfrak{P}_\alpha| = p^\alpha$, $\alpha \geqslant 2$, has order $p^{\alpha-1}$ and is normal in $\mathfrak{P}_\alpha$. Every proper subgroup $\mathfrak{P}_\lambda$ of $\mathfrak{P}_\alpha$ is a subgroup of a maximal subgroup $\mathfrak{P}_{\alpha-1}$.

6. If $\mathfrak{P}_\alpha$ contains a normal cyclic subgroup $\mathfrak{P}_1$, $|\mathfrak{P}_1| = p$, then $\mathfrak{P}_1 \leqslant \mathfrak{Z}(\mathfrak{P}_\alpha)$.

Hint: Notice that if $\mathfrak{P}_1 = \langle a \rangle_p$, the class $\mathscr{C}(a) = a$.

7. Using the corollary to Theorem 2 of Chap. II, §4, *b.* prove the following

THEOREM. (a) *If $\mathfrak{P}_\alpha$ is not abelian the index*

$$(\mathfrak{P}_\alpha : \mathfrak{Z}(\mathfrak{P}_\alpha)) \geqslant p^2.$$

(b) *Every group of order p^2 is abelian.*

(c) *If $|\mathfrak{G}| = p^3$ and $\mathfrak{G}$ is not abelian then $|\mathfrak{Z}(\mathfrak{G})| = p$ and $\mathfrak{Z}(\mathfrak{G}) = \mathfrak{G}'$, i.e. the commutator group of $\mathfrak{G}$.*

8. Making use of the definition of Chap. IV, §4, *b.* prove that every finite p-group $\mathfrak{P}_\alpha$ is nilpotent of class $r \leqslant \alpha$.

9. If $\mathfrak{G}$ is a finite group, $\mathfrak{H} \lhd \mathfrak{G}$, and $(\mathfrak{G} : \mathfrak{H}) = p^2$, then the commutator group $\mathfrak{G}' \trianglelefteq \mathfrak{H}$.

10. Let p, q, r denote three different prime numbers and $\mathfrak{G}$ a group of order pq or pqr. Show that $\mathfrak{G}$ cannot be a simple group.

11. Let $\mathfrak{P}_\lambda$ be a p-subgroup of the finite group $\mathfrak{G}$. Then

$$(\mathfrak{G} : \mathfrak{P}_\lambda) \equiv (\mathfrak{N}(\mathfrak{P}_\lambda) : \mathfrak{P}_\lambda) \pmod{p}.$$

Proof. Let $\mathfrak{H} < \mathfrak{G}$. Denote by $\mathbf{R}$ the transversal of the double cosets $\mathfrak{H}x\mathfrak{H}$ in $\mathfrak{G}$ and by $\mathbf{R}^*$ the subset of those $x \in \mathbf{R}$ for which $\mathfrak{H}x\mathfrak{H} \not\subseteq \mathfrak{N}(\mathfrak{H})$. By Chap. II, §6, (3) (taking $\mathfrak{A} = \mathfrak{B} = \mathfrak{H}$) one has

$$\mathfrak{G} = \mathfrak{N}(\mathfrak{H}) \cup \bigcup_{x\in\mathbf{R}^*} \mathfrak{H}x\mathfrak{H}$$

whence

$$(7) \qquad |\mathfrak{G}| = |\mathfrak{N}(\mathfrak{H})| + \sum_{x\in\mathbf{R}^*} \frac{|\mathfrak{H}|^2}{|\mathfrak{D}_x|}, \quad \mathfrak{D}_x = \mathfrak{H}^{x^{-1}} \cap \mathfrak{H},$$

and

$$(\mathfrak{G}:\mathfrak{H}) = (\mathfrak{N}(\mathfrak{H}):\mathfrak{H}) + \sum_{x\in\mathbf{R}^*} \frac{|\mathfrak{H}|}{|\mathfrak{D}_x|}.$$

Hence if $\mathfrak{H} = \mathfrak{P}_\lambda$, $|\mathfrak{P}_\lambda| = p^\lambda$, $|\mathfrak{D}_x| = p^{\delta_x}$ we have

$$(\mathfrak{G}:\mathfrak{P}_\lambda) = (\mathfrak{N}(\mathfrak{P}_\lambda):\mathfrak{P}_\lambda) + \Sigma p^{\lambda-\delta_x}.$$

It remains to be seen that all $\delta_x < \lambda$ if $x \notin \mathfrak{N}(\mathfrak{P}_\lambda)$, i.e. $x^{-1}\mathfrak{P}_\lambda x \neq \mathfrak{P}_\lambda$. This implies $\mathfrak{D}_x < \mathfrak{P}_\lambda$ and therefore $p^{\delta_x} < p^\lambda$ so that p divides every term of the sum $\Sigma p^{\lambda-\delta_x}$.

12. Let p be the smallest prime divisor of $\mathfrak{G}$. If $\mathfrak{H} < \mathfrak{G}$ and $(\mathfrak{G} : \mathfrak{H}) = p$ then $\mathfrak{H} \lhd \mathfrak{G}$.

Hint. Note that necessarily $(\mathfrak{G} : \mathfrak{N}(\mathfrak{H})) = 1$ or $= p$. In the first case $\mathfrak{G} = \mathfrak{N}(\mathfrak{H})$ and therefore $\mathfrak{H} \lhd \mathfrak{G}$. In the second case $\mathfrak{N}(\mathfrak{H}) = \mathfrak{H}$. Making use of (7) show that this case cannot occur. [Cf. *Amer. Math. Monthly* 70 (1963), 1016–17, Advanced Problem No. 5055.]

13. Let $\mathfrak{H} < \mathfrak{G}$ and p be the smallest prime divisor of $|\mathfrak{H}|$. Let $(\mathfrak{G}:\mathfrak{H}) \leqslant p$. Show that $\mathfrak{H} \lhd \mathfrak{G}$. [Cf. G. Frobenius "Über endliche Gruppen" (1895), Gesammelte Abhandlungen, Band II, No. 47, p. 640.]

14. Using the result of ex. 11 prove again that two arbitrary Sylow p-subgroups of a finite group are conjugate (cf. Theorem 2).

15. *The groups $\mathfrak{G}$ of the order pq where p and q are different primes.*

Suppose that $p > q$. By Sylow's Theorem 1, $\mathfrak{G}$ contains a Sylow p-subgroup $\mathfrak{P}$ and a Sylow q-subgroup $\mathfrak{Q}$. Since $|\mathfrak{P}| = p$ and $|\mathfrak{Q}| = q$, both are cyclic, $\mathfrak{P} = \langle a \rangle_p$ and $\mathfrak{Q} = \langle b \rangle_q$ and $\mathfrak{P} \cap \mathfrak{Q} = e$. Moreover $(\mathfrak{G} : \mathfrak{P}) = q < p$; hence by ex. 12: $\mathfrak{P} \lhd \mathfrak{G}$. Thus $\mathfrak{G} = \mathfrak{P}\mathfrak{Q} = \mathfrak{Q}\mathfrak{P}$ and $s_p = 1$ (cf. *d.* Corollary 3, and *e.* Theorem 3) whereas $s_q = (\mathfrak{G} : \mathfrak{N}(\mathfrak{Q})) = 1+kq$. Since $s_q | pq$ and $(s_q, q) = 1$ one has $s_q | p$ and therefore only the two possibilities:

(a) $s_q = 1$; (b) $s_q = p$.

In case (a) $\mathfrak{Q} \lhd \mathfrak{G}$ so that $\mathfrak{G} = \mathfrak{P} \times \mathfrak{Q}$. Thus $ab = ba$ and the cyclic group $\mathfrak{G} = \langle ab \rangle_{pq}$ is the only group of the order pq (cf. Chap. II, §5, ex. 6).

In case (b) we notice that

$$p = 1+kq, \quad \text{i.e.} \quad q|(p-1).$$

It will be shown that then, apart from the cyclic group $\langle ab \rangle_{pq}$, there exists exactly one non-abelian group of the order pq. As an example we mention the special case $p = 3$, $q = 2$ when $\mathfrak{G} \simeq \mathfrak{S}_3$, $\mathfrak{P} \simeq \mathfrak{A}_3$ and $\mathfrak{Q} \simeq \{I, (12)\}$; the 2-subgroup $\mathfrak{Q}$ has three conjugates.

In order to determine the non-cyclic group $\mathfrak{G}$ we make use of the fact that $\mathfrak{P} = \langle a \rangle_p \lhd \mathfrak{G}$; thus with any $b \in \mathfrak{Q}$, $b \notin e$, one has

(8) $\quad bab^{-1} = a^r \quad \text{where} \quad r \not\equiv 1 \pmod{p}$.

Indeed, if $r \equiv 1$ the group $\mathfrak{G}$ is abelian, hence cyclic.

The relation (8) makes it possible to construct the group table of $\mathfrak{G}$. Evidently all $x, y \in \mathfrak{G}$ can be written in the form

$$x = b^\lambda a^\kappa,\ y = b^\nu a^\mu,\ (\kappa, \mu = 0, 1, \ldots, p-1;\ \lambda, \nu = 0, 1, \ldots, q-1).$$

One has to find ρ, σ, modulo p, q respectively, for which

$$xy = b^\sigma a^\rho.$$

From (8)

$$ba^2b^{-1} = (bab^{-1})^2 = (a^r)^2 = a^{2r}, \ldots, ba^\kappa b^{-1} = a^{\kappa r},$$

$$b^2ab^{-2} = ba^r b^{-1} = (bab^{-1})^r = a^{r^2}, \ldots, b^\lambda ab^{-\lambda} = a^{r^\lambda}.$$

In particular $b^q a b^{-q} = a = a^{r^q}$ where

(9) $\quad r^q \equiv 1 \pmod p$.

Since $q \mid (p-1)$ this congruence has a root $r \not\equiv 1 \pmod p$. Indeed for every $t \not\equiv 0 \pmod p$ by Fermat's theorem: $t^{p-1} \equiv 1 \pmod p$, i.e. $t^{kq} \equiv 1$; thus $r = t^k$ for some t for which $t^k \not\equiv 1$. Further $b^\nu a^\mu b^{-\nu} = (b^\nu a b^{-\nu})^\mu = a^{r^\nu \mu}$ so that

$$\begin{aligned} xy &= b^\lambda a^\kappa b^\nu a^\mu = b^\lambda a^\kappa a^{r^\nu \mu} b^\nu = b^\lambda a^{r^\nu \mu + \kappa} b^\nu \\ &= b^{\lambda+\nu} b^{-\nu} a^{r^\nu \mu + \kappa} b^\nu = b^{\lambda+\nu} (b^{-\nu} a b^\nu)^{r^\nu \mu + \kappa} \\ &\qquad\qquad = b^{\lambda+\nu} a^{r^{-\nu}(r^\nu \mu + \kappa)} = b^{\lambda+\nu} a^{\mu + \kappa r^{-\nu}}. \end{aligned}$$

The composition law of the exponents therefore is given by

(10) $\quad \rho \equiv \mu + \kappa r^{-\nu} \pmod p, \quad \sigma \equiv \lambda + \nu \pmod q,$

and the group $\mathfrak{G}$ can be represented by the couples (λ, κ) which are multiplied $(\lambda, \kappa)(\nu, \mu) = (\sigma, \rho)$ according to (10).

So it appears that the group $\mathfrak{G}$ is a semi-direct product $\mathfrak{G} = (\mathfrak{Q}, \mathfrak{P})_\varphi$ (cf. Chap. III, §3, *a.*) with a homomorphism $\varphi : \mathfrak{G} \to \Gamma(\mathfrak{P})$. Since $\mathfrak{P} \simeq \mathbb{R}_q{}^+, \mathfrak{Q} \simeq \mathbb{R}_q{}^+$ we have $\Gamma(\mathfrak{P}) \simeq \mathfrak{R}_p$ and since $p-1 = kq$ we know that $\mathfrak{R}_p$ has a (unique) subgroup $\mathfrak{M}_q = \langle r \rangle_q$ of order q and for every fixed ν, $\nu = 1, 2, \ldots, q-1$, is $\alpha = \varphi(\nu) = r^{-\nu}$ an automorphism of $\mathbb{R}_p{}^+$: For $\mu \in \mathbb{R}_p{}^+$ we have $\alpha(\mu) = \mu r^{-\nu}$. Thus the formula (10) is indeed a special case of (1) in Chap. III, §3 *a.*

The group $\mathfrak{M}_q = \langle r \rangle_q$ is generated by every one of the $q-1$ roots r^i, $(i = 1, \ldots, q-1)$ of the congruence (9). To each of them corresponds another automorphism $\alpha_i = r^{i\nu}$ $(\nu \in \mathbb{R}_q{}^+)$. The corresponding groups $\mathfrak{G}_i$ are all isomorphic: Denote by (λ', κ') and (ν', μ') two elements of $\mathfrak{G}_i$. Their product (σ', ρ') then is defined by

$$\rho' \equiv \mu' + \kappa' r^{i\nu'} \pmod p, \quad \sigma' \equiv \lambda' + \nu' \pmod q.$$

Consider the mapping $\mathfrak{G} = \mathfrak{G}_1 \to \mathfrak{G}_i$:

$$\begin{aligned} \kappa' &\equiv r^{i-1}\kappa \pmod p, & \lambda' &\equiv \lambda i^{-1} \pmod q \\ \mu' &\equiv r^{i-1}\mu & \nu' &\equiv \nu i^{-1} \end{aligned}$$

where i^{-1} denotes the inverse of i in $\mathfrak{M}_q$. Since

$$\rho' \equiv r^{i-1}\rho \pmod{p}, \quad \sigma' \equiv \sigma i^{-1} \pmod{q}$$

the mapping represents an isomorphism $\mathfrak{G}_1 \to \mathfrak{G}_i$.

We summarize:

THEOREM. (E. Netto, 1882). *Let p and q be prime numbers, $p > q$. If $q \nmid (p-1)$ then the cyclic group $\langle c\rangle_{pq}$ is the only abstract group of order pq. If, however, $q \mid (p-1)$ then there is, apart from $\langle c\rangle_{pq}$, exactly one non-abelian group of order pq, namely*

$$\mathfrak{G} = \langle a, b \mid a^p = b^q = e,\ ba = a^r b,\ r \not\equiv 1 \pmod{p}\rangle.$$

Remark. The non-abelian group of order $2p$ is isomorphic to the dihedral group $\mathfrak{D}_p$, for every odd prime p.

By generalizing and refining the arguments used in the preceding discussion, O. Hölder has determined all the groups of orders p^3, pq^2, pqr, and p^4 where p, q, r are distinct prime numbers (*Math. Annalen* 43 (1893), 301–412).

§3 Finite nilpotent groups

A nilpotent group was defined as a group in which an upper (or a lower) central series exists. Some properties of the groups of this class have been discussed in Chap. IV, §4. The first object of the present section is a proof of the following theorem which to a certain extent describes what may be called the "structure" of a nilpotent group.

THEOREM 1. *A finite group $\mathfrak{G}$ of the order $g = p^\alpha q^\beta \ldots$ ($p^\alpha q^\beta \ldots$* denoting the prime factorization of g) *is nilpotent*:

(i) *if and only if its Sylow subgroups $\mathfrak{P}_\alpha, \mathfrak{Q}_\beta, \ldots, |\mathfrak{P}_\alpha| = p^\alpha$, $|\mathfrak{Q}_\beta| = q^\beta, \ldots$ are normal (and therefore unique) in $\mathfrak{G}$;*

(ii) *if and only if $\mathfrak{G}$ is the direct product of its Sylow subgroups.*

We conclude with some other characterizations of nilpotent groups.

a. The conditions of Theorem 1 are sufficient. First we show that *every p-group $\mathfrak{P}_\alpha$, $|\mathfrak{P}_\alpha| = p^\alpha$, is nilpotent* (cf. §2, ex. 8). Indeed, it is an easy consequence of the class equation that $\mathfrak{P}_\alpha$ has a non-trivial centre $\mathfrak{Z}_1 = \mathfrak{Z}(\mathfrak{P}_\alpha)$. Since the factor group $\mathfrak{P}_\alpha/\mathfrak{Z}_1$ is also a

p-group, it has a non-trivial centre $\mathfrak{Z}_2/\mathfrak{Z}_1$ and $\mathfrak{Z}_2 \lhd \mathfrak{P}_\alpha$. In the same way the p-group $\mathfrak{P}_\alpha/\mathfrak{Z}_2$ has the non-trivial centre $\mathfrak{Z}_3/\mathfrak{Z}_2$, etc. Thus $\mathfrak{e} \lhd \mathfrak{Z}_1 \lhd \mathfrak{Z}_2 \lhd \mathfrak{Z}_3 \lhd \ldots$ is the upper central chain of $\mathfrak{P}_\alpha$ which clearly ends with $\mathfrak{P}_\alpha = \mathfrak{Z}_r$, $r \leqslant \alpha$.

Now let p, q be two distinct prime numbers and $\mathfrak{Q}_\beta$ be a q-group of the order q^β. We shall show that the group $\mathfrak{G} = \mathfrak{P}_\alpha \times \mathfrak{Q}_\beta$ is nilpotent. With regard to Chap. II, §5, theorem 7 the group $\mathfrak{G}$ has the non-trivial centre $\mathfrak{Z}_1 = \mathfrak{Z}(\mathfrak{P}_\alpha) \times \mathfrak{Z}(\mathfrak{Q}_\beta)$ and $\mathfrak{G}/\mathfrak{Z}_1$ is isomorphic to the (external) direct product of a p-group and a q-group, namely $(\mathfrak{P}_\alpha/\mathfrak{Z}(\mathfrak{P}_\alpha), \mathfrak{Q}_\beta/\mathfrak{Z}(\mathfrak{Q}_\beta))$. In fact every x in $\mathfrak{G}$ appears in the form $x = a \cdot b$, $a \in \mathfrak{P}_\alpha$, $b \in \mathfrak{Q}_\beta$, and

$$a\mathfrak{Z}(\mathfrak{P}_\alpha) \cdot b\mathfrak{Z}(\mathfrak{Q}_\beta) = ab \cdot \mathfrak{Z}(\mathfrak{P}_\alpha) \cdot \mathfrak{Z}(\mathfrak{Q}_\beta) = x\mathfrak{Z}_1.$$

The same argument, applied to $\mathfrak{G}/\mathfrak{Z}_1$, yields from $\mathfrak{Z}_2/\mathfrak{Z}_1 = \mathfrak{Z}(\mathfrak{G}/\mathfrak{Z}_1)$ the second term $\mathfrak{Z}_2$ of the upper central series of $\mathfrak{G}$ and it can be repeated as often as required to obtain the central series of $\mathfrak{G}$.

In the same manner it can be shown that *the direct product of two or more nilpotent groups is a nilpotent group.*

b. The conditions of Theorem 1 *are necessary.* We shall first prove (i), the normality of the Sylow subgroups of a nilpotent group. The proof rests on the following two lemmas.

LEMMA 1. *Let $\mathfrak{P}_\alpha$ be a Sylow p-subgroup of a finite group $\mathfrak{G}$ and $\mathfrak{H} \leqslant \mathfrak{G}$. If $\mathfrak{N}(\mathfrak{P}_\alpha) \leqslant \mathfrak{H}$, then $\mathfrak{H} = \mathfrak{N}(\mathfrak{H})$, i.e. $\mathfrak{H}$ is its own normalizer in $\mathfrak{G}$.*

Proof. Let $x \in \mathfrak{N}(\mathfrak{H})$; then $x\mathfrak{P}_\alpha x^{-1} \leqslant \mathfrak{H}$ because

$$\mathfrak{P}_\alpha \leqslant \mathfrak{N}(\mathfrak{P}_\alpha) \leqslant \mathfrak{H} \leqslant \mathfrak{N}(\mathfrak{H}).$$

Since both, $\mathfrak{P}_\alpha$ and $x\mathfrak{P}_\alpha x^{-1}$, are p-Sylow subgroups in $\mathfrak{H}$ they are, in view of Sylow's theorem 2 (§2, *d.*), conjugate subgroups [not only in $\mathfrak{N}(\mathfrak{H})$ but] in $\mathfrak{H}$. Thus in $\mathfrak{H}$ there is an element y such that $y(x\mathfrak{P}_\alpha x^{-1})y^{-1} = \mathfrak{P}_\alpha$. Consequently $yx \in \mathfrak{N}(\mathfrak{P}_\alpha)$ and $yx \in \mathfrak{H}$. Hence $x \in \mathfrak{H}$ and $\mathfrak{N}(\mathfrak{H}) = \mathfrak{H}$.

LEMMA 2. *If $\mathfrak{G}$ is nilpotent and $\mathfrak{H} < \mathfrak{G}$, then $\mathfrak{H} < \mathfrak{N}(\mathfrak{H})$. ($\mathfrak{H}$ is proper subgroup in both cases.)*

This is essentially the content of Corollary 1 of Chap. IV, §4, *c*.

Now choose $\mathfrak{H} = \mathfrak{N}(\mathfrak{P}_\alpha)$; then by Lemma 1

(1) $\mathfrak{N}(\mathfrak{P}_\alpha) = \mathfrak{N}(\mathfrak{N}(\mathfrak{P}_\alpha))$.

On the other hand, if we assume that

(2) $\mathfrak{N}(\mathfrak{P}_\alpha) < \mathfrak{G}$,

then it follows from Lemma 2 that $\mathfrak{N}(\mathfrak{P}_\alpha) < \mathfrak{N}(\mathfrak{N}(\mathfrak{P}_\alpha))$. This stands in contradiction to the equality (1). Hence the assumption (2) is absurd and we conclude that $\mathfrak{N}(\mathfrak{P}_\alpha) = \mathfrak{G}$.

Now we show that $\mathfrak{G}$ is the direct product of its Sylow subgroups.

Slightly generalizing the situation let $\mathfrak{A} \lhd \mathfrak{G}$ and $\mathfrak{B} \lhd \mathfrak{G}$ and $(|\mathfrak{A}|, |\mathfrak{B}|) = 1$. Then $\mathfrak{A} \cap \mathfrak{B} = \mathfrak{e}$ since the order of $\mathfrak{A} \cap \mathfrak{B}$ must be a divisor of $|\mathfrak{A}|$ and of $|\mathfrak{B}|$. Hence $\mathfrak{A}\mathfrak{B} = \mathfrak{A} \times \mathfrak{B} \trianglelefteq \mathfrak{G}$. This can readily be extended to any number of normal subgroups with relatively prime orders, in particular to the system of the (normal) Sylow subgroups of $\mathfrak{G}$.

c. Maximal subgroups. In Chap. IV, §4, *c.*, Corollary 2, it has been shown that every subgroup $\mathfrak{H}$ of a nilpotent group $\mathfrak{G}$ is a member of a subnormal series, that is, $\mathfrak{H}$ is a subnormal subgroup of $\mathfrak{G}$. This implies that every maximal subgroup of $\mathfrak{G}$ is normal in $\mathfrak{G}$. Each of these conditions is not only necessary, but also sufficient for a finite group $\mathfrak{G}$ to be nilpotent.

To prove this we first show that if in a finite group $\mathfrak{G}$ every maximal subgroup is normal, then $\mathfrak{G}$ is nilpotent. In fact, suppose that $\mathfrak{H}$ is a maximal subgroup of $\mathfrak{G}$ and that $\mathfrak{H}$ contains the normalizer of some Sylow p-subgroup $\mathfrak{P}_\alpha$, i.e.

$$\mathfrak{N}(\mathfrak{P}_\alpha) \leqslant \mathfrak{H} \lhd \mathfrak{N}(\mathfrak{H}) = \mathfrak{G}.$$

Since by Lemma 1 we have $\mathfrak{N}(\mathfrak{H}) = \mathfrak{H}$ it follows that $\mathfrak{H} = \mathfrak{G}$. Thus for a given Sylow p-subgroup $\mathfrak{P}_\alpha$ there is no maximal subgroup $\mathfrak{H}$ which contains $\mathfrak{N}(\mathfrak{P}_\alpha)$. Now, in a finite group $\mathfrak{G}$ every proper subgroup either is a maximal subgroup or it is contained in some maximal subgroup of $\mathfrak{G}$; hence $\mathfrak{N}(\mathfrak{P}_\alpha) = \mathfrak{G}$, i.e. $\mathfrak{P}_\alpha \lhd \mathfrak{G}$. This being so for every Sylow subgroup of $\mathfrak{G}$, it follows by Theorem 1 that $\mathfrak{G}$ is nilpotent.

Further, if every subgroup of $\mathfrak{G}$ is subnormal, then so are the maximal subgroups of $\mathfrak{G}$. These are therefore normal in $\mathfrak{G}$ and $\mathfrak{G}$ is nilpotent.

We summarize:

THEOREM 2. (O. Schmidt, 1926). *A finite group $\mathfrak{G}$ is nilpotent* (i) *if and only if every maximal subgroup of $\mathfrak{G}$ is normal*; (ii) *if and only if every subgroup of $\mathfrak{G}$ is subnormal.*

d. The Frattini subgroup $\Phi(\mathfrak{G})$ of a group $\mathfrak{G}$ may be defined as the intersection of all maximal subgroups of $\mathfrak{G}$. Some of its elementary properties have been discussed in earlier parts of this book; in particular cf. Chap. III, §1, ex. 18. Its importance in the theory of nilpotent groups becomes evident from the following theorem:

THEOREM 3. (H. Wielandt, 1937). *A finite group $\mathfrak{G}$ is nilpotent if and only if its commutator subgroup is a subgroup of its Frattini subgroup.*

Proof. (i) Suppose that $\mathfrak{G}$ is a finite group and $\mathfrak{G}' \leqslant \Phi(\mathfrak{G})$. By definition every maximal subgroup $\mathfrak{H}$ of $\mathfrak{G}$ contains $\Phi(\mathfrak{G})$. Since $\mathfrak{G}' \trianglelefteq \mathfrak{G}$ and $\mathfrak{G}/\mathfrak{G}'$ is abelian, we have $\mathfrak{H}/\mathfrak{G}' \lhd \mathfrak{G}/\mathfrak{G}'$ whence $\mathfrak{H} \lhd \mathfrak{G}$ and by Theorem 2 the group $\mathfrak{G}$ is nilpotent.

(ii) Conversely suppose that $\mathfrak{G}$ is nilpotent. Let $\mathfrak{H} \underset{\max}{<} \mathfrak{G}$. By Lemma 2, $\mathfrak{H} < \mathfrak{N}(\mathfrak{H})$; thus $\mathfrak{H}$ being maximal in $\mathfrak{G}$ implies that $\mathfrak{H} \lhd \mathfrak{G}$ and that $\mathfrak{G}/\mathfrak{H}$ contains no proper subgroup. Therefore $\mathfrak{G}/\mathfrak{H}$ is cyclic of prime order and thus abelian; hence $\mathfrak{G}' \leqslant \mathfrak{H}$. So we see that $\mathfrak{G}'$ is contained in all maximal subgroups of $\mathfrak{G}$ and thus also in their intersection, i.e. $\Phi(\mathfrak{G})$.

LEMMA 3. *Let $\mathfrak{G}$ be a finite group and $\mathfrak{H} \leqslant \mathfrak{G}$. If $\mathfrak{G} = \Phi(\mathfrak{G})\mathfrak{H}$ then $\mathfrak{G} = \mathfrak{H}$.*

This follows immediately from the first definition of $\Phi(\mathfrak{G})$ as the set of all non-generators of $\mathfrak{G}$ (cf. Chap. II, §1, ex. 11). In fact, if $\mathfrak{G} = \Phi(\mathfrak{G})\mathfrak{H}$, then $\mathfrak{G} = \langle\Phi(\mathfrak{G}), \mathfrak{H}\rangle = \langle\mathfrak{H}\rangle = \mathfrak{H}$.

THEOREM 4. (G. Frattini, 1885). *The Frattini subgroup $\Phi(\mathfrak{G})$ of a finite group $\mathfrak{G}$ is nilpotent.*

Proof. Let $\mathfrak{P}$ be a Sylow p-subgroup of $\Phi(\mathfrak{G})$. Since $\Phi(\mathfrak{G}) \trianglelefteq \mathfrak{G}$ it follows that $x\mathfrak{P}x^{-1} \leqslant \Phi(\mathfrak{G})$ for every element x of $\mathfrak{G}$. Hence all subgroups conjugate to $\mathfrak{P}$ in $\mathfrak{G}$ are Sylow p-subgroups of $\Phi(\mathfrak{G})$ and therefore conjugate in $\Phi(\mathfrak{G})$. Thus there is an element y in $\Phi(\mathfrak{G})$ such that $x\mathfrak{P}x^{-1} = y\mathfrak{P}y^{-1}$ and therefore $y^{-1}x \in \mathfrak{N}(\mathfrak{P}) = \mathfrak{N}_{\mathfrak{G}}(\mathfrak{P})$. Consequently $x \in y \cdot \mathfrak{N}(\mathfrak{P}) \subseteq \Phi(\mathfrak{G})\mathfrak{N}(\mathfrak{P})$. (Cf. ex. 1.)

Now x was an arbitrary fixed element of $\mathfrak{G}$. Thus

$$\mathfrak{G} = \Phi(\mathfrak{G})\mathfrak{N}(\mathfrak{P})$$

and by Lemma 3

$$\mathfrak{G} = \mathfrak{N}(\mathfrak{P})$$

which implies that $\mathfrak{P}$ is normal in $\mathfrak{G}$ and therefore also normal in $\Phi(\mathfrak{G})$. This being so for all Sylow subgroups $\mathfrak{P}$ of $\Phi(\mathfrak{G})$, we derive from Theorem 1 that $\Phi(\mathfrak{G})$ is nilpotent.

COROLLARY. *If $\mathfrak{P}$ is a Sylow subgroup of $\Phi(\mathfrak{G})$ then $\mathfrak{P} \lhd \mathfrak{G}$.*

Examples and exercises

1. Show that if $\mathfrak{H} \trianglelefteq \mathfrak{G}$ and if $\mathfrak{P}$ is a Sylow subgroup of $\mathfrak{H}$, then $\mathfrak{G} = \mathfrak{H} \cdot \mathfrak{N}_{\mathfrak{G}}(\mathfrak{P})$.

Hint. Use the argument employed in the proof of Theorem 4 (as well as in that of Lemma 1) namely: For every x in $\mathfrak{G}$ there is an element y of $\mathfrak{H}$ such that $x\mathfrak{P}x^{-1} = y\mathfrak{P}y^{-1}$.

2. If $\mathfrak{H} \lhd \mathfrak{G}$ and $\mathfrak{H} \trianglelefteq \Phi(\mathfrak{G})$, then $\mathfrak{G}$ does not split over $\mathfrak{H}$. (Cf. Chap. III, §3, ex. 6.)

3. THEOREM (W. Gaschütz, 1953). *If* $\mathfrak{L} \trianglelefteq \mathfrak{G}$, $\mathfrak{H} \leqslant \mathfrak{G}$, $\mathfrak{L} \trianglelefteq \mathfrak{H} \cap \Phi(\mathfrak{G})$ *and if* $\mathfrak{H}/\mathfrak{L}$ *is nilpotent, then* $\mathfrak{H}$ *is nilpotent.*

Proof. Let $\mathfrak{P}$ be a Sylow p-subgroup of $\mathfrak{H}$. By §2, Theorem 4, $\mathfrak{P}\mathfrak{L}/\mathfrak{L}$ is a Sylow p-subgroup of $\mathfrak{H}/\mathfrak{L}$; as such it is unique because $\mathfrak{H}/\mathfrak{L}$ is nilpotent. It is normal and characteristic in $\mathfrak{H}/\mathfrak{L}$, thus

$$\mathfrak{P}\mathfrak{L}/\mathfrak{L} \trianglelefteq \mathfrak{G}/\mathfrak{L} \quad \text{whence} \quad \mathfrak{P}\mathfrak{L} \trianglelefteq \mathfrak{G}.$$

Now $\mathfrak{L} \trianglelefteq \mathfrak{H}$; therefore $\mathfrak{P}$ is a Sylow subgroup of $\mathfrak{P}\mathfrak{L}$ and by ex. 1

$$\mathfrak{G} = \mathfrak{P} \cdot \mathfrak{L} \cdot \mathfrak{N}_{\mathfrak{G}}(\mathfrak{P}) = \mathfrak{L} \cdot \mathfrak{N}_{\mathfrak{G}}(\mathfrak{P})$$

because $\mathfrak{P} \trianglelefteq \mathfrak{N}_{\mathfrak{G}}(\mathfrak{P})$. Since moreover $\mathfrak{L} \trianglelefteq \Phi(\mathfrak{G})$ we conclude that $\mathfrak{G} = \mathfrak{N}_{\mathfrak{G}}(\mathfrak{P})$ and thus $\mathfrak{P} \trianglelefteq \mathfrak{G}$ and $\mathfrak{P} \trianglelefteq \mathfrak{H}$.

There are two immediate consequences of this theorem:

(i) Theorem 4: $\Phi(\mathfrak{G})$ is nilpotent for every finite group $\mathfrak{G}$.

(ii) If $\mathfrak{G}$ is finite and $\mathfrak{G}/\Phi(\mathfrak{G})$ is nilpotent, then $\mathfrak{G}$ is itself nilpotent.

§4 The structure of finite abelian groups

Let $\mathfrak{G}$ be a finite abelian group of order $|\mathfrak{G}| = p^\alpha q^\beta \ldots$ where $p, q, \ldots$ are distinct prime numbers, $\alpha, \beta, \ldots$ natural numbers. Since $\mathfrak{G}$ is nilpotent it is, by §3, theorem 1, the direct product of its Sylow subgroups $\mathfrak{P}_\alpha, \mathfrak{Q}_\beta, \ldots$, the so-called *primary factors* of $\mathfrak{G}$. If $\mathfrak{P}_\alpha$ is a cyclic p-group, it cannot be split into direct factors. Indeed let $\mathfrak{P}_{\alpha_1}, \mathfrak{P}_{\alpha_2}$ be two cyclic p-groups such that $\mathfrak{P}_\alpha = \mathfrak{P}_{\alpha_1} \times \mathfrak{P}_{\alpha_2}$ and $\alpha_1 \geqslant \alpha_2$; then the highest order of an element of the product $\mathfrak{P}_{\alpha_1}\mathfrak{P}_{\alpha_2}$ is p^{α_1} which is smaller than $|\mathfrak{P}_\alpha| = p^{\alpha_1+\alpha_2}$ [cf. Chap. II, §5, ex. 6(c)].

The structure of a finite abelian group is described by the so-called *Basis Theorem*:

THEOREM 1. *Every finite abelian group $\mathfrak{G}$ is either a cyclic p-group or a direct product of cyclic p-subgroups $\mathfrak{P}^{(j)}$, $j = 1, \ldots, n$, where $|\mathfrak{P}^{(j)}| = p_j^{\alpha(j)}$ and $p_1, \ldots, p_n$ are (not necessarily distinct) prime divisors of $|\mathfrak{G}|$. The orders of the cyclic factors $\mathfrak{P}^{(j)}$ are uniquely determined by $\mathfrak{G}$.*

Remark. Although the cyclic factors $\mathfrak{P}^{(j)}$ are not, in general, unique, their orders $p_j^{\alpha(j)}$ are uniquely determined by the group $\mathfrak{G}$; they are called the *invariants of* $\mathfrak{G}$.

a. Evidently it is sufficient to prove the theorem for a non-cyclic finite abelian p-group $\mathfrak{P}_\alpha = \mathfrak{P}$, $|\mathfrak{P}| = p^\alpha$.

Choose in $\mathfrak{P}$ an element a_1 of the highest possible order p^{α_1}. Then $\alpha_1 < \alpha$ because $\mathfrak{P}$ is not cyclic. Let $\langle a_1 \rangle = \mathfrak{A}_1$. Then $\mathfrak{A}_1 \lhd \mathfrak{P}$ and $\mathfrak{P}/\mathfrak{A}_1 = \mathfrak{B}$ is an abelian p-group of order $p^{\alpha-\alpha_1} < p^\alpha = |\mathfrak{P}|$. Let us assume the theorem to be correct for all abelian p-groups of order smaller than p^α. Thus

$$\mathfrak{B} = \mathfrak{B}_2 \times \mathfrak{B}_3 \times \ldots \times \mathfrak{B}_k$$

where the $\mathfrak{B}_\kappa$ ($\kappa = 2, 3, \ldots, k$) are cyclic p-groups generated by some of the cosets of $\mathfrak{A}_1$ in $\mathfrak{P}$. Say $\mathfrak{B}_\kappa = \langle x_\kappa \mathfrak{A}_1 \rangle$, $x_\kappa \in \mathfrak{P}$. Let $|\mathfrak{B}_\kappa| = p^{\alpha_\kappa}$,

$\alpha_2+\alpha_3+\ldots+\alpha_k = \alpha-\alpha_1$ so that $(x_\kappa\mathfrak{A}_1)^{p^{\alpha_\kappa}} = \mathfrak{A}_1$ and therefore

$$x_\kappa^{p^{\alpha_\kappa}} = y_\kappa \in \mathfrak{A}_1, \text{ i.e. } y_\kappa = a_1^{n_\kappa} \; (0 \leqslant n_\kappa \leqslant p^{\alpha_1}).$$

The order of y_κ is obviously a power of p, say p^{β_κ}. We shall show that $\beta_\kappa = \alpha_1-\gamma_\kappa$ where p^{γ_κ} is the highest power of p dividing n_κ, i.e. $n_\kappa = m_\kappa p^{\gamma_\kappa}$, $p \nmid m_\kappa$. In fact, if p^{β_κ} is the order of y_κ we have

$$y_\kappa^{p^{\beta_\kappa}} = a_1^{n_\kappa p^{\beta_\kappa}} = a_1^{m_\kappa p^{\beta_\kappa+\gamma_\kappa}} = e$$

and since $a_1^{m_\kappa}$ $(p \nmid m_\kappa)$ has the same order as a_1, namely p^{α_1}, it follows that $\beta_\kappa+\gamma_\kappa = \alpha_1$.

Now let p^{ξ_κ} be the order of x_κ. Since $\mathfrak{B}_\kappa = \langle x_\kappa\mathfrak{A}_1\rangle$ has the order p^{α_κ} in $\mathfrak{P}/\mathfrak{A}_1$, we conclude that $\alpha_\kappa \leqslant \xi_\kappa$ (cf. Chap. II, §3, ex. 11). Let $\xi_\kappa-\alpha_\kappa = \delta_\kappa$. Then

$$x_\kappa^{p^{\xi_\kappa}} = e = (x_\kappa^{p^{\alpha_\kappa}})^{p^{\delta_\kappa}} = y_\kappa^{p^{\delta_\kappa}}.$$

Thus $\delta_\kappa = \beta_\kappa$ and $\xi_\kappa = \alpha_\kappa+\beta_\kappa$.

By supposition p^{α_1} is the largest order of an element of $\mathfrak{P}$. Hence it follows that $\xi_\kappa = \alpha_\kappa+\beta_\kappa \leqslant \alpha_1 = \beta_\kappa+\gamma_\kappa$ whence $\alpha_\kappa \leqslant \gamma_\kappa$.

The next step is to show that the coset $x_\kappa\mathfrak{A}_1$ contains an element a_κ of the order p^{α_κ}. In fact, the element

$$a_\kappa = x_\kappa a_1^{-m_\kappa p^{\gamma_\kappa-\alpha_\kappa}}$$

satisfies this condition. For verification notice that $a_1^{m_\kappa p^{\gamma_\kappa}} = y_\kappa$, thus

$$a_\kappa^{p^{\alpha_\kappa}} = y_\kappa y_\kappa^{-1} = e$$

and since $a_\kappa\mathfrak{A}_1 = x_\kappa\mathfrak{A}_1$ and has the order p^{α_κ} in $\mathfrak{P}/\mathfrak{A}_1$, the order of a_κ cannot be smaller than p^{α_κ}.

To conclude the proof of the first part of the theorem we put $\langle a_\kappa\rangle = \mathfrak{A}_\kappa$ and $\mathfrak{A}^* = \mathfrak{A}_2\mathfrak{A}_3\ldots\mathfrak{A}_k < \mathfrak{P}$. Consider the natural homomorphism $\varphi : \mathfrak{P} \to \mathfrak{P}/\mathfrak{A}_1$. Under φ the image of each element $a_\kappa a_1^{\mu}$ $(\mu = 0, 1, \ldots, p^{\alpha_\kappa}-1)$ of the coset $a_\kappa\mathfrak{A}_1$ is the coset $a_\kappa\mathfrak{A}_1$ itself, an element of $\mathfrak{P}/\mathfrak{A}_1$; in particular $\varphi(a_\kappa) = a_\kappa\mathfrak{A}_1$. Hence

$$\langle\varphi(a_\kappa)\rangle = \varphi(\langle a_\kappa\rangle) = \varphi(\mathfrak{A}_\kappa) = \langle a_\kappa\mathfrak{A}_1\rangle = \mathfrak{B}_\kappa \leqslant \mathfrak{P}/\mathfrak{A}_1.$$

Consequently $\varphi(\mathfrak{A}^*) = \mathfrak{B}$ and

$$p^{\alpha-\alpha_1} = |\mathfrak{B}| = |\varphi(\mathfrak{A}^*)| \leqslant |\mathfrak{A}^*| \leqslant p^{\alpha_2} \dots p^{\alpha_k} = p^{\alpha-\alpha_1}.$$

Hence $|\mathfrak{A}^*| = |\mathfrak{B}|$ and it follows that $\varphi : \mathfrak{A}^* \to \mathfrak{B}$ is an isomorphism. Since $\varphi(\mathfrak{A}_k) = \mathfrak{B}_k$ we conclude that $\mathfrak{A}^* = \mathfrak{A}_2 \times \mathfrak{A}_3 \times \dots \times \mathfrak{A}_k$.

The isomorphism $\mathfrak{A}^* \simeq \mathfrak{P}/\mathfrak{A}$ finally entails that $|\mathfrak{P}| = |\mathfrak{A}_1|\,|\mathfrak{A}^*|$ whence $|\mathfrak{A}_1 \cap \mathfrak{A}^*| = 1$ and

$$\mathfrak{P} = \mathfrak{A}_1 \times \mathfrak{A}^* = \mathfrak{A}_1 \times \mathfrak{A}_2 \times \dots \times \mathfrak{A}_k.$$

b. A system of generators $a_1, \dots, a_k$ of $\mathfrak{A}_1, \dots, \mathfrak{A}_k$ respectively is called a *basis* of $\mathfrak{P}$. A basis of $\mathfrak{P}$ is not unique and simple examples show that even the direct factors $\mathfrak{A}_1, \dots, \mathfrak{A}_k$ of $\mathfrak{P}$ are, in general, not uniquely defined. According to the theorem, however, *the exponents $\alpha_1, \alpha_2, \dots, \alpha_k$ are uniquely defined by the group $\mathfrak{P}$.*

To prove this we make the following arrangement. Let, as before, $\mathfrak{A}_\kappa = \langle a_\kappa \rangle$, $|\mathfrak{A}_\kappa| = p^{\alpha_\kappa}$ $(\kappa = 1, \dots, k)$,

$$(1) \qquad \mathfrak{P} = \mathfrak{A}_1 \times \dots \times \mathfrak{A}_k, \quad \alpha_1 \geqslant \alpha_2 \geqslant \dots \geqslant \alpha_k \geqslant 1.$$

With a second system of cyclic p-subgroups $\mathfrak{B}_\lambda = \langle b_\lambda \rangle$, $|\mathfrak{B}_\lambda| = p^{\beta_\lambda}$ $(\lambda = 1, \dots, l)$ let

$$(2) \qquad \mathfrak{P} = \mathfrak{B}_1 \times \dots \times \mathfrak{B}_l, \quad \beta_1 \geqslant \beta_2 \geqslant \dots \geqslant \beta_l \geqslant 1.$$

We may assume that for a certain m, $1 \leqslant m \leqslant k$, we have

$$\alpha_1 = \beta_1, \dots, \alpha_{m-1} = \beta_{m-1}, \quad \text{but} \quad \alpha_m < \beta_m.$$

To show that this is impossible we form the subgroup $\mathfrak{P}^{(m)} < \mathfrak{P}$, consisting of the p^{α_m}-th powers of the elements of $\mathfrak{P}$ (cf. Chap. I, §2, ex. 14). According to (1)

$$(1') \qquad \mathfrak{P}^{(m)} = \mathfrak{A}_1^{(m)} \times \mathfrak{A}_2^{(m)} \times \dots \times \mathfrak{A}_{m-1}^{(m)}, \quad \mathfrak{A}_\kappa^{(m)} = \langle a_\kappa^{p^{\alpha_m}} \rangle.$$

Indeed $\mathfrak{A}_\kappa^{(m)} = e$ if $\kappa \geqslant m$. According to (2)

$$(2') \qquad \mathfrak{P}^{(m)} = \mathfrak{B}_1^{(m)} \times \mathfrak{B}_2^{(m)} \times \dots \times \mathfrak{B}_{m-1}^{(m)} \times \mathfrak{B}_m^{(m)} \times \dots, \qquad \mathfrak{B}_\lambda^{(m)} = \langle b_\lambda^{p^{\alpha_m}} \rangle.$$

Now we compute the order of $\mathfrak{P}^{(m)}$. From (1′)

$$|\mathfrak{P}^{(m)}| = p^{\alpha_1-\alpha_m}p^{\alpha_2-\alpha_m}\dots p^{\alpha_{m-1}-\alpha_m},$$

and from (2′)

$$|\mathfrak{P}^{(m)}| = p^{\beta_1-\alpha_m}p^{\beta_2-\alpha_m}\dots p^{\beta_{m-1}-\alpha_m}p^{\beta_m-\alpha_m}\dots.$$

Since $\beta_1 = \alpha_1, \dots, \beta_{m-1} = \alpha_{m-1}$ we conclude that

$$|\mathfrak{P}^{(m)}| = p^{\alpha_1-\alpha_m}p^{\alpha_2-\alpha_m}\dots p^{\alpha_{m-1}-\alpha_m}p^{\beta_m-\alpha_m}\dots$$

and therefore $\beta_m = \alpha_m$. Thus our assumption was incorrect and hence $l = k$ and $\beta_\kappa = \alpha_\kappa$ ($\kappa = 1, \dots, k$). This completes the proof of the basis theorem.

c. Application. The theorem we have just proved enables us to determine all finite abstract abelian groups $\mathfrak{G}$ of a given order g. Each of these groups is uniquely determined by g and its system of invariants $p_1{}^{\alpha^{(1)}}, \dots, p_n{}^{\alpha^{(n)}}$ whose product equals g.

On the other hand, if $g = \prod_{p|g} p^\alpha$ is the factorization of g with *distinct* prime numbers p then the factors p^α are the orders $|\mathfrak{P}_\alpha|$ of the Sylow subgroups $\mathfrak{P}_\alpha$ of $\mathfrak{G}$. These are normal in $\mathfrak{G}$ and therefore unique. For every additive decomposition of α, namely

$$\alpha = \alpha_1 + \alpha_2 + \dots + \alpha_k, \quad \alpha_1 \geqslant \alpha_2 \geqslant \dots \geqslant \alpha_k \geqslant 1$$

there is exactly one abelian p-group $\mathfrak{P}_\alpha$ with these exponents. To the decomposition $\alpha = 1 + \dots + 1$ corresponds the elementary abelian p-group of the order p^α (cf. Chap. III, §1, *d.*), and to the exponent α corresponds the cyclic group of order p^α.

In general, the symbol

$$(p_1{}^{\alpha^{(1)}}, \dots, p_n{}^{\alpha^{(n)}}), \quad \prod_{\nu=1}^{n} p_\nu{}^{\alpha^{(\nu)}} = |\mathfrak{G}|,$$

(using the notation employed in theorem 1) gives a complete description of the finite abelian group $\mathfrak{G}$; it is called the *type* of $\mathfrak{G}$. Thus

$(p, p, \ldots, p)$ is the type of an elementary abelian p-group, and (p^α) the type of a cyclic group of order p^α.

Remark. It is difficult to establish who first has stated and proved the basis theorem for finite abelian groups. According to the paper of G. Frobenius and L. Stickelberger (*Journal für die reine und angewandte Mathematik* 86 (1879) or Frobenius' Gesammelte Abhandlungen, Band I, p. 545–590) the theory has been founded by Euler and Gauss in arithmetical investigations (without using the word "group") and by Lagrange and Abel in the theory of algebraic equations (referring to systems of permutations of the roots).

A basis theorem using a basis different from the one chosen above has been developed by Frobenius and Stickelberger. For a detailed exposition of this theory reference may be made to their paper or to the work by A. Chatelet, *Les groupes abéliens finis et les modules des points entiers*, Paris et Lille 1925.

Examples and exercises

1. Determine the types of all abelian p-groups of order p^5.

2. Given the type of a finite abelian group $\mathfrak{G}$, determine the types of the two subgroups $\mathfrak{H}_n$ and $\mathfrak{L}_n$ as introduced in Chap. I, §2, ex. 14.

3. Show that the additive group of a finite field is an elementary abelian p-group.

4. One of the most important abelian groups is the multiplicative group $\mathfrak{R}_m$ $(m > 1)$ of the residues (mod m) relatively prime to m (cf. Chap. I, §3, *b*.). Its order was found to be given by Euler's function $\varphi(m)$.

(a) If $m = m_1 m_2$, $(m_1, m_2) = 1$, show that $\mathfrak{R}_m$ contains two subgroups $\mathfrak{R}_{m_1}{}^*$, $\mathfrak{R}_{m_2}{}^*$ such that

$$\mathfrak{R}_m = \mathfrak{R}_{m_1}{}^* \times \mathfrak{R}_{m_2}{}^*, \quad \mathfrak{R}_{m_1}{}^* \simeq \mathfrak{R}_{m_1}, \quad \mathfrak{R}_{m_2}{}^* \simeq \mathfrak{R}_{m_2}.$$

(b) Determine the types of the groups $\mathfrak{R}_{30}$, $\mathfrak{R}_{60}$, $\mathfrak{R}_{32}$.

5. (a) The number of subgroups of order p in a finite abelian p-group which splits into k direct cyclic factors equals

$$f_k(p) = p^{k-1} + p^{k-2} + \ldots + p + 1.$$

(b) Let $\mathfrak{G}$ be a finite abelian group and $p_1, \ldots, p_m$ the distinct prime divisors of $|\mathfrak{G}|$. If k_j is the number of direct cyclic factors of the p_j-primary subgroup of $\mathfrak{G}$, then the number of subgroups of prime order in $\mathfrak{G}$ equals $\sum_{j=1}^{m} f_{k_j}(p_j)$.

Appendix

HINTS OR SOLUTIONS TO SOME OF THE EXERCISE PROBLEMS

Chapter I

§1. *Ex.* 5(a). Replace x and y by their prime factor products (cf. §3, *a.*, VI) and express g_1, g_2 in terms of the functions f_1, f_2 [cf. ex. 2(b)].

Ex. 10(b). Find a function $x = g(\xi)$ so that $x \circ y = g(\xi + \eta)$.

Ex. 12. Using the uniqueness of the pseudo-inverse show that a is the pseudo-inverse of a^* and note that for two arbitrary elements a and b of **S** the element $c = a(ba)^*b$ has aa^* as well as b^*b as pseudo-inverse. Hence $aa^* = a^*a = e$ is a right identity in $(\mathbf{S} \mid \cdot)$ and a^* is a right inverse of a. (Cf. *Amer. Math. Monthly* 79 (1972), p. 1138.)

§2. *Ex.* 13. To a fixed element $a \in \mathfrak{G}$ associate the mapping $\varphi_a : x \to ax, x \in \mathfrak{G}$; these mappings φ_a form a group $\mathfrak{G}$ and the function $f(a) : a \to \varphi_a$ is an isomorphism: To $f(b)f(a)$ corresponds the mapping $\varphi_b(\varphi_a(x)) = \varphi_b(ax) = b(ax) = (ba)x = \varphi_{ba}(x)$ which corresponds to $f(ba)$.

Ex. 15. The equation $|xy^{-1}| = |x^{-1}y|$ implies either $xy^{-1} = y^{-1}x$ or $xy^{-1} = x^{-1}y$. In the first case x and y commute. For non-commuting x, y we conclude in the second case that $x^2 = y^2 = x^{-2} = y^{-2}$. Since x, xy and y, xy are two non-commuting couples we have $(xy)^2 = (yx)^2 = x^2 = x^{-2} = y^2 = y^{-2}$. The squares play the role of e' (or -1): $x^4 = y^4 = (xy)^4 = e$ and also $yxyx = x^2$ whence $yxy = x$ or $yx = xy^{-1} = xyy^{-2} = xyy^2$. Thus we may identify x with i, y with j, xy with k.

§3. *Ex.* 7(b). Write the condition $\alpha\beta = 1$ in the form of a system of linear equations in b_1, b_2:

$$\begin{aligned} a_1b_1 - a_2b_2 &= 1 + mz_1, \\ a_2b_1 + a_1b_2 &= mz_2, \end{aligned} \qquad z_1, z_2 \in \mathbb{Z}.$$

Chapter II

§1. *Ex.* 7. If $a \in \mathbf{A}$ then $(\mathbf{A} \cap \mathbf{B})c = \{ac \mid a \in \mathbf{B}\} = \{ac \mid ac \in \mathbf{B}c\} = \mathbf{A}c \cap \mathbf{B}c$.

Ex. 11. Note that if $x \in \Phi$ then $\langle \mathbf{A}, x\rangle = \langle \mathbf{A}, x^{-1}\rangle$. If also $y \in \Phi$ then $\langle \mathbf{A}, x, y\rangle = \langle \mathbf{A}, x\rangle = \langle \mathbf{A}\rangle$ and since $\langle \mathbf{A}\rangle \leqslant \langle \mathbf{A}, xy\rangle \leqslant \langle \mathbf{A}, x, y\rangle$ it follows that $xy \in \Phi$.

§2. *Ex.* 11. Let $m\mu + n\nu = 1$; then $c = c^{m\mu}\, c^{n\nu}$ and $c^{n\nu}$ has order m. If $c = ab = ba$ one has $c^{n\nu} = a^{n\nu}\, b^{n\nu} = a^{n\nu} = a^{m\mu+n\nu} = a$.

Ex. 15. $\Phi(\langle a\rangle_p) = \mathfrak{e}$, $\Phi(\langle a\rangle_{p^\alpha}) = \langle a^p\rangle_{p^{\alpha-1}}$. If $m = p^\alpha q^\beta$ (p, q different prime numbers):

$$\Phi(\langle a\rangle_m) = \langle a^p\rangle_{m/p} \cap \langle a^q\rangle_{m/q} = \langle a^{pq}\rangle_{m/pq}.$$

Ex. 16. Let $z_1, \ldots, z_h$ be the elements of $\mathfrak{H}_n : z_i = x_i^{\,n}$, $x_i \in \mathfrak{G}$. The elements $x_1, \ldots x_h$ form a transversal for $\mathfrak{L}_n$ in $\mathfrak{G}$. Indeed, if x_i and x_j lie in the same coset of $\mathfrak{L}_n$, i.e. $x_j = yx_i$, $y \in \mathfrak{L}_n$, then $z_j = z_i$. On the other hand, every $x \in \mathfrak{G}$ lies in one of the cosets $x_i\mathfrak{L}_n$. For x^n equals one of the elements z_i of $\mathfrak{H}_n$, $z_i = x_i^{\,n}$, thus $x_i^{-n}x^n = (x_i^{-1}x)^n = e$. Hence $x_i^{-1}x \in \mathfrak{L}_n$.

§4. *Ex.* 12. Apply the method described in subsection *e*.

Ex. 17(b). Consider the reflexions $\sigma_1, \sigma_2, \sigma_3$ with respect to the coordinate planes x_2x_3, x_3x_1, x_1x_2 of the system Σ_0 respectively, and the corresponding matrices B_1, B_2, B_3 (cf. ex. 14 of §2).

§5. *Ex*. 10. Use Theorem 3 and the result of §4, ex. 17(a).

§6. *Ex*. 4. For every $x \in \mathfrak{G}$, $a \in \mathfrak{H}$ one has $\mathfrak{H}x = \mathfrak{H}ax$ and by supposition also $x\mathfrak{H} = ax\mathfrak{H}$ (J. Fischer).

Ex. 6. By supposition, every element c of $\mathfrak{A} \circ \mathfrak{B}$ is contained in each of the products $\mathfrak{A}^x \mathfrak{B}$; $x \in \mathfrak{G}$. Hence for every c there is an $x \in \mathfrak{G}$ and elements $a \in \mathfrak{A}$, $b \in \mathfrak{B}$ such that $c = xax^{-1}b$. Another $c' \in \mathfrak{A} \circ \mathfrak{B}$ may be written in the form $c' = b^{-1}xa'(b^{-1}x)^{-1}b'$. Hence $cc' \in \mathfrak{A} \circ \mathfrak{B}$.

Chapter III

§1. *Ex*. 17(f). Apply (e) for $\mathfrak{L} = \mathfrak{K}(\mathfrak{G})$.

Ex. 19. For each $tat^{-1} \in \mathscr{C}(a)$ consider $tat^{-1}a^{-1}$.

§2. *Ex*. 14(a). Let $\mathfrak{G} = \mathfrak{Z}(\mathfrak{G}) \cup \mathscr{F}(a)$, $a \notin \mathfrak{Z}(\mathfrak{G})$. Since $\mathfrak{Z}(\mathfrak{G}) \leqslant \mathfrak{C}(a)$ and $\mathscr{F}(a) \subseteq \mathfrak{C}(a)$ also $\mathfrak{G} = \mathfrak{C}(a)$, thus $a \in \mathfrak{Z}(\mathfrak{G})$.

Ex. 14(b). Let $\mathfrak{G} = \mathfrak{Z}(\mathfrak{G}) \cup \mathscr{F}(a) \cup \mathscr{F}(b)$. Repeat the argument of ex. 14(a) to derive that $\mathfrak{G} = \mathfrak{C}(a) \cup \mathfrak{C}(b)$ and apply the corollary of Theorem 3 of Chap. II, §1.

AUTHOR INDEX

SUBJECT INDEX